# Methods in Molecular Biology

AF786591

*Series Editor*
**John M. Walker**
**School of Life and Medical Sciences**
**University of Hertfordshire Hatfield,**
**Hertfordshire, AL10 9AB, UK**

For further volumes:
http://www.springer.com/series/7651

# Cryopreservation of Mammalian Gametes and Embryos

## Methods and Protocols

Edited by

**Zsolt Peter Nagy**

*Reproductive Biology Associates*
*Atlanta, GA, USA*

**Alex C. Varghese**

*ASTRA Fertility Group*
*Mississauga, ON, Canada*

**Ashok Agarwal**

*Cleveland Clinic Foundation*
*Cleveland, OH, USA*

Humana Press

*Editors*
Zsolt Peter Nagy
Reproductive Biology Associates
Atlanta, GA, USA

Alex C. Varghese
ASTRA Fertility Group
Mississauga, ON, Canada

Ashok Agarwal
Cleveland Clinic Foundation
Cleveland, OH, USA

ISSN 1064-3745 ISSN 1940-6029 (electronic)
Methods in Molecular Biology
ISBN 978-1-4939-6826-8 ISBN 978-1-4939-6828-2 (eBook)
DOI 10.1007/978-1-4939-6828-2

Library of Congress Control Number: 2017934237



Printed on acid-free paper

This Humana Press imprint is published by Springer Nature
The registered company is Springer Science+Business Media LLC
The registered company address is: 233 Spring Street, New York, NY 10013, U.S.A.

# Preface

Cryopreservation of oocytes, sperm, and ovarian and testicular tissues, as well as embryos, is one of the most critical procedures to preserve the reproductive capacity of individuals. To grasp its importance it is essential to know that in the United States alone, there are over forty thousand women younger than 40 years old who develop malignant tumors every year—many of whom do not have children/families and whose fertility capacity is at risk because of the disease or the medical therapy applied against the disease. Their best chance to have children in the future is to cryopreserve their gametes or reproductive tissues/organs. Medical conditions can also affect adult men and adolescents, where sperm or testicular tissue cryopreservation serves as the best option to maintain their future fertility. In addition to men and women with medical conditions healthy individuals may also consider preserving their fertility at a younger age to ensure that when they are ready to have a family, they would be able to have their biological offspring. It is a well-known fact that most countries are facing an infertility "epidemic" because social conditions are changing and individuals are postponing marriage and/or having children. As a result, over 20 % of women at the age of 45 are childless, mostly involuntarily. In the past, cryopreserving ovarian tissue and oocytes were a challenge, but today, with recent breakthroughs in technology, female reproductive cells and tissues can be preserved much more efficiently. Age-related decline in female fertility is well established; however, there are indications that age (ageing) may affect men also. There is an increasing number of studies that show a correlation between paternal age (above 45) and higher incidence of different disorders, including autism and certain mental illnesses. For these reasons, cryopreservation of reproductive cells and tissues is fast becoming a critical part of modern reproductive medicine.

Furthermore, it is for these reasons that we aimed to bring a book to readers of the medical profession that provides a comprehensive and technically detailed presentation on all aspects of cryopreservation of reproductive cells and tissues. This book presents current, well-established procedures, as well as novel techniques with the latest innovations described in detail. This book is extremely valuable because each topic is written by a world-renowned author, who is a leader in their field.

We want to express our sincere gratitude to all of the authors who helped write these outstanding chapters for this book. Our special appreciation goes to the development editor, Mike Koy, for his outstanding support from the beginning to the end. The editors are also thankful to their families for their love and support in this important endeavor.

We are truly honored to be able to bring this book to you.

*Atlanta, GA, USA* — *Zsolt Peter Nagy*
*Mississauga, ON, Canada* — *Alex C. Varghese*
*Cleveland, OH, USA* — *Ashok Agarwal*

# Contents

# Contributors

MOHAMED ABDELGAWAD • *Department of Mechanical Engineering, Assiut University, Assiut, Egypt*
ASHOK AGARWAL • *American Center for Reproductive Medicine, Cleveland Clinic, Cleveland, OH, USA*
NAIF H. ALHARBI • *REIM Department, Women's Specialized Hospital, King Fahad Medical City, Riyadh, Kingdom of Saudi Arabia*
NAFISA ALI • *School of Sociology, University of New South Wales, Sydney, NSW, Australia*
JAFFAR ALI • *Faculty of Medicine, Department of Obstetrics and Gynaecology, University of Malaya, Kuala Lumpur, Malaysia*
AMIR ARAV • *Fertilesafe Ltd., Ness Ziona, Israel*
JANELL ARCHER • *Reproductive Services/Melbourne IVF, Royal Women's Hospital, Parkville, VIC, Australia*
KIRI BEILBY • *Monash University, Melbourne, VIC, Australia*
A. BORINI • *9.baby Family and Fertility Center, Bologna, Italy*
SUSANNA BRANDI • *Genea Biomedx, Sydney, NSW, Australia*
AMBER BREWER • *Arizona Reproductive Medicine Specialists, Phoenix, AZ, USA*
RI-CHENG CHIAN • *Department of Obstetrics and Gynecology, McGill University, Montreal, Canada*
ANA COBO • *Cryo Preservation Laboratory, IVI Valencia, Valencia, Spain*
DAVID H. EDGAR • *Reproductive Services/Melbourne IVF and Department of Obstetrics & Gynaecology, Royal Women's Hospital, University of Melbourne, Parkville, VIC, Australia*
ANN-SOFIE FORSBERG • *Product Development, Nidacon International AB, Mölndal, Sweden*
GORAL GANDHI • *IVF Department, Rotunda—The Center For Human Reproduction, Mumbai, Maharashtra, India*
DEBRA A. GOOK • *Reproductive Services/Melbourne IVF and Department of Obstetrics & Gynaecology, Royal Women's Hospital, University of Melbourne, Parkville, VIC, Australia*
JAMES GRAHAM • *Shady Grove Fertility Reproductive Science Center, Rockville, MD, USA*
THORIR HARDARSSON • *Fertility Center, Carlanderska Hospital, Gothenburg, Sweden*
PAUL V. HOLMES • *Product Development, Nidacon International AB, Mölndal, Sweden*
VLADIMIR ISACHENKO • *Department of Obsterics and Gynecology, Medical Faculty, University Maternal Hospital, Cologne University, Cologne, Germany*
EVGENIA ISACHENKO • *Department of Obsterics and Gynecology, Medical Faculty, University Maternal Hospital, Cologne University, Cologne, Germany*
SAKINA KAGALWALA • *IVF Department, Rotunda—The Center For Human Reproduction, Mumbai, Maharashtra, India*
MOHAN S. KAMATH • *Reproductive Medicine Unit, Christian Medical College, Vellore, Tamil Nadu, India*
DANA KEILTY • *Department of Obstetrics and Gynecology, McGill University, Montreal, Canada*
SEUL KI KIM • *Department of Obstetrics and Gynecology, Seoul National University Bundang Hospital, Seongnam, South Korea*

S. SAMUEL KIM • *Center for Fertility and Reproduction, American-Sino Women's and Children's Hospital, Shanghai, China*
MASASHIGE KUWAYAMA • *Repro-Support Medical Research Centre, REPROLIFE Inc., Tokyo, Japan*
ANNA NILäNG LAESSKER • *Product Development, Nidacon International AB, Mölndal, Sweden*
MARK G. LARMAN • *Research, Vitrolife, Englewood, CO, USA*
JUNG RYEOL LEE • *Department of Obstetrics and Gynecology, Seoul National University Bundang Hospital, Seongnam, South Korea*
JANA LIEBENTHRON • *Cryobank at the Department of Gynecological Endocrinology and Reproductive Medicine, University Clinic of Bonn, Bonn, Germany*
JUERGEN LIEBERMANN • *Fertility Centers of Illinois, Chicago, IL, USA*
JOSH LIM • *Shady Grove Fertility Center, Rockville, MD, USA*
JUN LIU • *Department of Mechanical and Industrial Engineering, University of Toronto, Toronto, ON, Canada*
G. MAL • *ICAR-Indian Veterinary Research Institute, Palampur, Himachal Pradesh, India*
PETER MALLMANN • *Department of Obsterics and Gynecology, Medical Faculty, University Maternal Hospital, Cologne University, Cologne, Germany*
MARKUS MONTAG • *ilabcomm GmbH, St. Augustin, Germany*
TETSUNORI MUKAIDA • *Hiroshima Hart Clinic, Hiroshima, Japan*
K. MUTHUKUMAR • *Reproductive Medicine, Christian Medical College, Vellore, Tamil Nadu, India*
JULIA C. NEUBAUER • *Fraunhofer Institute for Biomedical Engineering, Sulzbach, Germany*
PRIYANKA PANGERKAR • *IVF Department, Rotunda—The Center For Human Reproduction, Mumbai, Maharashtra, India*
TEIJA T. PEURA • *Genea Biomedx, Sydney, NSW, Australia*
DEREK G. PYNE • *Department of Mechanical and Industrial Engineering, University of Toronto, Toronto, ON, Canada*
PATRICK QUINN • *Quinn's IVF Consulting, Apache Junction, AZ, USA*
GOHAR RAHIMI • *Department of Obsterics and Gynecology, Medical Faculty, University Maternal Hospital, Cologne University, Cologne, Germany*
LAURA RIENZI • *Genera Centers for Reproductive Medicine, Clinical Valle Giulia, Rome, Italy*
MARTA F. RIESCO • *Department Molecular Biology, Cell Biology Area, University of León, León, Spain*
VANESA ROBLES • *Department Molecular Biology, Cell Biology Area, University of León, León, Spain*
TAMMIE K. ROY • *Genea Biomedx, Sydney, NSW, Australia*
RAUL SANCHEZ • *Department of Preclinical Science, Center for Traslation Medicine, Facultad de Medicina, Universidad de la Frontera, Temuco, Chile*
SHERMAN SILBER • *Infertility Center of St. Louis, St. Louis, MO, USA*
B. SINGH • *Regional Station, ICAR-Indian Veterinary Research Institute, Palampur, Himachal Pradesh, India*
SARABPREET SINGH • *Reproductive Medicine, Artemis Health Institute, Gurgaon, Haryana, India*
S. K. SINGLA • *Animal Biotechnology Centre, ICAR-National Dairy Research Institute, Karnal, Haryana, India*
FRANK STRACKE • *Fraunhofer Institute for Biomedical Engineering, Sulzbach, Germany*

Chang Suk Suh • *Department of Obstetrics and Gynecology, Seoul National University Bundang Hospital, Seongnam, South Korea*
Yu Sun • *Department of Mechanical and Industrial Engineering, University of Toronto, Toronto, ON, Canada*
Pankaj Talwar • *ART Centre, Army Hospital (Research and Referral), New Delhi, India*
Michael Tucker • *Shady Grove Fertility Center, Rockville, MD, USA*
Eva Tvrda • *American Center for Reproductive Medicine, Cleveland Clinic, Cleveland, OH, USA; Department of Animal Physiology, Slovak University of Agriculture, Nitra, Slovak Republic*
Filippo Maria Ubaldi • *Genera Center for Reproductive Medicine, Clinical Valle Giulia, Rome, Italy*
David G. Valcarce • *Department Molecular Biology, Cell Biology Area, University of León, León, Spain*
Matthew VerMilyea • *Ovation Fertility, Austin, TX, USA*
Yixin Xu • *Department of Obstetrics and Gynecology, McGill University, Montreal, Canada*
Yehudit Natan • *Fertilesafe Ltd, Ness Ziona, Israel*
Hye Won Youm • *Department of Obstetrics and Gynecology, Seoul National University Bundang Hospital, Seongnam, South Korea*
C. Zacà • *9.baby Family and Fertility Center, Bologna, Italy*
Heiko Zimmermann • *Fraunhofer Institute for Biomedical Engineering, Sulzbach, Germany*

# Part I

## Introduction

# Chapter 1

# Historical Background on Gamete and Embryo Cryopreservation

Jaffar Ali, Naif H. AlHarbi, and Nafisa Ali

## Abstract

This chapter describes the development of the science of cryopreservation of gametes and embryos of various species including human. It attempts to record in brief the main contributions of workers in their attempts to cryopreserve gametes and embryos. The initial difficulties faced and subsequent developments and triumphs leading to present-day state of the art are given in a concise manner. The main players and their contributions are mentioned and the authors' aim is to do justice to them. This work also attempts to ensure that credit is correctly attributed for significant advances in gamete and embryo cryopreservation. In general this chapter has tried to describe the historical development of the science of cryopreservation of gametes and embryos as accurately as possible without bias or partiality.

**Key words** Assisted reproduction, Cryopreservation, Cryoprotectants, Embryos, Frozen-storage, Gametes, Infertility, Oocytes, Spermatozoa, Vitrification

## 1 Background

Early humans learnt to preserve food and seeds of plants. Ability to preserve food made possible travel over vast lands and oceans. More importantly the ability to preserve food made it possible to store food in preparation for leaner times that helped humankind survive famines, a concept that led us, in recent years, to the development of techniques for the frozen storage of biological materials for future use. Frozen storage or cryopreservation is useful for food production, for treatment of infertility, and for maintenance of genetic diversity in both domesticated and wild animals. The key elements in the preservation of food were dehydration; the use of protectants or preservatives such as vinegar, salt, or sugar that also served as desiccants; and the use of exposure to low or high temperatures. The lessons learnt in preserving food led humankind to consider preserving other materials of importance such as organs, blood, blood products, and a large number of biological materials, spermatozoa, eggs, and embryos in a number of species. In very

Zsolt Peter Nagy et al. (eds.), *Cryopreservation of Mammalian Gametes and Embryos: Methods and Protocols*, Methods in Molecular Biology, vol. 1568, DOI 10.1007/978-1-4939-6828-2_1, © Springer Science+Business Media LLC 2017

simplistic terms, precisely the same elements used in food preservation such as dehydration and preservatives and low temperature were utilized to cryopreserve gametes and embryos. After many years of research and experimentation, we appear to have succeeded in this quest.

Cryobiology supposedly originated in ancient Egypt where cold temperatures were used to stop bleeding and reduce inflammation and as local anesthesia. One of the earliest reports of an attempt at preservation of biological material related to animal reproduction was that of stallion spermatozoa by the Arabs in 1322 [1]. As the story goes, an Arab chieftain got his spy to steal semen from a prize stallion belonging to a rival tribe. The spy collected the stallion semen in a sponge to which he had added camel milk as a preservative. After his fruitful escapade, the spy successfully inseminated his chieftain's mare using the stolen stallion's semen. The inseminated mare later became pregnant and delivered a normal foal. The actual nature of the concoction of protectants used by the spy to preserve the viability of the spermatozoa except for milk is not known. However, it is of interest to note that centuries later, reproductive biologists used milk in cryopreservation solutions for the cryopreservation of mammalian spermatozoa. In more recent times, cryobiology was investigated in detail by Luyet and coworkers beginning in the late 1930s till about the 1960s [2].

### 1.1 Accuracy of Historical Accounts

Historical accounts should be accurate and devoid of bias or discrimination on the basis of language or origin. Polge et al. [3] have been credited with being the first to cryopreserve spermatozoa in 1949 using the cryoprotectant glycerol although Bernstein and Petropavlovski demonstrated much earlier in 1937 the cryoprotective effects of glycerol for the cryopreservation of spermatozoa (cited in ( [4])). Oftentimes when recording historical facts, there appears to be a tendency to cite only work published in mainstream journals in the English language. Indeed it is seldom mentioned that Smirnov had also cryopreserved spermatozoa in 1949 [4] in the same year as Polge, probably because Smirnov's work was published in Russian. Likewise, work published in national journals of developing countries is often not accorded any recognition whatsoever. For instance, the work of Mukerji and coworkers [5] in India, published in the *Indian Journal of Cryogenics* in 1978, was even to this date not accorded any recognition. Rather, it was dismissed as a hoax, not only by the rest of the world but even by their own compatriots. Mukerji's group reported the first live birth of a baby on 3 October 1978 from an IVF embryo that was successfully cryopreserved at the 8-cell stage using 1.5 M DMSO, a cryoprotectant. The frozen embryo was cryostored for about 50 days before transfer. Mukerji's team consisted of a Cornell-trained cryobiologist which suggests the team had skilled personnel who had the ability to cryopreserve embryos. Indeed a scrutiny of their lab records favors the conclusion that Mukerji's group succeeded in cryopreserving the human embryo at the 8-cell stage in 1978 [6].

To this day, however, the world does not acknowledge the fact that Mukerji's group reported the first live birth of a human from a frozen embryo, a mere 2 months and 9 days after the birth of the first IVF baby on 25 July 1978 reported by Steptoe and Edwards [7]. The extreme ostracism faced by Mukerji following his report and his subsequent unfair chastisement by the authorities, ultimately, led to his untimely and unnatural demise [6]. The historical account of Mukerji's predicament and subsequent chastisement was clearly recorded by Anand Kumar in 1997 [6] and is freely available. Anand Kumar knew of and lost his own place in history as the architect of India's first test tube baby when he revealed the truth about Mukerji. The fraternity must reflect on the misfortune which befell Mukerji. This was the direct consequence of according recognition to only those reports which appear in mainstream publications. Interestingly both Robert Edwards and Subhas Mukerji obtained their doctorate at the University of Edinburgh, Scotland.

## 2 Cryopreservation of Spermatozoa

About four and a half centuries after the Arabs demonstrated the ability to retain viability of stallion spermatozoa in a concoction containing camel milk, Lazzaro Spallanzani, in a landmark development, demonstrated that spermatozoa frozen in snow did not die but were merely temporarily inactivated under cold conditions and brought to life after the semen was thawed [8]. This was in 1776. Spallanzani was an Italian priest. Spallanzani may be thought of as the Father of Gamete Cryobiology. He was born in Scandiano, in the province of Reggio Emilia, Italy, in 1729. He obtained his education at the University of Bologna where he came under the influence of an accomplished kinswoman, Laura Bassi, who was thought to have influenced his subsequent scientific career and discoveries. Spallanzani's work would not have been possible without the use of the microscope which was invented by Leeuwenhoek about a century earlier. Indeed Leeuwenhoek was the first to describe human spermatozoa using his microscope in 1677. In Greek "cryos" means "cold." Most of the early work in cryobiology was performed in spermatozoa because they were readily available and their motility can be used as an index of viability post-thaw.

The cryoprotective properties of glycerol were discovered quite by chance. As a research student in Alan Parkes' Laboratory at the Medical Research Council, UK, Christopher Polge, together with Audrey Smith, was trying to cryopreserve fowl spermatozoa using a levulose (sugar) solution without much success. The work on spermatozoa cryopreservation was an extension of the work of CS Schaffner (1942) on vitrification [9]. Polge temporarily gave up freezing fowl spermatozoa for more productive endeavors. Later on, when Polge decided to have one last try, the frozen-thawed fowl spermatozoa survived and retained their motility, but when the experiment was repeated with freshly prepared cryoprotectant solution, the spermatozoa did not survive. It would have been baffling if the laboratory technician had not mentioned that the labels had previously come off and been refixed. This revelation raised the possibility that the labels could have been switched. Glycerol solution was used in their laboratory to mount microtome section. The bottles were tested. The presence of glycerol was confirmed by its characteristic smell when burnt. The test revealed the solution that gave the better result contained about 25 % glycerol instead of the intended cryoprotectant, levulose. This serendipitous discovery of the cryoprotective property of glycerol laid the foundations for future scientific developments in gamete and embryo cryopreservation. The observation that followed was that the frozen-thawed fowl spermatozoa though motile appeared to have lost their fertility. What probably happened was the presence of glycerol prevented the loss of motility in the frozen-thawed spermatozoa but had caused the loss of its fertilizing potential. The problem was resolved when the glycerol was removed from the suspension prior to insemination. Parkes immediately foresaw the immense potential of this discovery for the cattle industry. He proceeded to organize artificial insemination using frozen-thawed spermatozoa in cattle. However they obtained only a single success. Polge called the calf "Frosty." Frosty was the first calf to be produced from frozen-thawed spermatozoa. Polge moved to Cambridge where he joined Tim Rowson. At Cambridge both Polge and Rowson modified the frozen-thawed spermatozoa preparation protocol to successfully perform artificial insemination in the cattle. This work was reported in 1952 [9].

The artificial insemination program in farm animals and the treatment of infertility in human patients provided the impetus to seek the means to effectively cryostore spermatozoa. Following on the work of Polge and Rowson, Bratton et al. also demonstrated in 1955 the successful cryopreservation of bull spermatozoa which achieved high fertility following insemination. It was Sherman and Bunge in 1953 who reported the first successful cryopreservation of human spermatozoa and their use to achieve pregnancies and live births (cited in ( [9])) following insemination with frozen-thawed sperm. It is noteworthy that the use of liquid nitrogen for cold storage of spermatozoa was first reported by Sherman. He demonstrated the superior qualities of liquid nitrogen storage at −196 °C over the then prevailing convention of −75 °C. Subsequently, in 1963, Sherman was able to establish that there was no loss of motility after storage of spermatozoa at the lower temperature after 1 year of storage, but, on the contrary, motility diminished with time when stored at −75 °C. In 1964 Perloff et al. [10] demonstrated for the first time that spermatozoa can be stored for more than 5 months and indeed for long periods of time, without loss of viability. It is now well recognized that spermatozoa can retain their capability to fertilize eggs and give rise to normal pregnancies in the human even after frozen storage for 28 years [11].

Apart from artificial insemination, frozen-thawed spermatozoa have been used to successfully achieve pregnancies in intrauterine insemination [IUI] in 1990 [12], in vitro fertilization [IVF] in 1984 [13], subzonal insemination [SUZI] in 1992 [14], and intracytoplasmic sperm injection [ICSI] with frozen-thawed ejaculatory spermatozoa in 1994 [15]. The first pregnancy following ICSI with cryopreserved testicular sperm was reported by Romeros et al. in 1996 [16]. The successful cryopreservation of ejaculatory and testicular human spermatozoa with establishment of a clinical pregnancy in a synthetic cryoprotectant medium devoid of added donor proteins was demonstrated in 2010 [17].

## 3 Cryopreservation of Embryos

The history of embryo cryopreservation is far more extensive than that of oocyte cryopreservation. Efforts and successes in embryo cryopreservation predate oocyte cryopreservation. For this reason the history of embryo cryopreservation shall precede that of oocyte cryopreservation in this chapter; however, advances in sperm, oocyte, and embryo cryopreservation have been interrelated. The earliest scientifically formulated attempts to achieve cryopreservation point to B.J. Luyet as the pioneer of this area of endeavor that began in the 1930s. Ferdows and coworkers (1958) attempted to cryopreserve rabbit zygotes with little success (cited in ( [18])). The report of Whittingham in 1971 [19] on the successful survival of mouse embryos in polyvinylpyrrolidone (PVP) is the first successful cryopreservation of embryos, but others were unable to repeat Whittingham's successes.

Real success in embryo cryopreservation ensued when Whittingham combined forces with Mazur and Leibo where the skills of Whittingham in embryo handling were combined with the cryobiological knowledge of Mazur and Leibo. Leibo calculated that, to avoid intracellular ice formation in cells the size of mouse embryos, the freezing should not be faster than 1 °C per minute. The experiments that followed in which the effects of the composition of the suspending medium, consisting of glycerol and DMSO, together with the cooling and warming rates showed Leibo's calculations to be accurate and true. This collaboration led to successes in embryo cryopreservation in which 65 % of surrogates became pregnant and 40 % of them proceeded to term [20]. In the same year, Wilmut [21] then working in Cambridge came to the same conclusion independently of Whittingham and coworkers. He too reported successful mouse embryo cryopreservation in 1972. Two years later, Whittingham and Whitten [22] demonstrated the survival of frozen mouse embryos after transatlantic transport in liquid nitrogen. Most workers reported high embryonic death when embryos were exposed to cryoprotectants in a single step so Willadsen and coworkers developed a stepwise exposure to cryoprotectant to avoid loss of embryos [23].

Most of the early work on embryo cryopreservation was on mouse embryos, but other species were successfully cryopreserved within a short span of time in the ensuing years. These included successful embryo cryopreservation in cattle [24–26], rabbits [27, 28], rats [29], sheep [30], goats [31], horses [32], baboons [33], marmoset monkeys [34], cats [35], and many more.

The history of embryo cryopreservation in the human is somewhat complicated. The first report of a live birth from a frozen-thawed embryo was reported by Mukerji and coworkers from India in 1978 [5], but this report is largely unacknowledged by the scientific community because it was published in nonmainstream literature. Later in 1983, Trounson and Mohr of Monash University, Australia, reported a stillbirth from a frozen-thawed human embryo which many consider to be the first report of a successful cryopreservation of a human embryo [36]. The first human live births were reported by the group of Zeilmaker in 1984 from Holland [37].

By the mid-1980s, the slow controlled-rate freezing technique was well documented with reasonable success in most laboratories worldwide for embryos of numerous species. The methodology for slow freezing was however tedious, very expensive, and demanding in time of laboratory workers. Typically, each cycle of slow controlled-rate cooling could take 3–4 h to complete. Naturally investigators searched for methods that are less demanding. This search led to investigations of rapid and ultrarapid freezing. This has yet to meet with universal acceptance although early studies of rapid and ultrarapid freezing of embryos of numerous species using glycerol and sucrose or propylene glycol and sucrose did show promising results with high survival post-thaw [38–44]. It is noteworthy that sucrose was used to achieve partial dehydration initially by Mazur, Miller, and Leibo [45] in 1974 for bovine blood cells and applied to embryos by Kasai, Niwa, and Iritani [38] and in almost all subsequent work on cryopreservation of embryos and gametes. Using rapid freezing methods, post-thaw survival has been as high as 95 % [44] in the mouse. Likewise in the mouse, a post-thaw survival of 92 % and in vivo survival similar to that observed with fresh embryos were obtained when DMSO and sucrose were used as cryoprotectants [46]. DMSO proved more efficacious than glycerol for ultrarapid freezing procedures. Early studies on the ultrarapid freezing of human embryos using DMSO and sucrose were also encouraging with live births [47, 48], but ultrarapid freezing did not gain much foothold probably because its development was overshadowed by the reemergence of the vitrification technique of cryopreservation. Around the first quarter of the 1990s, work on ultrarapid freezing ceased, and attention shifted to vitrification which appeared to promise excellent survival and outcome. This is likely to be attributable to the fact that, in vitrification, both intracellular and extracellular ice formation can be avoided [49].

In the mid-1980s, the report by Rall and Fahy [50, 51] of successful cryopreservation by vitrification of mouse embryos with live births created considerable interest in this unique technique of cryopreservation. This led to the revival of vitrification as a strategy for ultrarapid cryopreservation. Vitrification is a curious state in which a highly concentrated solution, when cooled, becomes a very viscous solid called a "glass." This viscous solid has the molecular structure of a liquid but the mechanical properties of a solid. Vitrification had been previously abandoned due to chemo-toxicity associated with the high solute concentration of VS's needed to achieve glass formation. Later Kono et al. [52] successfully vitrified rat embryos using the VS1 solution formulated by Rall and coworkers [50, 51]. Soon after this report, Kasai et al. [53] reported the vitrification of mouse day 4 morulae with almost complete survival using an ethylene glycol-based VS. This was a major development, and it fulfilled the promise of almost total survival envisioned by many cryobiologists for vitrification. However the result of vitrification in other mammalian species was not quite as spectacular. This is probably because, according to this author's (JA) opinion, the VS's used were not well researched for a given developmental stage of the embryo of a given species for which it was used. In any event, the report of Kasai et al. [53] indicated without any doubt that very high survival and viability could be obtained following vitrification if the conditions were right. There was a need to find the "right" conditions.

By the time the Kasai report came out, Ali and Shelton had completed a systematic and extensive study of 3044 ternary cryoprotectant solutions to identify cryoprotectant solutions possessing glass-forming properties. Of these 3044 cryoprotectant solutions, 66 ternary cryoprotectants could form glass when cooled [54]. Of these, three VS's appeared less toxic based on simultaneous toxicology studies undertaken by the same group. Of these three VS's, VS14 was the least embryotoxic [54–56]. Ali and Shelton successfully vitrified all developmental stages of the mouse and day 6 stages of the sheep embryo using VS14 (5.5 M ethylene glycol and 1.0 M sucrose) with no appreciable loss of viability in vitro and in vivo followed by live births. Later, Ali used the VS14 technique to successfully vitrify human day 2 embryos with insignificant interference with survival in vitro [57–59]. In the mid-1990s, Ali moved to the Middle East where he was not permitted to perform research on vitrification due to regulations. This, plus the lack of an animal facility and the retirement of Shelton at around this time effectively ensured Ali did no more work on vitrification until 2015. However, many workers from around the world managed to successfully vitrify oocytes and embryos of various species, some followed by live births, between the 1990s and 2010s [60–85] using the VS14 formulated by Ali and Shelton [54–56], in some instances with minor modifications [62, 86–89]. The successful application of VS14 in a number of species including humans proved its versatility. It is noteworthy out of all reports that successfully utilized the VS14 formulated by Ali and Shelton, only one group [84, 85] attributed the VS14 to the inventors. As a consequence of this inconsistency in reporting, the credit for the pioneering contributions of Ali and Shelton to vitrification was not universally known or acknowledged but instead erroneously attributed to other workers. Even as recently as 2010 and 2011, some workers who were not involved in its development claimed VS14 to be "their method." Similar parallels can be seen in all areas of scientific endeavor.

## 4 Cryopreservation of Oocytes

Historical records indicate Picket attempted to cryopreserve oocytes in 1893 [90]. It is well documented that early attempts to freeze oocytes failed. In the 1950s, the realization that glycerol [3] could serve as a cryoprotectant led some workers to use it for the cryopreservation of oocytes in the mouse and sheep [91, 92], but they were not successful. It is well recognized that oocytes of almost all species are prone to chilling injury which made oocyte cryopreservation a challenge [93]. Considerable groundwork on cryobiology during the subsequent two and half decades by numerous workers worldwide provided the knowledge and clues about the physiological and physicochemical factors which determine the survival or death of cells when challenged with a variety of chemicals (cryoprotectants) or physical conditions such as temperature, osmotic pressure, pH, atmospheric pressure, membrane permeability, chemical toxicity, and so forth which are likely to affect cells during cryopreservation (cited in ( [18])).

These studies enabled workers to devise techniques of cryopreservation for gametes, embryos, organs, tissues, and cells. Prior to the 1990s, attempts to cryopreserve oocytes were not successful. Studies during this period discovered that cooling led to disruption of the meiotic spindle [94, 95] which would reconstitute itself in an abnormal configuration with the chromatids strewn in the ooplasm in a haphazard manner when the temperature was brought back to body temperature [95]. Indeed studies by Kola et al. in 1988 [96] lent credence to this assumption. Nevertheless, two studies published 1 year previously and in the same year did not demonstrate the same effect [97, 98]. Notwithstanding these discordant observations, the telltale appearance of digynic polyploidy in the latter studies suggested that, while the effect on the meiotic spindle may not be obvious, cooling and exposure to cryoprotectant chemicals impact negatively on the oocyte meiotic process by inducing parthenogenesis. Before drawing this conclusion, account should be taken of the likelihood that, when human oocytes have been used experimentally, these have usually been aged. Consequently, the higher incidence of polyploidy reported [99, 100] may reflect oocyte aging rather than the cryopreservation procedure [101]. A number of physical factors are known to activate oocytes resulting in parthenogenesis.

Al-Hasani et al. [100] could obtain fertilization and cleavage but only one pregnancy from cryopreserved human oocytes. The following year, there was a report of a birth following the fertilization of eggs obtained from frozen-thawed eggs that previously failed to fertilize during conventional IVF cycles during a subsequent treatment cycle in the same patient [102]. In both instances, the poor outcome was probably related to oocyte aging rather than to the efficacy of the cryopreservation methodology. Contemporaneously, notwithstanding recurrent negative observations on oocyte cryopreservation, Chen [103], a Singaporean working in Australia, reported birth of a set of twins from cryopreserved human oocytes. Chen's report heralded a new era in oocyte cryopreservation and appeared to suggest the problem of oocyte cryopreservation that had plagued researchers in the field for so long, and which seemed insurmountable previously, had been finally solved. The ensuing jubilation was short-lived as it soon became clear Chen's success could not be repeated by other workers. This failure suggested that further research would be required before oocyte cryopreservation could be applied on a routine basis. One possible explanation for the inconsistent outcomes could relate to differences in the quality of individual oocytes. While some oocytes may be able to withstand the physical and chemical challenges imposed by the cryopreservation procedures, most others may not do so. There was a need to develop cryopreservation techniques which were benign for the target cells. Indeed, another 11 years of research was required before oocyte freezing could be successfully applied on a routine basis. The first births of babies from frozen oocytes following the use of ICSI were reported by Eleonora Porcu and her coworkers at the University of Bologna, Italy. Subsequently Porcu's slow controlled-rate freezing oocyte cryopreservation technique utilizing propanediol [104–108] was reported to be reproducible in other laboratories, and a number of groups elsewhere established pregnancies from oocyte cryopreservation especially from 1998 onward [109–112].

Even as the challenge of slow controlled-rate freezing of the human oocyte was being resolved in the 1990s, reports of successful oocyte cryopreservation using vitrification appeared. In 1989 one worker, Nakagata [113], in Japan reported the successful vitrification of mature mouse oocytes which were subsequently fertilized resulting in the birth of live young. The technique of vitrification of mouse oocytes as performed by Nakagata was technically challenging, and most workers could not reproduce his work partly because of the extremely short duration (20 s) of exposure to cryoprotectants before ultrarapid cooling. This was not technically feasible for other workers. At around the time that Nakagata [113] reported his work on mouse oocyte vitrification, van Blerkom [114] reported vitrification of germinal vesicle (GV) stage mouse oocytes. About 90 % of the GV oocytes survived vitrification to resume meiosis and normal maturation but, on examination, were found to have sustained considerable cytoplasmic and nuclear perturbations. These perturbations appeared to be corrected during culture. The matured oocytes were capable of fertilization; nevertheless, van Blerkom warned of possible nuclear defects due to the potential for deletion of segments of DNA. Reflecting on the technical challenges and the risk of serious developmental issues post-rehydration and/or implantation, research on human oocyte cryopreservation almost ceased.

At a similar time, Porcu et al. [104–108] reported their successes with oocyte slow controlled-rate freezing; Alex Martino [60] then working in the laboratory of Stanley Leibo at the University of Guelph in Canada demonstrated the successful cryopreservation of bovine oocytes held in electronic microscope grids using a vitrification solution (called VS14) formulated by Ali and Shelton [54–56] some 3 years earlier. Martino called the VS14, devised by Ali et al., "EG5.5" which led to some confusion in other workers on the origins of the VS14 [115, 116]. VS14 (EG5.5) consists of 5.5 M ethylene glycol and 1.0 M sucrose. About 6 years prior to Martino's report, Ali joined Jim Shelton at the John Curtin School of Medical Research of the Australian National University (ANU) as a graduate student. The VS14 (or EG5.5) was developed by Ali and Shelton [54], and the details of its preparation were published in 1993. This vitrification solution was the product of an extensive evaluation undertaken by Ali and Shelton in which 3044 cryoprotectant solutions were investigated. VS14 was the least toxic of all the VS's investigated. Following the report of Martino et al. of successful vitrification of bovine oocytes using VS14, Chung et al. and Hong et al. (cited in ([61])) reported the successful vitrification of human oocytes during the annual conference of the American Society of Reproductive Medicine in 1998. They also used the VS14 developed by Ali and Shelton [54]. Both reports appear to be from the same group based in Korea. In the following year, the same group [61] reported the first pregnancy after vitrification of oocytes with VS14. Again, in the same year, a case study by Kulosheva et al. [62] reported a live birth from vitrified oocytes using a similar ethylene glycol-based VS that consisted of 40 % (approximately 6 M) ethylene glycol and 0.6 M sucrose. For the next decade or so, workers affiliated to the Korean group based at Cha General Hospital, Seoul, achieved pregnancies using VS14-vitrified oocytes [61, 63–67] and embryos. It is noteworthy that a large number of workers who utilized the VS14, some with minor modifications, failed to acknowledge the origins of VS14 [115, 116]. The lack of formulation data on most VS's employed in clinical investigations casts doubts on originator claims other than those relating to VS14 since its original description by Ali and Shelton [54–56].

The lack of formulation and toxicological data could also suggest the VS's may be of whimsical origin rather than products of painstaking research. Whimsically conceived VS's could have the potential for devastating long-term impact on health of individuals who are derived from embryos vitrified in such less researched VS's. A French study warns of nonlethal damage during the cryopreservation process which could have long-term negative effects on the individual. These workers noted violent behavior in mice derived from frozen- thawed embryos and have warned of possible similar consequences in the human [117]. They also predicted delayed detrimental effects of cryopreservation on the affected individual.

Subsequent to the emergence of VS's that employed single permeating cryoprotectants, a number of workers began using ethylene glycol and sucrose in combination with other permeating cryoprotectants such as DMSO [118–121] or propylene glycol [122, 123]. Present-day commercially available VS's appear to consist of combinations of permeating cryoprotectants plus sucrose. It is noteworthy that vitrification appears to have become the preferred technique for the preservation of oocytes by the late 2000s. A 5-year review of oocyte cryopreservation reports which described the use of slow cool methods versus vitrification showed the latter to be more efficacious [124]. By the end of the 2000s, oocyte cryopreservation by vitrification has become routine in some laboratories with excellent pregnancy rates almost comparable to those achieved using fresh oocytes [125, 126]. It now seems the problem of oocyte cryopreservation especially with the use of vitrification has been finally put to rest [127, 128] although it remains to be routinely applied universally. However, the long-term impact of oocyte vitrification on offspring remains to be elucidated. It is noteworthy that Delphine Parrott had obtained live young from frozen-grafted ovarian tissue some 30 years earlier [129]. Her work is seldom mentioned although it augured attempts to cryopreserve ovarian tissue many decades later.

From the mid-1990s through to 2015, vitrification as a strategy for embryo cryopreservation has been firmly established especially in the human with excellent outcomes similar to those achieved with fresh embryos. Present-day VS's utilize at least two permeating plus one non-permeating cryoprotectants. This practice has resulted in excellent outcomes. Most of the current commercially available VS's for human embryos appear to be of this kind. The proportion of pregnancies from cryopreserved embryos is now similar to that for fresh embryos [128, 130]. The problems faced by workers in the past in the cryopreservation of embryos now appear to have been resolved although any long-term effects of embryo cryopreservation remain to be elucidated conclusively. A recent review [128] of publications on human embryo and oocyte cryopreservation, beginning from 1980 through to 2013, indicated improved pregnancy rates from vitrified embryos. This was especially so in recent years with outcomes almost the same as those observed with fresh embryos, albeit with a higher rate of Cesarean section. Significant increases in risks of DNA damage, spindle configuration, embryonic aneuploidy, and genomic imprinting were not observed when fresh and slow-freezing procedures were compared.

Even though the significant and tangible advances made in the area of cryopreservation of gametes and embryos over the last three decades promise well for the future of vitrification in healthcare and other industries, one lingering problem remains to be resolved. This relates to the use of micro-vehicles in vitrification. In this situation, asepsis is difficult to attain. Consequently, it is known as the *open* system in which human gametes, embryos, and other tissues in the healthcare and other industries are subjected to non-asepsis, a practice that is not in keeping with Good Laboratory Practice/Good Clinical Practice (GLP/GCP). This situation has persisted to the present time since micro-carriers were first utilized in the 1990s [131, 132]. Newly innovated safe aseptic methods of cryopreservation can be adopted or have to be innovated to institute GLP-/GCP-compliant techniques in healthcare.

In conclusion the quest for the cryopreservation of gametes and embryos has been largely resolved, but there remains scope for further improvement.

## Acknowledgment

The authors are grateful to Dr. Peter J McCullagh and Dr. James N Shelton of the John Curtin School of Medical Research, the Australian National University, Canberra, Australia, for their valuable assistance in copy editing the manuscript and valuable comments that helped improve the quality of the manuscript.

## References


1. de Oliveira RA, Wolf CA, Viu MAO, Gambarini ML (2015) Cooling of equine semen at 16 °C for 36 h with the addition of cysteine in different concentrations. Pferdeheilkunde 31:27–32
2. Schmidt PJ, Basile J (2006) Luyet and the beginnings of transfusion cryobiology. Transfus Med Rev 20(3):242–246
3. Polge C, Smith AU, Parkes AS (1949) Revival of spermatozoa after vitrification and dehydration at low temperatures. Nature 164 (4172):666
4. Isachenko V, Isachenko E, Katkov II, Montag M, Dessole S, Nawroth F, Van Der Ven H (2004) Cryoprotectant-free cryopreservation of human spermatozoa by vitrification and freezing in vapor: effect on motility, DNA integrity, and fertilization ability. Biol Reprod 71(4):1167–1173
5. Mukerji S, Mukherjee S, Bhattacharya SK (1978) The feasibility of long term cryogenic freezing of viable human embryos - a brief pilot study report. Indian J Cryogenics 3 (1):80
6. Anand Kumar TC (1997) Historical notes: architect of India's first test tube baby: Dr. Subhas Mukerji (16 January 1931 to 19 July 1981). Curr Sci 72:526–531
7. Steptoe PC, Edwards RG (1978) Birth after the reimplantation of a human embryo. Lancet 2:366
8. Clarke GN (2006) A.R.T. and history, 1678-1978. Hum Reprod 21(7):1645–1650
9. Watson PF. Chris Polge, CBE FRS (16 August 1926–17 August 2006): a personal appreciation. Cryo Letters 2008;29(1):1–6.
10. Perloff WH, Steinberger E, JK S (1964) Conception with human spermatozoa frozen by nitrogen vapor technic. Fertil Steril 15:501–504
11. Feldschuh J, Brassel J, Durso N, Levine A (2005) Successful sperm storage for 28 years. Fertil Steril 84(4):1017
12. Byrd W, Bradshaw K, Carr B, Edman C, Odom J, Ackerman G (1990) A prospective randomized study of pregnancy rates following intrauterine and intracervical insemination using frozen donor sperm. Fertil Steril 53 (3):521–527
13. Schill WB, Trotnow S (1984) Use of frozen sperm for in vitro fertilization. A case report. Hautarzt 35(6):313–315
14. Levron J, Lightman A, Stein DW, Brandes JM, Itskovitz-Eldor J (1992) Pregnancy after subzonal insertion of cryopreserved spermatozoa from a patient with testicular seminoma. Fertil Steril 58(4):839–840
15. Payne D, Flaherty SP, Jeffrey R, Warnes GM, Matthews CD (1994) Successful treatment of severe male factor infertility in 100 consecutive cycles using intracytoplasmic sperm injection. Hum Reprod 9(11):2051–2057
16. Romero J, Remohí J, Mínguez Y, Rubio C, Pellicer A, Gil-Salom M (1996) Fertilization after intracytoplasmic sperm injection with cryopreserved testicular spermatozoa. Fertil Steril 65(4):877–879
17. Ali J., Proceeding of the Workshops on the WHO 2010 Semen Analysis Format; total quality management in ART laboratory and ultra micro-drop embryo culture. Department of obstetrics and gynaecology, University of Malaya (UM), Kuala Lumpur, Malaysia.15th &16th-February-2014
18. Ali J. (1992) Factors affecting ultra rapid vitrification and cryopreservation of embryos. The Australian National University. PhD Thesis
19. Whittingham DG (1971) Survival of mouse embryos after freezing and thawing. Nature 233(5315):125–126
20. Whittingham DG, Leibo SP, Mazur P (1972) Survival of mouse embryos frozen to −196 degrees and −269 degrees C. Science 178 (4059):411–414
21. Wilmut I (1972) The low temperature preservation of mammalian embryos. J Reprod Fertil 31(3):513–514
22. Whittingham DG, Whitten WK (1974) Long-term storage and aerial transport of frozen mouse embryos. J Reprod Fertil 36(2):433–435
23. Willadsen SM, Polge C, Rowson LEA, Moor RM Preservation of sheep embryos in liquid nitrogen. Cryobiology 11:560
24. Wilmut I, Rowson LE (1973) Experiments on the low-temperature preservation of cow embryos. Vet Rec 92(26):686–690
25. Wilmut I, Rowson LE (1973) The successful low-temperature preservation of mouse and cow embryos. J Reprod Fertil 33(2):352–353
26. Willadsen SM, Polge C, Rowson LEA (1978) The viability of deep-frozen cow embryos. J Reprod Fertil 52(2):391–393
27. Bank H, Maurer RR (1974) Survival of frozen rabbit embryos. Exp Cell Res 89(1):188–196
28. Whittingham DG, Adams CE (1976) Low temperature preservation of rabbit embryos. J Reprod Fertil 47(2):269–274
29. Whittingham DG (1975) Survival of rat embryos after freezing and thawing. J Reprod Fertil 43(3):575–578

30. Willadsen SM, Polge C, Rowson LEA, Moor RM 1974 Preservation of sheep embryos in liquid nitrogen. Cryobiology 11:560
31. Bilton RJ, Moore NW (1977) Frozen storage of embryos of farm animals: progress and implications. Ciba Found Symp Jan 18–20 (52):203–219 Cited In: Ali J. Factors affecting ultra rapid vitrification and cryopreservation of embryos. The Australian National University, 1992
32. Yamamoto Y, Oguri N, Tsutsumi Y, Hachinohe Y (1982) Experiments in the freezing and storage of equine embryos. J Reprod Fertil Suppl 32:399–403
33. Pope CE, Pope VZ, Beck LR (1984) Live birth following cryopreservation and transfer of a baboon embryo. Fertil Steril 42(1):143–145
34. Summers PM, Shephard AM, Taylor CT, Hearn JP (1987) The effects of cryopreservation and transfer on embryonic development in the common marmoset monkey, *Callithrix jacchus*. J Reprod Fertil 79(1):241–250
35. Dresser BL, Gelwicks EJ, Wachs KB, Keller GL (1988) First successful transfer of cryopreserved feline (Felis catus) embryos resulting in live offspring. J Exp Zool 246 (2):180–186
36. Trounson A, Mohr L (1983) Human pregnancy following cryopreservation, thawing and transfer of an eight-cell embryo. Nature 305(5936):707–709
37. Zeilmaker GH, Alberda AT, van Gent I, Rijkmans CM, Drogendijk AC (1984) Two pregnancies following transfer of intact frozen–thawed embryos. Fertil Steril 42:293–296
38. Kasai M, Niwa K, Iritani A (1980) Survival of mouse embryos frozen and thawed rapidly. J Reprod Fertil 59(1):51–56
39. Wood MJ, Farrant J (1980) Preservation of mouse embryos by two-step freezing. Cryobiology 17(2):178–180
40. Massip A, Van der Zwalmen P, Leroy F (1984) Effect of stage of development on survival of mouse embryos frozen–thawed rapidly. Cryobiology 21(5):574–577
41. Bui-Xuan-Nguyen N, Heyman Y, Renard JP (1984) Direct freezing of cattle embryos after partial dehydration at room temperature. Theriogenology 22(4):389–399
42. Takeda T, Elsden RP, Seidel GE Jr. Use of sucrose during removal of cryoprotectants after thawing eight-cell mouse embryos. Theriogenology 1987;28(1):101–108.
43. Chupin D, De Reviers MM (1986) Quick freezing of rat embryos. Theriogenology 26 (2):157–166
44. Széll A, Shelton JN (1987) Osmotic and cryoprotective effects of glycerol-sucrose solutions on day-3 mouse embryos. J Reprod Fertil 80 (1):309–316
45. Mazur P, Miller RH, Leibo SP (1974) Survival of frozen-thawed bovine red cells as a function of the permeation of glycerol and sucrose. J Membr Biol 15(2):137–158
46. Shaw JM, Trounson AO (1989) Effect of dimethyl sulfoxide and protein concentration on the viability of two-cell mouse embryos frozen with a rapid freezing technique. Cryobiology 26(5):413–421
47. Gordts S, Roziers P, Campo R, Noto V (1990) Survival and pregnancy outcome after ultrarapid freezing of human embryos. Fertil Steril 53(3):469–472
48. Feichtinger W, Hochfellner C, Ferstl U (1991) Clinical experience with ultra-rapid freezing of embryos. Hum Reprod 6(5):735–736
49. Mazur P (1984) Freezing of living cells: mechanisms and implications. Am J Physiol 247(3 Pt 1):C125–C142 Review
50. Rall WF, Fahy GM (1985) Ice-free cryopreservation of mouse embryos at −196 degrees C by vitrification. Nature 313 (6003):573–575
51. Rall WF, Wood MJ, Kirby C, Whittingham DG (1987) Development of mouse embryos cryopreserved by vitrification. J Reprod Fertil 80(2):499–504
52. Kono T, Suzuki O, Tsunoda Y (1988) Cryopreservation of rat blastocysts by vitrification. Cryobiology 25(2):170–173
53. Kasai M, Komi JH, Takakamo A, Tsudera H, Sakurai T, Machida T (1990) A simple method for mouse embryo cryopreservation in a low toxicity vitrification solution, without appreciable loss of viability. J Reprod Fertil 89 (1):91–97
54. Ali J, Shelton JN (1993) Design of vitrification solutions for the cryopreservation of embryos. J Reprod Fertil 99(2):471–477
55. Ali J, Shelton JN (1993) Vitrification of preimplantation stages of mouse embryos. J Reprod Fertil 98(2):459–465
56. Ali J, Shelton JN (1993) Successful vitrification of day-6 sheep embryos. J Reprod Fertil 99(1):65–70
57. Ali J and Shelton JN Development of vitrification solutions. In: M. Tucker, J. Liebermann, (Eds.) Vitrification in assisted reproduction. A User's Manual and Troubleshooting Guide. Informa Healthcare Medical Books, UK, p45–63
58. Ali J (1996) Developmental competence of unipronuclear and triploid day-2 human

embryos after vitrification with VS14. Med Sci Res 24:377–378

59. Ali J, Bongso A, Ratnam SS (1995) Chromosomal analysis of day-2 human embryos vitrified with VS14. Med Sci Res 23:539–540
60. Martino A, Songsasen N, Leibo SP (1996) Development into blastocysts of bovine oocytes cryopreserved by ultrarapid cooling. Biol Reprod 54:1059–1069
61. Cha KY, Hong SW, Chung HM, Choi DH, Ko JJ, Yoon TK. 1999 Pregnancy and implantation from vitrified oocytes following In vitro fertilization (IVF) and In Vitro Culture (IVC). Proceedings of the 54th Annual Meeting of the American Society for Reproductive Medicine, Toronto, CA, September 25–30
62. Kuleshova L, Gianaroli L, Magli C, Ferraretti A, Trounson A (1999) Birth following vitrification of a small number of human oocytes: case report. Hum Reprod 14(12):3077–3079
63. Cha KY, Hong SW, Chung HM, Choi DH, Ko JJ, Yoon TK. 1999 Pregnancy and implantation from vitrified oocytes following In vitro fertilization (IVF) and In Vitro Culture (IVC). Proceedings of the 55th Annual Meeting of the American Society for Reproductive Medicine, Toronto, CA, September 25–30, Fertil Steril, p.S2
64. Hong SW, Chung HM, Lim JM, Ko JJ, Yoon TK, Yee B, Cha KY (1999) Improved human oocyte development after vitrification: a comparison of thawing methods. Fertil Steril 72(1):142–146
65. Chung HM, Hong SW, Lim JM, Lee SH, Cha WT, Ko JJ, Han SY, Choi DH, Cha KY (2000) In vitro blastocyst formation of human oocytes obtained from unstimulated and stimulated cycles after vitrification at various maturational stages. Fertil Steril 73(3):545–551
66. Kim TJ, Hong SW (2011) Successful live birth from vitrified oocytes after 5 years of cryopreservation. J Assist Reprod Genet 28(1):73–76. doi:10.1007/s10815-010-9487-9493
67. Kim TJ, Laufer LR, Hong SW (2010) Vitrification of oocytes produces high pregnancy rates when carried out in fertile women. Fertil Steril 93:467–474
68. Hong SW, Hyung MS, Chung HM et al (1999) Improved human oocyte development after vitrification: a comparison of thawing methods. Fertil Steril 72:142–146
69. Chen SU, Lien YR, Chen HF et al (2000) Open pulled straws for vitrification of mature mouse oocytes preserve patterns of meiotic spindles and chromosomes better than conventional straws. Hum Reprod 15:2598–2603
70. Chen SU, Lien YR, Chao KH et al (2000) Cryopreservation of mature human oocytes by vitrification with ethylene glycol in straws. Fertil Steril 74:804–808
71. Choi DH, Chung HM, Lim JM et al (2000) Pregnancy and delivery of healthy infants developed from vitrified blastocysts in an IVFET program. Fertil Steril 74:838–839
72. Chung HM, Seung WH, Hong MS et al (2000) In vitro blastocyst formation of human oocytes obtained from unstimulated and stimulated cycles after vitrification at various maturational stages. Fertil Steril 73:545–551
73. Yoon TK, Chung HM, Lim JM et al (2000) Pregnancy and delivery of healthy infants developed from vitrified oocytes in a stimulated in vitro fertilization-embryo transfer program. Fertil Steril 74:180–181
74. Yoon TK, Kim TJ, Park SE et al (2003) Live births after vitrification of oocytes in a stimulated in vitro fertilization-embryo transfer program. Fertil Steril 79:1323–1326
75. Yin H, Kim SS, Fisher J et al (2001) Investigation of optimal conditions for equilibrating ovarian tissue with ethylene glycol prior to vitrification. Fertil Steril 76(Suppl 1):S101 (abstr)
76. Kim TJ, Hong SW, Park SE et al (2003) Pregnancy after vitrification of human oocytes and blastocysts using same cryoprotectant solution, ethylene glycol, and sucrose. Fertil Steril 80(Suppl 3):143 abstr.
77. Kim T, Hong S, Cha K (2005) Pregnancies from cryopreserved oocytes using vitrification protocol. Fertil Steril 84(Suppl. 1):S179 (abstr)
78. Kim SH, Ku SY, Sung KC et al (2006) Simplified EM grid vitrification is a convenient and efficient method for mouse mature oocyte cryopreservation. Yonsei Med J 47:399–404
79. Kim TJ, Hong SW, Chung HM et al (2005) Pregnancy and delivery after vitrification of human oocytes. Fertil Steril 83(Suppl. 5): S13 (abstr)
80. Park SE, Chung HM, Cha KY et al (2001) Cryopreservation of ICR mouse oocytes: improved post-thawed preimplantation development after vitrification using Taxol, a cytoskeleton stabilizer. Fertil Steril 75:1177–1184
81. Park SE, Kim TJ, Hong SW et al (2003) Vitrification of human mature oocytes in a straw to prevent the risk of liquid nitrogen contamination during storage. Fertil Steril 80(Suppl. 3):64–65 (abstr)

82. Hong S, Kim T, Lee S et al (2005) Cryopreserved blastocysts using vitrification protocol give excellent pregnancy and implantation rates after thawing. Fertil Steril 84(Suppl.1): S178–S179 (abstr)
83. Martins RD, Costa EP, Chagas JSC et al (2005) Effects of vitrification of immature bovine oocytes on in vitro maturation. Anim Reprod 2:128–134
84. Papis K, Avery H, Holm P et al (1995) The effect of vitrification solution, equilibration time, and direct dilution method on survivability of equilibrated or vitrified bovine in vitro matured oocytes. Theriogenology 43:293
85. Papis K, Shimizu M, Izaike Y (2000) Factors affecting the survivability of bovine oocytes vitrified in droplets. Theriogenology 54 (5):651–658
86. El-Danasouri I, Selman HA (2001) Successful pregnancies and deliveries after a simple vitrification protocol for day 3 human embryos. Fertil Steril 76:400–402
87. Selman HA, El-Danasouri I (2002) Pregnancies derived from vitrified human zygotes. Fertil Steril 77:422–423
88. Rama Raju GA, Haranath GB, Krishna KM, Prakash GJ, Madan K (2005) Vitrification of human 8-cell embryos, a modified protocol for better pregnancy rates. Reprod Biomed Online 11(4):434–437
89. Rama Raju GA, Jaya Prakash G, Murali Krishna K, Madan K (2009) Neonatal outcome after vitrified day 3 embryo transfers: a preliminary study. Fertil Steril 92(1):143–148
90. Hafez ES (1969) Superovulation and preservation of mammalian egg. Acta Endocrinol Suppl 140:5–44
91. Sherman JK, Lin TP (1959) Temperature shock and cold-storage of unfertilized mouse eggs. Fertil Steril 10(4):384–396
92. Avrill RLW, Rowson LES (1959) Attempts at storage of sheep ova at low temperatures. J Agri Sci 52:392–395 Cited In: Loi P, Ptak G, Daattena M, Ledda S, Naitana S, Cappai P. Embryo transfer and related technologies in sheep reproduction. Reprod Nutr Dev 1998;38:615–628
93. Vincent C, Johnson MH (1992) Cooling, cryoprotectants, and the cytoskeleton of the mammalian oocyte. Oxf Rev Reprod Biol 14:73–100
94. Magistrini M, Szöllösi D (1980) Effects of cold and of isopropyl-N-phenylcarbamate on the second meiotic spindle of mouse oocytes. Eur J Cell Biol 22(2):699–707
95. Pickering SJ, Johnson MH (1987) The influence of cooling on the organization of the meiotic spindle of the mouse oocyte. Hum Reprod 2(3):207–216
96. Kola I, Kirby C, Shaw J, Davey A, Trounson A (1988) Vitrification of mouse oocytes results in aneuploid zygotes and malformed fetuses. Teratology 38(5):467–474
97. Glenister PH, Wood MJ, Kirby C, Whittingham DG (1987) Incidence of chromosome anomalies in first-cleavage mouse embryos obtained from frozen-thawed oocytes fertilized in vitro. Gamete Res 16(3):205–216
98. Van den Abbeel E, Van der Elst J, Van Waesberghe L, Camus M, Devroey P, Khan I, Smitz J, Staessen C, Wisanto A, Van Steirteghem A (1988) Hyperstimulation: the need for cryopreservation of embryos. Hum Reprod 3(Suppl 2):53–57
99. Al-Hasani S, Diedrich K, van der Ven H, Reinecke A, Hartje M, Krebs D (1987) Cryopreservation of human oocytes. Hum Reprod 2(8):695–700
100. Al-Hasani S, Diedrich K, van der Ven H, Krebs D. [Initial results of the cryopreservation of human oocytes]. Geburtshilfe Frauenheilkd. 1986;46(9):643–644. [Article in German]
101. Gook DA, Osborn SM, Bourne H, Johnston WI (1994) Fertilization of human oocytes following cryopreservation; normal karyotypes and absence of stray chromosomes. Hum Reprod 9(4):684–691
102. van Uem JF, Siebzehnrübl ER, Schuh B, Koch R, Trotnow S, Lang N (1987) Birth after cryopreservation of unfertilized oocytes. Lancet 1(8535):752–753
103. Chen C (1988) Pregnancies after human oocyte cryopreservation. Ann N Y Acad Sci 541:541–549
104. Porcu E, Fabbri R, Seracchioli R, Ciotti PM, Magrini O, Flamigni C (1997) Birth of a healthy female after intracytoplasmic sperm injection of cryopreserved human oocytes. Fertil Steril 68(4):724–726
105. Fabbri R, Porcu E, Marsella T, Primavera MR, Seracchioli R, Ciotti PM, Magrini O, Venturoli S, Flamigni C (1998) Oocyte cryopreservation. Hum Reprod 13(Suppl 4):98–108
106. Porcu E (1999) Freezing of oocytes. Curr Opin Obstet Gynecol 11(3):297–300 Review
107. Porcu E, Fabbri R, Ciotti PM, Petracchi S, Seracchioli R, Flamigni C (1999) Ongoing pregnancy after intracytoplasmic sperm injection of epididymal spermatozoa into cryopreserved human oocytes. J Assist Reprod Genet 16(5):283–285
108. Porcu E, Fabbri R, Petracchi S, Ciotti PM, Flamigni C (1999) Ongoing pregnancy

after intracytoplasmic injection of testicular spermatozoa into cryopreserved human oocytes. Am J Obstet Gynecol 180 (4):1044–1045

109. Nawroth F, Kissing K (1998) Pregnancy after intracytoplasmatic sperm injection (ICSI) of cryopreserved human oocytes. Acta Obstet Gynecol Scand 77(4):462–463

110. Polak de Fried E, Notrica J, Rubinstein M, Marazzi A, Gómez GM (1998) Pregnancy after human donor oocyte cryopreservation and thawing in association with intracytoplasmic sperm injection in a patient with ovarian failure. Fertil Steril 69(3):555–557

111. Young E, Kenny A, Puigdomenech E, Van Thillo G, Tiverón M, Piazza A (1998) Triplet pregnancy after intracytoplasmic sperm injection of cryopreserved oocytes: case report. Fertil Steril 70(2):360–361

112. Würfel W, Schleyer M, Krüsmann G, Hertwig IV, Fiedler K (1999) Fertilization of cryopreserved and thawed human oocytes (Cryo-Oo) by injection of spermatozoa (ICSI)–medical management of sterility and case report of a twin pregnancy. Zentralbl Gynakol 121(9):444–448 [Article in German]

113. Nakagata N (1989) High survival rate of unfertilized mouse oocytes after vitrification. J Reprod Fertil 87(2):479–483

114. Van Blerkom J (1989) Maturation at high frequency of germinal-vesicle-stage mouse oocytes after cryopreservation: alterations in cytoplasmic, nuclear, nucleolar and chromosomal structure and organization associated with vitrification. Hum Reprod 4(8):883–898

115. Ali J (2001) Source of vitrification solution. Fertil Steril 76:214–215

116. Ali J (2001) Vitrification of embryos and oocytes with 5.5 mol/l ethylene glycol and 1.0 mol/l sucrose. Hum Reprod 16:1777–1778

117. Dulioust E, Toyama K, Busnel MC, Moutier R, Carlier M, Marchaland C, Ducot B, Roubertoux P, Auroux M (1995) Vitrification of embryos and oocytes with 5.5 mol/l ethylene glycol and 1.0 mol/l sucrose. Proc Natl Acad Sci U S A 92(2):589–593

118. Vanderzwalmen, P. Bertin G., 2000 1-Ch. Debauche, 1 V. Standaart, 1E.Schoysman. "In Vitro" Survival of metaphase II oocytes (MII) and blastocysts after vitrification in a hemi-straw (HS) system. Proceedings of the 56th annual meeting of the American society for reproductive medicine, San Diego, CA, October 21–26 Fertil Steril 74: p. S215

119. Stoop D, De Munck N, Jansen E, Platteau P, Van den Abbeel E, Verheyen G, Devroey P (2012) Clinical validation of a closed vitrification system in an oocyte-donation programme. Reprod Biomed Online 24 (2):180–185

120. Boyer P, Montjean D, Tourame P, Gervoise-Boyer M (2013) Oocyte vitrification in an ART laboratory. Gynecol Obstet Fertil 41 (9):551–553

121. Chang CC, Elliott TA, Wright G, Shapiro DB, Toledo AA, Nagy ZP (2013) Prospective controlled study to evaluate laboratory and clinical outcomes of oocyte vitrification obtained in in vitro fertilization patients aged 30 to 39 years. Fertil Steril 99 (7):1891–1897

122. Chian RC, Kuwayama M, Tan L, Tan J, Kato O, Nagai T (2004) High survival rate of bovine oocytes matured in vitro following vitrification. J Reprod Dev 50(6):685–696

123. Zhang Z, Liu Y, Xing Q, Zhou P, Cao Y (2011) Cryopreservation of human failed-matured oocytes followed by in vitro maturation: vitrification is superior to the slow freezing method. Reprod Biol Endocrinol 9:156

124. Levi Setti PE, Porcu E, Patrizio P, Vigiliano V, de Luca R, d'Aloja P, Spoletini R, Scaravelli G (2014) Human oocyte cryopreservation with slow freezing versus vitrification. Results from the National Italian Registry data, 2007-2011. Fertil Steril 102(1):90–95

125. Noyes N, Porcu E, Borini A (2009) Over 900 oocyte cryopreservation babies born with no apparent increase in congenital anomalies. Reprod Biomed Online 18(6):769–776

126. Noyes N, Knopman J, Labella P, McCaffrey C, Clark-Williams M, Grifo J (2010) Oocyte cryopreservation outcomes including pre-cryopreservation and post-thaw meiotic spindle evaluation following slow cooling and vitrification of human oocytes. Fertil Steril 94(6):2078–2082

127. Paramanantham J, Talmor AJ, Osianlis T, Weston GC (2015) Cryopreserved oocytes: update on clinical applications and success rates. Obstet Gynecol Surv 70(2):97–114

128. Zhang L, Yan LY, Zhi X, Yan J, Qiao J (2015) Female fertility: is it safe to "Freeze?". Chin Med J (Engl) 128(3):390–397

129. Parrott DMV (1960) The fertility of mice with orthotopic ovarian grafts derived from frozen tissue. J Reprod Fertil 1:230–241

130. Wong KM, Mastenbroek S, Repping S (2014) Cryopreservation of human embryos and its contribution to in vitro fertilization success rates. Fertil Steril 102(1):19–26

131. Riha J, Landa V, Kneissl J, Matus J, Jindra J, Kloucek Z (1991) Vitrification of cattle embryos by direct dropping into liquid nitrogen and embryo survival after nonsurgical transfer. Zivoc Vir 36:113–120 Cited In: Vajta G, Kuwayama M. Improving cryopreservation systems. Theriogenology 65:236–244 (2006).

132. Elnahas A, Alcolak E, Abu Marar E, Elnahas T, Elnahas K, Palapelas V, Diedrich K, Al-Hasani S (2010) Vitrification of human oocytes and embryos of different development stages. J Reprod Stem Cell Biotechnol 1(2):181–192

# Part II

## Basics and Advanced Biology

# Chapter 2

# Utility of Animal Models for Human Ovarian Tissue Cryopreservation

**Seul Ki Kim, Jung Ryeol Lee, and S. Samuel Kim**

## Abstract

Success in cryopreservation of ovarian tissue (OT) in animal models has led to develop efficient cryotechnologies for human ovarian tissue. In this chapter, cryopreservation protocols developed for animal experiments are described.

**Key words** Ovarian tissue, Cryopreservation, Slow freezing, Vitrification, Fertility preservation

## 1 Introduction

Although the technology of cryopreservation of ovarian tissue (OT) is still at a relatively experimental stage, it seems to be a promising strategy for fertility preservation with encouraging results. Indeed, it is the only available option for fertility preservation in pre-pubertal girls. In the last decade, a wealth of information has been collected through studies using different animal models, and this has led to the development of new technologies and optimization of existing ones. Success in cryopreservation of OT in animal models has allowed its application in humans. Today, the results of those studies still continue to form the basis of OT cryobiology.

The most commonly used animal models in developing the principles of OT cryopreservation for application in humans are sheep and bovine models. Especially bovine ovaries have become useful due to the similarities in the ovary size and composition as well as length of folliculogenesis. Nevertheless, the rodent model has contributed more than any other animal models to scientific experiments. Although establishing the optimal cryopreservation protocol is one of the most important steps to improve OT survival and many laboratories have developed various protocols, cryopreservation of OT has not been fully optimized (especially in humans).

Zsolt Peter Nagy et al. (eds.), *Cryopreservation of Mammalian Gametes and Embryos: Methods and Protocols*, Methods in Molecular Biology, vol. 1568, DOI 10.1007/978-1-4939-6828-2_2, © Springer Science+Business Media LLC 2017

Cryopreservation of OT can be done by either slow freezing or vitrification. To date, the slow freezing method has been considered a standard technique for OT cryopreservation in clinical settings. Recently, vitrification of OT has been studied by multiple investigators; however, the results are still controversial. In theory, vitrification of OT should provide better results than those of slow freezing; however, it will take some time to perfect the vitrification technology for OT.

Cryoprotective agents (CPAs), by increasing the total concentration of solutes in the system, reduce the amount of ice formation at any given temperature; however, to be biologically acceptable, they must be able to penetrate into the cells and have low toxicity. CPAs commonly used in OT vitrification are dimethyl sulfoxide (DMSO), propanediol (PrOH), and ethylene glycol (EG) as permeating CPAs and sucrose, trehalose, and Ficoll as nonpermeating CPAs. Some studies suggested that the combination of two or more CPAs can reduce the toxicity and may be more effective than a single type of CPA. Nevertheless, the efficacy of different types and concentrations of CPA should be further investigated.

Here, we describe slow freezing and vitrification protocols for mouse, sheep, and bovine OTs.

## 2 Materials

### *2.1 Mouse Ovarian Tissue Cryopreservation [1, 2]*

From 4- to 7-week-old B6D2F1 female mice were housed under controlled conditions as follows: lighting (12-h light/dark cycle), temperature (20–22 °C), humidity (40–60 %) and fed ad libitum.

#### *2.1.1 Vitrification*

1. Basic medium (BM): 20 % fetal bovine serum (FBS) in Dulbecco's phosphate buffered saline (D-PBS).
2. Equilibration solution (ES): 7.5 % (v:v) Ethylene glycol (EG), 7.5 % (v:v) Dimethylsulfoxide (DMSO) in BM.
3. Vitrification solution (VS): 20 % (v:v) EG, 20 % (v:v) DMSO, 0.5 M sucrose in BM.
4. Warming solution: 1 M, 0.5 M, 0.25 M, and 0 M sucrose in BM.
5. Four-well dish.
6. Sterilized filter paper or gauze.
7. Liquid nitrogen (LN2).
8. Cryovial.
9. Cane.
10. Electron microscopic grid (IGC400)
11. Sterile hood.

#### 2.1.2 Slow Freezing

1. Basic medium (BM): 20 % FBS in D-PBS.
2. Slow freezing solution: 1.5 M DMSO, 0.1 M sucrose in BM.
3. Warming solution: 0.5 M, 0.25 M, 0 M sucrose in BM.
4. Four-well dish.
5. LN2.
6. Cryovial.
7. Forceps/cotton tips.
8. Sterile hood.
9. Slow freezing machine: Kryo 360, Cryologic etc.

### 2.2 Sheep Ovarian Tissue Cryopreservation [3–5]

Sheep ovaries immersed in cold (0–4 °C) D-PBS were transported from a local slaughterhouse to laboratory within 2 h on ice (*see* **Note 1**). Prepare the ovarian cortex into thin cortical slices (1 mm thick).

#### 2.2.1 Vitrification

1. Basic medium (BM): 20 % FBS in L-15 medium.
2. Equilibration solution (ES): 7.5 % (v:v) EG, 7.5 % (v:v) DMSO in BM.
3. Vitrification solution (VS): 20 % (v:v) EG, 20 % (v:v) DMSO, 0.5 M sucrose in BM.
4. Warming solution: 1 M, 0.5 M, 0.25 M, and 0 M Sucrose in BM.
5. Sterilized filter paper or gauze.
6. LN2.
7. Curved scissors.
8. Forceps.
9. 35 mm and 60 mm cell culture dishes.
10. Sterile hood.
11. Water bath.

#### 2.2.2 Slow Freezing

1. Basic medium (BM): 20 % FBS in L-15 medium.
2. Slow freezing solution: 1.5 M DMSO + 0.1 M sucrose in BM.
3. Thawing solution: TS1: 1.0 M DMSO + 0.1 M sucrose, TS2: 0.5 M DMSO + 0.1 M sucrose, TS3: 0.1 M sucrose
4. Sterilized filter paper or gauze.
5. LN2.
6. Cryovial.
7. Cane.
8. Curved scissors.
9. Forceps/cotton tips.
10. 35 mm and 60 mm cell culture dishes.

11. Sterile hood.
12. Water bath.
13. Slow freezing machine: Kryo 360, Cryologic etc.

### 2.3 Bovine Ovarian Tissue Cryopreservation [6, 7]

#### 2.3.1 Vitrification

1. Basic medium (BM): 20 % FBS in L-15 medium.
2. Equilibration solution (ES): 7.5 % (v:v) EG, 7.5 % (v:v) DMSO in BM.
3. Vitrification solution (VS): 20 % (v:v) EG, 20 % (v:v) DMSO, 0.5 M sucrose in BM.
4. Warming solution: 1 M, 0.5 M, 0.25 M, and 0 M Sucrose in BM.
5. Sterilized filter paper or gauze.
6. LN2.
7. Curved scissors.
8. Forceps.
9. 35 mm and 60 mm cell culture dishes.
10. Sterile hood.
11. Water bath.

#### 2.3.2 Slow Freezing

1. Basic medium (BM): 20 % FBS in L-15 medium.
2. Slow freezing solution: 1.5 M DMSO + 0.1 M sucrose in BM.
3. Thawing solution: TS1: 1.0 M DMSO + 0.1 M sucrose, TS2: 0.5 M DMSO + 0.1 M sucrose, TS3: 0.1 M sucrose
4. Sterilized filter paper or gauze.
5. LN2.
6. Cryovial.
7. Cane.
8. Curved scissors.
9. Forceps/cotton tips.
10. 35 mm and 60 mm cell culture dishes.
11. Sterile hood.
12. Water bath.
13. Slow freezing machine: Kryo 360, Cryologic etc.

## 3 Methods

Carry out all procedures at room temperature unless otherwise specified and use the sterile hood when processing.

### 3.1 Mouse Ovarian Tissue Cryopreservation [2]

The whole ovary of the BDF1 female mice (*see* **Note 2**).

#### 3.1.1 Vitrification of Mouse Ovarian Tissue

Vitrification procedure

1. Expose intact ovaries to the equilibration solution for 10 min.
2. Expose to a vitrification solution for 3–5 min (*see* **Note 3**).
3. Load each of the ovaries onto the EM grid and excess vitrification solution be removed using sterilized filter papers (*see* **Note 4**).
4. Immediately plunge the ovaries on the EM grids into LN2 for 30 s (*see* **Note 5**).
5. Place the vitrified ovaries attached to the EM grids into 1.5 mL cryovials filled with LN2 and store.

Warming procedure

1. Hold cryovials for 20 s in air at room temperature and then fill with warming solution (1.0 M sucrose in BM) at 37 °C for 30 s. Move the ovaries with the grids to the 4-well dishes and wash in warming solutions in a stepwise manner (0.5 M, 0.25 M, and 0 M sucrose in BM, 3 min each).
2. Further incubate the ovaries detached from the grids for 10 min in BM.

#### 3.1.2 Slow Freezing of Mouse Ovarian Tissue

Freezing procedure

1. After washing in the isolation medium, transfer ovarian tissue to the dish containing the slow freezing solution.
2. Place each ovary into the cryovial containing a small volume of slow freezing solution (*see* **Note 6**).
3. After 20 min exposure to the cryoprotectant at room temperature for equilibration, place the cryovials in a programmable freezer precooled to 4 °C.
4. Cool at 2 °C/min to −7 °C and hold for 5 min (*see* **Note 7**).
5. Induce an ice formation (seeding) manually by touching the cryovial at the level of the meniscus with cotton tip/precooled scissors/forceps. Hold 5 min after seeding before futher cooling.
6. Further cool down to −40 °C at a rate of −0.3 °C/min and then cool to −100 °C at −10 °C/min.
7. Plunge directly into LN2 for storage.

Warming procedure

1. Hold cryovials for 20 s in air at room temperature and place the cryovials in a water bath at 37 °C for 30–60 s.

2. Remove the cryoprotectant by serial dilutions of the cryoprotectants in medium stepwise, which could avoid rapid osmotic changes in the cryopreserved tissue.
3. After removing cryoprotectant, wash the thawed ovaries repeatedly in L-15 medium containing 10 % (v:v) FBS.

### 3.2 Sheep Ovarian Tissue Cryopreservation [8, 9]

#### 3.2.1 Vitrification of Sheep Ovarian Tissue

Vitrification procedure

1. Equilibrate prepared ovarian cortical tissue (5 × 5 × 1 mm) in equilibration solution for 15-min at room temperature.
2. After the equilibration period, transfer tissues into the vitrification solution and equilibrate for 10–15 min at room temperature.
3. Place the tissues on a piece of gauze to remove the residual vitrification medium.
4. Place each ovarian tissue in a cryovial.
5. Plunge the cryovial with ovarian tissue into LN2. And then, close the cover, and place in a LN2 tank for storage.

Warming procedure

1. Hold cryovials for 20 s in air at room temperature and fill with 37 °C pre-warming solution (1.0 M sucrose in BM) for 2 min.
2. Move the ovaries to the 35 mm dishes and wash in warming solutions for 3 min each in a stepwise manner (0.5 M, 0.25 M, and 0 M sucrose in BM).
3. Wash in BM (x2–3).

#### 3.2.2 Slow Freezing of Sheep Ovarian Tissue

Freezing procedure

1. After washing in the isolation medium, transfer ovarian tissue to the dish containing slow freezing solution. Place each section of ovarian tissue into the cryovial containing a small volume of slow freezing solution (1 ml).
2. After 30 min exposure to the cryoprotectant at room temperature for equilibration, place the cryovials into the programmable freezer precooled to 4 °C.
3. Cool at 2 °C/min to −7 °C and hold for 5 min.
4. Induce seeding manually by touching the cryovial at the level of the meniscus with precooled cotton tips/forceps. Hold 5 min after seeding before futher cooling.
5. Further cool down to −40 °C at a rate of −0.3 °C/min and then cool to −100 °C at −10 °C/min.
6. Plunge directly into LN2 for storage.

Warming procedure

1. Hold cryovials for 20–30 s in air at room temperature.
2. Then place into 37 °C water bath for 2–3 min and agitate vigorously.
3. Thaw stepwise using prepared thawing solutions (TS1: 1.0 mol/L DMSO with 0.1 mol/L sucrose, TS2: 0.5 mol/L DMSO with 0.1 mol/L sucrose, TS3: 0.1 mol/L sucrose), 3 min each.
4. Wash thawed tissue in L-15 medium (x2–3).

### 3.3 Bovine Ovarian Tissue Cryopreservation

#### 3.3.1 Vitrification of Bovine Ovarian Tissue

Vitrification procedure

1. Equilibrate prepared ovarian tissue (5 × 5 × 1mm cortex) in equilibration solution for 15-min at room temperature.
2. After equilibration period, transfer ovarian tissue into the vitrification solution and equilibrate for 10–15 min at room temperature.
3. Place ovarian tissue on a piece of gauze to remove the residual vitrification medium.
4. Place each ovarian tissue in a cryovial.
5. Plunge the cryovial with ovarian tissue into LN2. And then, close the cover and place in a LN2 tank for storage.

Warming procedure

1. Hold cryovials for 20 s in air at room temperature and then fill with 37 °C pre-warming solution (1.0 M sucrose in BM) for 2 min.
2. Move ovarian tissue to the 35 mm cell culture dishes and wash in warming solutions for 3 min each in a stepwise manner (0.5 M, 0.25 M, and 0 M sucrose in BM).
3. Wash in BM (x2–3).

#### 3.3.2 Slow Freezing of Bovine Ovarian Tissue

Freezing procedure

1. After washing in the isolation medium, transfer ovarian tissue to culture dishes containing slow freezing solution.
2. Place each section of ovarian tissue into the cryovial containing a small volume of slow freezing solution (1 ml).
3. After 30 min exposure to the cryoprotectant at room temperature for equilibration, place the cryovials in a programmable freezer precooled to 4 °C.
4. Cool at 2 °C/min to −7 °C and hold for 5 min.
5. Induce seeding manually by touching the cryovial at the level of the meniscus with precooled cotton tips/forceps and hold 5 min more.

6. Further cool down to −40 °C at a rate of −0.3 °C/min and then cool to −100 °C at −10 °C/min.
7. Plunge directly into LN2 for storage.

Warming procedure

1. Hold cryovials for 20 s in air at room temperature.
2. Place the cryovials in a water bath at 37 °C for 2–3 min and agitate vigorously.
3. Wash tissue stepwise using prepared thawing solutions for 3 min each (TS1: 1.0 mol/L DMSO with 0.1 mol/L sucrose, TS2: 0.5 mol/L DMSO with 0.1 mol/L sucrose, TS3: 0.1 mol/L sucrose).
4. Wash thawed ovarian tissue (x2–3) in L-15 medium.

## 4 Notes

1. The effect of transport temperature on ovaries is unclear and different temperatures (0–4 °C, 22–25 °C and 35–38 °C) are used. And D-PBS or 0.9 % saline can be used as a storage solution during transportation. We prefer transporting tissue at 4 °C in D-PBS or L-15 solution.
2. Because of the small dimension (~2 × 2 × 2 $mm^3$) of the mouse ovary, we vitrified whole organs instead of ovarian slices.
3. Keep ES and VS vials at the room temperature (25 °C–27 °C) at least 1 hour before vitrification.
4. Only cryovials or cryotubes without EM grids can be used. However, EM grids offer good performance for cooling due to their very high thermal conductivity.
5. The transparent glassy appearance should be maintained during cooling and warming.
6. Various kinds of carriers are being used (plastic straw, cryovial, etc.).
7. Different cooling rates are used depending on the laboratory.

### References

1. Kim GA, Kim HY, Kim JW, Lee G, Lee E, Ahn JY, Park JH, Lim JM (2011) Effectiveness of slow freezing and vitrification for long-term preservation of mouse ovarian tissue. Theriogenology 75:1045–1051
2. Youm HW, Lee JR, Lee J, Jee BC, Suh CS, Kim SH (2014) Optimal vitrification protocol for mouse ovarian tissue cryopreservation: effect of cryoprotective agents and in vitro culture on vitrified-warmed ovarian tissue survival. Hum Reprod 29:720–730
3. Bordes A, Lornage J, Demirci B, Franck M, Courbiere B, Guerin JF, Salle B (2005) Normal gestations and live births after orthotopic autograft of vitrified-warmed hemi-ovaries into ewes. Hum Reprod 20:2745–2748
4. Oskam IC, Asadi BA, Santos RR (2010) Histologic and ultrastructural features of

cryopreserved ovine ovarian tissue: deleterious effect of 1,2-propanediol applying different thawing protocols. Fertil Steril 93:2764–2766

5. Santos RR, Rodrigues AP, Costa SH, Silva JR, Matos MH, Lucci CM, Bao SN, van den Hurk R, Figueiredo JR (2006) Histological and ultrastructural analysis of cryopreserved sheep preantral follicles. Anim Reprod Sci 91:249–263
6. Kim SS, Yang HW, Kang HG, Lee HH, Lee HC, Ko DS, Gosden RG (2004) Quantitative assessment of ischemic tissue damage in ovarian cortical tissue with or without antioxidant (ascorbic acid) treatment. Fertil Steril 82:679–685
7. Wang YS, Zhao X, Su JM, An ZX, Xiong XR, Wang LJ, Liu J, Quan FS, Hua S, Zhang Y (2011) Lowering storage temperature during ovary transport is beneficial to the developmental competence of bovine oocytes used for somatic cell nuclear transfer. Anim Reprod Sci 124:48–54
8. Bedaiwy MA, Jeremias E, Gurunluoglu R, Hussein MR, Siemianow M, Biscotti C, Falcone T (2003) Restoration of ovarian function after autotransplantation of intact frozen-thawed sheep ovaries with microvascular anastomosis. Fertil Steril 79:594–602
9. Maffei S, Pennarossa G, Brevini TA, Arav A, Gandolfi F (2014) Beneficial effect of directional freezing on in vitro viability of cryopreserved sheep whole ovaries and ovarian cortical slices. Hum Reprod 29:114–124

# Chapter 3

# Current Challenges in Immature Oocyte Cryopreservation


## Ri-Cheng Chian, Yixin Xu, and Dana Keilty



## Abstract

Current freezing technology, especially the vitrification method, has markedly improved oocyte survival rate after warming, and the pregnancy rate is comparable to that achieved with fresh oocytes. However, most groups report using oocytes matured in vivo for vitrification. Although immature oocytes can be vitrified successfully, clinical outcomes do not reach that of vitrification of matured oocytes. The current literature suggests that oocytes should be vitrified at mature metaphase II (M-II) stage following IVM rather than at the immature germinal vesicle (GV) stage, because the potential for oocyte maturation is reduced when vitrification is performed on immature oocytes at the GV stage.



**Key words** Immature oocytes, Metaphase II stage, GV stage, Cryopreservation, Vitrification


## 1 Introduction

Initial attempts to freeze oocytes employed the same slow-freezing methods that were considered the gold standard for embryo cryopreservation. However, the slow-freezing oocytes were met with very low survival and pregnancy rates [1–3], and this intervention was thus considered experimental [4]. With time, the efficacy of the slow freezing of oocytes has been improved by increasing the sucrose concentration in the freezing solution [5–8]. With the recent advancement of the vitrification technique, there has been a significant improvement in the efficacy of oocyte cryopreservation [9–18].

A recent guideline issued by the Practice Committees of the American Society for Reproductive Medicine (ASRM) and the Society for Assisted Reproductive Technology (SART) indicated that mature oocyte vitrification and warming should no longer be considered only an experimental procedure [19]: this technology is recommended in cases of gonadotoxic therapies when there is a lack of alternative options for fertility preservation. Several groups worldwide have reported high survival and pregnancy rates as well as high live birth rates using vitrified oocytes [20–27]. However, it


Zsolt Peter Nagy et al. (eds.), *Cryopreservation of Mammalian Gametes and Embryos: Methods and Protocols*, Methods in Molecular Biology, vol. 1568, DOI 10.1007/978-1-4939-6828-2_3, © Springer Science+Business Media LLC 2017

appears that the majority of the live births were from egg donor programs [28] rather than from a clinic-specified population seeking to use this technology for fertility cryopreservation.

In addition, to date, most of the literature analyzes oocyte cryopreservation in the context of ovarian stimulation for in vitro fertilization (IVF) treatment. Ovarian stimulation with gonadotropins may not be suitable for many women who are seeking fertility preservation, especially patients with breast cancer and other hormone-dependent cancers, and those who require immediate chemotherapy. Immature oocyte retrieval followed by in vitro maturation (IVM) from unstimulated ovaries and then cryopreservation of those of in vitro matured oocytes is indeed an attractive strategy for these women. Immature oocyte retrieval followed by IVM has proven effective, with this treatment resulting in pregnancies and many healthy live births [29].

## 2 Cooling and Warming Rates for Oocyte Cryopreservation

The cryopreservation protocols for oocytes can be divided into two categories: slow freezing/rapid thawing (cooling rates of 0.3–2 °C/min) and rapid cooling/warming or vitrification (it was previously believed that the cooling rate should be at least 20,000 °C/min) [30]. Freezing induces the precipitation of water into ice, leading to the separation of water from the dissolved substances. The presence of both intracellular ice crystals and a high concentration of solute can be lethal to the oocytes or embryos during the freezing procedure. Therefore, a slow cooling rate has been proposed in order to minimize the damage from ice crystallization, osmosis, and chilling [31].

The typical slow-freezing method uses a cooling rate of 1 °C/min from −5 to −9 °C. Ice crystallization is then induced by a process known as "seeding," which results in a heterogeneous ice nucleation that is more stable than a supercooling or homogenous nucleation. After seeding, the cooling rate is reduced to 0.3–0.5 °C/min until a lower temperature is reached (usually between −30 and −150 °C), and then the cells are stored in liquid nitrogen (LN2). Thawing is done rapidly, with temperature changes that can exceed 360 °C/min [32]. This procedure prevents the occurrence of recrystallization, where water enters the oocytes or embryos and transforms into a solid state around previously formed small ice crystals.

Alternatively, rapid freezing/thawing techniques can be used to prevent the formation of ice crystals by transforming the oocytes or embryos into a vitreous- or glasslike state [33]. While this approach improves the viability of the cells, a high concentration of cryoprotectant is required in order to prevent ice crystal formation [34].

"Vitrification" refers to a glasslike solidification of a solution at a low temperature, importantly the *ice-free* solidification of an aqueous solution: living cells can be successfully frozen because they are cooled so rapidly that ice crystallization does not occur [34]. Strictly speaking, a glasslike state also occurs when slow-freezing procedures are applied at the glass transition temperature (−130 °C); to avoid the intra- and extracellular ice formation at this temperature, vitrification is done using a high concentration of cryoprotectant and extremely rapid cooling/warming rates so that the oocytes or embryos move quickly through the glass transition temperature [35–39].

The majority of experts believe that the cooling rate is the most critical factor for successful oocyte vitrification. A strategy of direct contact of various devices with $LN_2$ has thus been adopted, which has been able to achieve a cooling rate of >20,000 °C/min [30]. However, many concerns about this approach have been raised [40–44]. In addition to the contamination risk from direct contact, the necessity of direct contact with LN2 in order to achieve the essential cooling rate for vitrification continues to be debated.

Until recently, the optimal cooling and warming rates with a very small amount (<1.0 μL) of solution and samples were unknown [45]. Mouse models indicate that a slow-warming process proves lethal as it allows the growth of small intracellular ice crystals by recrystallization [45, 46], indicating that the cooling rate is of less consequence than the warming rate during vitrification-warming procedures [47]. Interestingly, the same authors have also demonstrated that a very high warming rate, as opposed to a high cooling rate, was the essential element to survival of the oocyte in this method [48]. The data also indicated that oocytes could be vitrified successfully with the JY Straw vitrification system (Fig. 1), even with relatively slow cooling (Fig. 2) and warming (Fig. 3) rates. These approaches have yielded high survival and pregnancy rates [27, 49], and so the necessity of direct contact with $LN_2$ to achieve the optimal cooling rate requires further verification. The literature suggests that the optimal cooling and warming rates with a very small amount (<1.0 μL) of solution and samples do not have to reach >20,000 °C/min, indicating that further understanding of the principles of vitrification is required in the field.

## 3 Cryopreservation of Immature Oocytes

With the conventional slow-freezing method, an alternative strategy to avoid spindle depolymerization is to cryopreserve immature germinal vesicle (GV) stage oocytes. These immature oocytes are arrested at the diplotene state of prophase I. Theoretically, the use

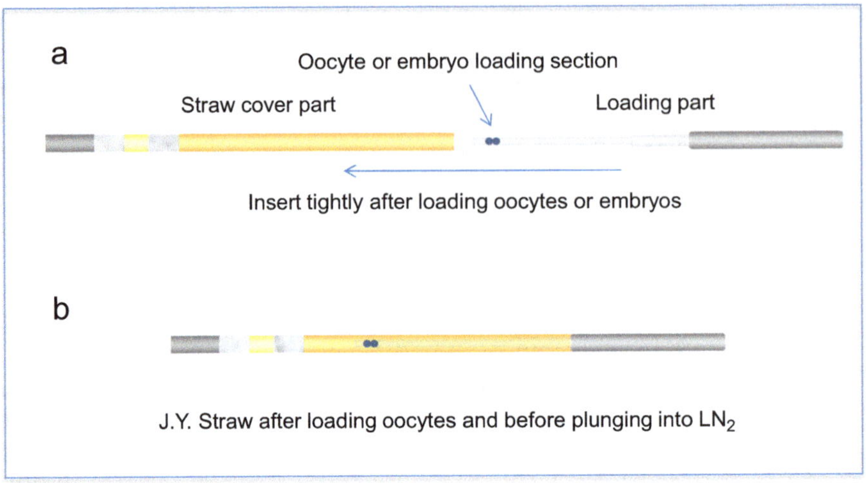


**Fig. 1** Structure of JY Straw vitrification system. (**a**) JY Straw has two parts: an oocyte loading part and a straw cover part; (**b**) after fitting the straw cover tightly, JY Straw is gently plunged in LN2. This device does not allow the loading part to contact LN2 directly. As there is space between the loading part and the straw cover, the cooling rate can be expected to be reduced compared to a direct-contact $LN_2$ method. Reproduced from Wang et al. [49] with modification

of immature GV stage oocytes circumvents the risk of polyploidy and aneuploidies, because the chromatin is diffuse and surrounded by a nuclear membrane [50, 51]. Although the oocyte survival rate seems improved, the maturation and fertilization rates are poor, and embryonic development is stunted in immature oocyte freezing [52–54] by the slow-freezing method. So far, only a single live birth has been reported following the slow freezing of immature oocytes at the GV stage [55]. Therefore, cryopreservation of mature oocytes (M-II stage) using the slow-freezing method seems more efficient than immature oocytes. Recently, the modified slow-freezing methods have been introduced for cryopreservation of oocytes [5, 6, 56–66]. Although the survival and fertilization rates seem improved, the clinical outcomes need to be confirmed.

There is limited information on the oocyte survival, fertilization rates, and early embryonic development after vitrification at the immature GV stage. While survival rates have not proven significantly different between oocytes vitrified at the GV and mature metaphase II (M-II) stages (Fig. 4) [67, 68], oocyte maturation rates were found to be significantly reduced in oocytes vitrified at the immature GV stage followed by IVM compared to those cultured without vitrification (Fig. 5) [67, 68]. However, following insemination by ICSI, there were no significant differences in fertilization, cleavage, or blastocyst development rates between these two groups. These results suggest that vitrification of mature oocytes is more efficient than immature oocytes.

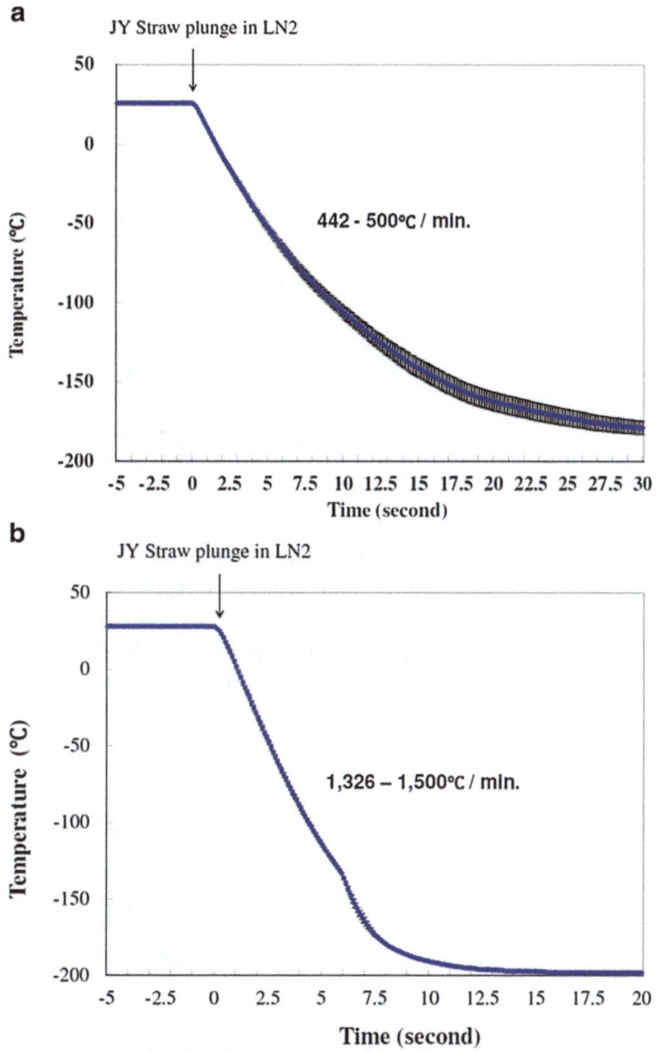


**Fig. 2** Cooling curve of JY Straw for oocyte vitrification. (**a**) The cooling rate of the interior of JY Straw after plunging into LN2. The temperature drop from 25 °C to −196 °C required approximately 26–30 s, and thus a cooling rate of approximately 442–500 °C/min was achieved; (**b**) in contrast, the cooling rate was approximately 1326–1500 °C/min when the loading part was inserted into LN2 with direct contact, i.e., without the straw cover. Reproduced from Wang et al. [49] with permission

## 4 Vitrification of In Vitro Matured Oocytes

Immature oocyte retrieval without preceding gonadotropin stimulation to the ovaries, followed by IVM culture and oocyte fertilization, is a safe and effective treatment option for a large population of infertile patients and has yielded clinical pregnancy rates of 35 % [69–71], which are comparable to those achieved with standard

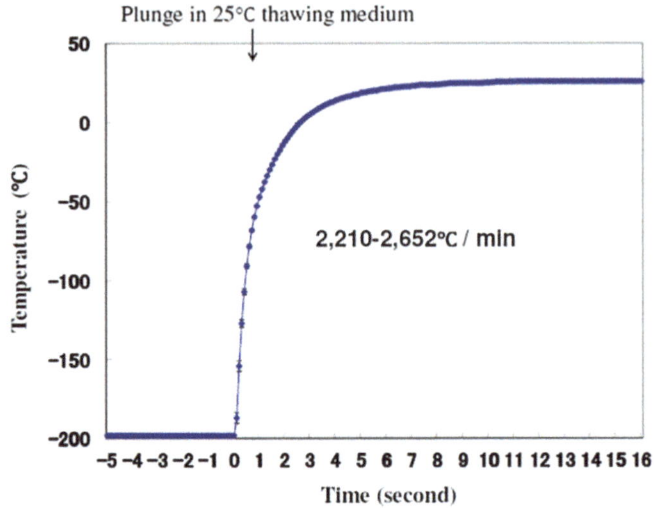


**Fig. 3** Warming curve of JY Straw for oocyte thawing. The loading part of JY Straw was inserted directly into 25 °C thawing solution. An increase in temperature from −196 °C to 25 °C required approximately 4–5 s, and thus a warming rate of approximately 2210–2652 °C/min was achieved. Reproduced from Wang et al. [49] with permission

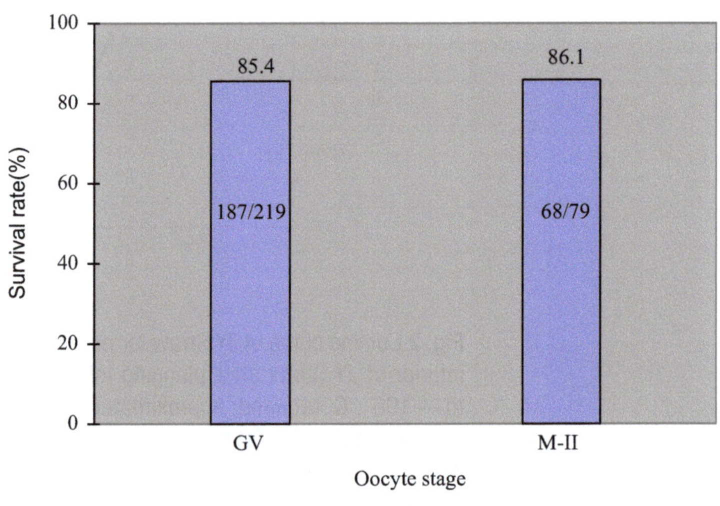


**Fig. 4** Survival rates for human oocytes vitrified at immature GV stage and at mature M-II stage. A total of 298 oocytes (219 immature oocytes and 79 mature oocytes) were vitrified and thawed. There were no differences in the survival rates between the two groups. Reproduced from Cao et al. [67] with permission

IVF treatment cycles reported by several national registries [72–74]. It also has been recently demonstrated that babies born via IVM do not show increased fetal abnormality rates compared to those born via regular IVF or spontaneous conceptions in fertile women [75, 76]. Immature oocyte retrieval without ovarian

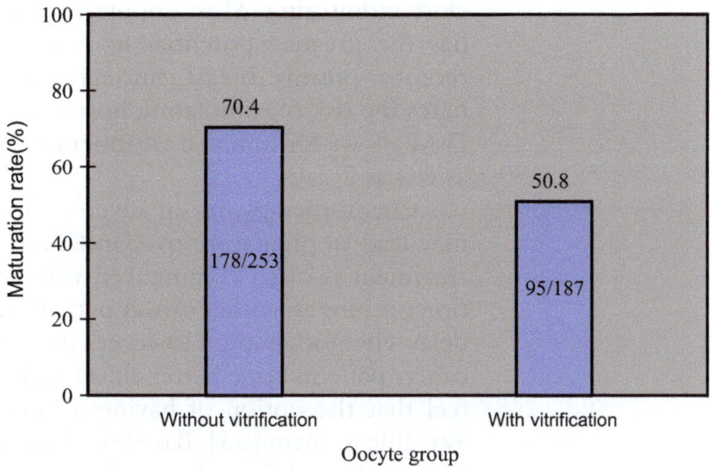


**Fig. 5** In vitro maturation rates of immature human oocytes with or without vitrification for IVM. A total of 440 immature oocytes (253 immature oocytes without vitrification and 187 immature oocytes with vitrification) were cultured in vitro for maturation. There was a significant difference ($P < 0.05$) between these two groups. Reproduced from Cao et al. [67] with permission

stimulation allows the avoidance of expensive medications and ovarian hyperstimulation syndrome (OHSS). The incidence of severe OHSS is 0.6–1.9 % but may be as high as 6 % in high-risk groups, such as young women with polycystic ovaries [77]. Although a number of strategies have been proposed to predict and prevent this potentially life-threatening complication, the only preventive strategy that is certain to eliminate OHSS is the avoidance of ovarian stimulation altogether.

At present, there are limited options for fertility preservation applicable to young cancer patients, many of whom do not have a male partner, must start chemotherapy quickly, and are concerned about the long-term effects of gonadotropin stimulation. A standard ovarian stimulation treatment protocol includes a 4–6-week delay to allow downregulation of the pituitary, followed by gonadotropin stimulation and oocyte retrieval [78]. Even in short protocols, a delay of at least 2–3 weeks is suggested. The IVM oocyte cryopreservation protocol, on the other hand, takes only 2–10 days. Women may also wish to avoid repeated gonadotropin stimulation because of the substantial cost of the drugs and the long-term effects of repeated doses of gonadotropins on the risk of developing ovarian, endometrial, and breast cancers [79, 80].

Although it has been demonstrated that healthy live births can be achieved from the combination of IVM and oocyte vitrification [81], vitrification of in vitro matured oocytes is currently less effective than vitrification of in vivo matured oocytes. However, the combination of IVM and oocyte vitrification has several advantages, including the elimination of the high cost of drugs and the need for close monitoring. Most importantly, this novel treatment protocol has the greatest potential to help cancer patients. In estrogen receptor-positive breast cancer patients, the IVM protocol eliminates the risk of stimulating hormone-sensitive tumors. Moreover, IVM allows for minimal treatment delay and may be completed in as few as 2 days.

Chemotherapy has an adverse effect on ovarian reserve, which may lead to premature ovarian failure and infertility [82]. Cancer treatment is often commenced without pursing fertility preservation options as neither cancer patients nor their oncologists wish to delay chemotherapy. However, there is evidence to suggest that cancer patients cope better emotionally with their treatment if they feel that the option of having a biological child in the future is available to them [83]. Based on the results of a series of healthy live births achieved from the combination of IVM and oocyte vitrification, this fertility-preserving approach should be considered for some cancer patients before chemotherapeutic treatment [84].

The application of vitrification techniques to human oocytes has been questioned due to the lack of basic biological studies addressing its effects on oocyte viability [4]. Indeed, reviewing current vitrification protocols found that relatively high concentrations of cryoprotectant have been used to vitrify oocytes and the cytotoxicity of cryoprotectant and osmotic changes of vitrification solutions are the main concern [85]. The clinical efficiency of, and safety issues surrounding, vitrified oocytes cannot be precisely assessed because of the lack of well-controlled clinical trials [83]. While it is important to assess the viability and any abnormalities of vitrified oocytes using embryological parameters [86], the most important tool for evaluation is the rate of healthy live births from vitrified oocytes [87].

## 5 Expert Commentary

There are no differences in the survival rates between oocytes vitrified at the immature GV stage and at the mature M-II stage. However, the potential for oocyte maturation is reduced in immature oocytes vitrified at the GV stage. This suggests that oocytes should be vitrified at mature metaphase II (M-II) stage following IVM rather than at the immature GV stage.

## 6 Key Issues

Vitrification of in vitro matured oocytes represents a novel option for fertility preservation. The most important issue remains the evaluation of healthy live births produced from in vitro matured oocytes followed by vitrification in the future.

## References


1. Tucker M, Wright G, Morton P et al (1996) Preliminary experience with human oocyte cryopreservation using 1,2-propanediol and sucrose. Hum Reprod 11:1513–1515
2. Porcu E, Fabbri R, Seracchioli R et al (1997) Birth of a healthy female after intracytoplasmic sperm injection of cryopreserved human oocytes. Fertil Steril 68:724–726
3. Tucker MJ, Morton PC, Wright G et al (1998) Clinical application of human egg cryopreservation. Hum Reprod 13:3156–3159
4. Gook DA, Edgar DH (2007) Human oocyte cryopreservation. Hum Reprod Update 13:591–605
5. Borini A, Bonu MA, Coticchio G et al (2004) Pregnancies and births after oocyte cryopreservation. Fertil Steril 82:601–605
6. Borini A, Sciajno R, Bianchi V et al (2006) Clinical outcome of oocyte cryopreservation after slow cooling with a protocol utilizing a high sucrose concentration. Hum Reprod 21:512–517
7. Bianchi V, Lappi M, Bonu MA, Borini A (2012) Oocyte slow freezing using a 0.2–0.3 M sucrose concentration protocol: is it really the time to trash the cryopreservation machine? Fertil Steril 97:1101–1107
8. Edgar DH, Gook DA (2012) A critical appraisal of cryopreservation (slow cooling versus vitrification) of human oocytes and embryos. Hum Reprod Update 18:536–554
9. Yoon TK, Kim TJ, Park SE et al (2003) Live births after vitrification of oocytes in a stimulated in vitro fertilization-embryo transfer program. Fertil Steril 79:1323–1326
10. Katayama KP, Stehlik J, Kuwayama M et al (2003) High survival rate of vitrified human oocytes results in clinical pregnancy. Fertil Steril 80:223–224
11. Chian RC, Son WY, Huang JYJ et al (2005) High survival rates and pregnancies of human oocytes following vitrification: preliminary report. Fertil Steril 84(Suppl. 1):S36
12. Kim TJ, Hong SW, Chung HM et al (2005) Pregnancy and delivery after vitrification of human oocytes. Fertil Steril 83(Suppl. 1):S13
13. Kim T, Hong S, Cha K (2005) Pregnancies from cryopreserved oocytes using vitrification protocol. Fdrtil Steril 84(Suppl. 1):S179
14. Okimura T, Kato K, Zhan Q et al (2005) Update on clinical efficiency of the vitrification method for human oocytes in an in vitro fertilization program. Fertil Steril 84(Suppl 1):S174
15. Lucena E, Bernal DP, Lucena C et al (2006) Successful ongoing pregnancies after vitrification of oocytes. Fertil Steril 85:108–111
16. Chian RC, Huang JYJ, Tan SL et al (2008) Obstetric and perinatal outcome in 200 infants conceived from vitrified oocytes. Reprod Biomed Online 16:608–610
17. Chian RC, Gilbert L, Huang JYJ et al (2009) Live birth after vitrification of in vitro matured human oocytes. Fertil Steril 91:372–376
18. Chian RC, Gilbert L, Huang JYJ et al (2009) Obstetrical outcomes following vitrification of oocytes matured in-vivo or in-vitro. Fertil Steril 91:2391–2398
19. The Practice Committees of the American Society for Reproductive Medicine and the Society for Assisted Reproductive Technology (2013) Mature oocyte cryopreservation: a guideline. Fertil Steril 99:37–43
20. Cobo A, Kuwayama M, Perez S et al (2008) Comparison of concomitant outcome achieved with fresh and cryopreserved donor oocytes vitrified by the Cryotop method. Fertil Steril 89:1657–1664
21. Cobo A, Meseguer M, Remohí J, Pellicer A (2010) Use of cryo-banked oocytes in an ovum donation programme: a prospective, randomized, controlled, clinical trial. Hum Reprod 25:2239–2246
22. Cobo A, Remohí J, Chang CC, Nagy ZP (2011) Oocyte cryopreservation for donor egg banking. Reprod Biomed Online 23:341–346
23. Cobo A, Diaz C (2011) Clinical application of oocyte vitrification: a systematic review and meta-analysis of randomized controlled trials. Fertil Steril 96:277–285
24. Smith GD, Serafini PC, Fioravanti J et al (2010) Prospective randomized comparison of human oocyte cryopreservation with slow-rate freezing or vitrification. Fertil Steril 94:2088–2095
25. Parmegiani L, Cognigni GE, Bernardi S et al (2011) Efficiency of aseptic open vitrification and hermetical cryostorage of human oocytes. Reprod Biomed Online 23:505–512
26. Rienzi L, Cobo A, Paffoni A et al (2012) Consistent and predictable delivery rates after oocyte vitrification: an observational longitudinal cohort multicentric study. Hum Reprod 27:1606–1612
27. Cai LB, Qian XQ, Wang W et al (2012) Oocyte vitrification technology has made egg-sharing donation easier in China. Reprod Biomed Online 24:186–190

28. Quaas AM, Melamed A, Chung K et al (2013) Egg banking in the United States: current status of commercially available cryopreserved oocytes. Fertil Steril 99:827–831
29. Chian RC, Lim JH, Tan SL (2004) State of the art in in-vitro oocyte maturation. Curr Opin Obstet Gynecol 16:211–219
30. Kuwayama M, Vajta G, Kato O, Leibo SP (2005) Highly efficient vitrification method for cryopreservation of human oocytes. Reprod Biomed Online 11:300–308
31. Mazur P (1990) Equilibrium, quasi-equilibrium, and nonequilibrium freezing of mammalian embryos. Cell Biophys 17:53–92
32. Hunter JE (1995) Cryopreservation of mammalian embryos. Slow cooling. Methods Mol Biol 38:199–209
33. Rall WF, Fahy GM (1985) Ice-free cryopreservation of mouse embryos at −196 °C by vitrification. Nature 313:573–575
34. Kasai M, Zhu SE, Pedro PB et al (1996) Fracture damage of embryos and its prevention during vitrification and warming. Cryobiology 33:459–464
35. MacFarlane DR, Forsyth M (1990) Resent insights on the role of the cryoprotection against in vitrification. Cryobiology 27:345–358
36. Nakagata N (1989) High survival rate of unfertilized mouse oocytes after vitrification. J Reprod Fertil 87:479–483
37. Pensis M, Loumaye E, Psalti I (1989) Screening of conditions for rapid freezing of human oocytes: preliminary study toward their cryopreservation. Fertil Steril 52:787–794
38. Kono T, Kwon OY, Nakahara T (1991) Development of vitrified mouse oocytes after in vitro fertilization. Cryobiology 28:50–54
39. Shaw PW, Bernard AG, Fuller BJ et al (1992) Vitrification of mouse oocytes using short cryoprotectant exposure: effects of varying exposure times on survival. Mol Reprod Dev 33:210–214
40. Chian RC, Kuwayama M, Tan L et al (2004) High survival rate of bovine oocytes matured in vitro following vitrification. J Reprod Dev 50:685–696
41. Bielanski A, Nadin-Davis S, Sapp T, Lutze-Wallace C (2000) Viral contamination of embryos cryopreserved in liquid nitrogen. Cryobiology 40:110–116
42. Bielanski A, Bergeron H, Lau PC, Devenish J (2003) Microbial contamination of embryos and semen during long term banking in liquid nitrogen. Cryobiology 46:146–152
43. Bielanski A, Vajta G (2009) Risk of contamination of germplasm during cryopreservation and cryobanking in IVF units. Hum Reprod 24:2457–2467
44. Cobo A, Romero JL, Pérez S et al (2010) Storage of human oocytes in the vapor phase of nitrogen. Fertil Steril 94:1903–1907
45. Leibo SP, Pool TB (2011) The principal variables of cryopreservation: solutions, temperatures, and rate changes. Fertil Steril 96:269–276
46. Seki S, Mazur P (2008) Effect of warming rate on the survival of vitrified mouse oocytes and on the recrystallization of intracellular ice. Biol Reprod 79:727–737
47. Seki S, Mazur P (2009) The dominance of warming rate over cooling rate in the survival of mouse oocytes subjected to a vitrification procedure. Cryobiology 59:75–82
48. Seki S, Mazur P (2012) Ultra-rapid warming yields high survival of mouse oocytes cooled to −196 °C in dilutions of a standard vitrification solution. PLoS One 7:6058
49. Wang Y, Okitsu O, Zhao XM et al (2014) The effect of minimal concentration of ethylene glycol (EG) combined with polyvinylpyrrolidone (PVP) on mouse oocyte survival and subsequent embryonic development following vitrification. J Assist Reprod Genet 31:55–63
50. Cooper A, Paynter SJ, Fuller BJ et al (1998) Differential effects of cryopreservation on nuclear or cytoplasmic maturation in vitro in immature mouse oocytes from stimulated ovaries. Hum Reprod 13:971–978
51. Isachenko EF, Nayudu PL (1999) Vitrification of mouse germinal vesicle oocytes: effect of treatment temperature and egg yolk on chromosomal normality and cumulus integrity. Hum Reprod 14:400–408
52. Toth TL, Lanzendorf SE, Sandow BA et al (1994) Cryopreservation of human prophase I oocytes collected from unstimulated follicles. Fertil Steril 61:1077–1082
53. Toth TL, Baka SG, Veeck LL et al (1994) Fertilization and in vitro development of cryopreserved human prophase I oocytes. Fertil Steril 61:891–894
54. Son WY, Park SE, Lee KA et al (1996) Effects of 1,2-propanediol and freezing-thawing on the in vitro developmental capacity of human immature oocytes. Fertil Steril 66:995–999
55. Tucker MJ, Wright G, Morton PC et al (1998) Birth after cryopreservation of immature oocytes with subsequent in vitro maturation. Fertil Steril 70:578–579
56. Yang DS, Blohm PL, Winslow KL et al (1998) A twin pregnancy after microinjection of human cryopreserved oocyte with a specially

developed oocyte cryopreservation regime. Fertil Steril 70:S86

57. Stachecki JJ, Cohen J, Willadsen SM (1998) Cryopreservation of unfertilized oocytes: the effect of replacing sodium with Choline in the freezing medium. Cryobiology 37:346–355
58. Fabbri R, PorcuE MT et al (2001) Human oocyte cryopreservation: new perspectives regarding oocyte survival. Hum Reprod 16:411–416
59. Quintans CJ, Donaldson MJ, Bertolino MV et al (2002) Birth of two babies using oocytes that were cryopreserved in a choline-based freezing medium. Hum Reprod 17:3149–3152
60. Boldt J, Cline D, McLaughlin D (2003) Human oocyte cryopreservation as an adjunct to IVF-embryo transfer cycles. Hum Reprod 18:1250–1255
61. Boldt J, Tidswell N, Sayers A et al (2006) Human oocyte cryopreservation: 5-year experience with a sodium-depleted slow freezing method. Reprod Biomed Online 13:96–100
62. Stachecki JJ, Cohen J (2004) An overview of oocyte cryopreservation. Reprod Biomed Online 9:152–163
63. Azambuja R, Badalotti M, Teloken C et al (2005) Successful birth after injection of frozen human oocytes with frozen epididymal spermatozoa. Reprod Biomed Online 11:449–451
64. Bianchi V, Coticchio G, Fava L et al (2005) Meiotic spindle imaging in human oocytes frozen with a slow freezing procedure involving high sucrose concentration. Hum Reprod 20:1078–1083
65. Coticchio G, Bonu MA, Bianchi V et al (2005) Criteria to assess human oocyte quality after cryopreservation. Reprod Biomed Online 11:421–427
66. Levi-Setti PE, Albani E, Novara PV et al (2006) Cryopreservation of supernumerary oocytes in IVF/ICSI cycles. Hum Reprod 21:370–375
67. Cao YX, Xing Q, Zhang ZG et al (2009) Cryopreservation of immature and in-vitro matured human oocytes by vitrification. Reprod Biomed Online 19:369–373
68. Cao YX, Chian RC (2009) Fertility preservation with immature and in vitro matured oocytes. Semin Reprod Med 27:456–464
69. Chian RC (2004) In-vitro maturation of immature oocytes for infertile women with PCOS. Reprod Biomed Online 8:547–552
70. Le Du A, Kadoch IJ, Bourcigaux N et al (2005) In vitro oocyte maturation for the treatment of infertility associated with polycystic ovarian syndrome: the French experience. Hum Reprod 20:420–424
71. Jurema MW, Nogueira D (2006) In vitro maturation of human oocytes for assisted reproduction. Fertil Steril 86:1277–1291
72. Gunby J, Daya S (2007) IVF Directors Group of the Canadian Fertility and Andrology Society. Assisted reproductive technologies (ART) in Canada: 2003 results from the Canadian ART Register. Fertil Steril 88:550–559
73. Centers for Disease Control and Prevention (www.cdc.gov). 2003. Assisted Reproductive Technology (ART) Report: Section 2—ART Cycles Using Fresh, Nondonor Eggs or Embryos (Updated on March 27, 2006; cited on October 19, 2006). Available from: http://www.cdc.gov/ART/ART2003/sect2_fig3-13.htm#Figure%2010. Accessed on March 29, 2007.
74. Andersen AN, Gianaroli L, Felberbaum R, de Mouzon J, Nygren KG (2006) The European IVF-monitoring programme (EIM) for the European Society of Human Reproduction and Embryology (ESHRE); Assisted reproductive technology in Europe, 2002. Results generated from European registers by ESHRE. Hum Reprod 21:1680–1697
75. Buckett WM, Chian RC, Holzer H et al (2007) Obstetric outcomes and congenital abnormalities after in vitro maturation, in vitro fertilization, and intracytoplasmic sperm injection. Obstet Gynecol 110:885–891
76. Chian RC, Xu CL, Huang JYJ et al (2014) Obstetric outcomes and congenital abnoralities in infants conceived with oocytes matured in vitro. Facts Views Vis Obgyn 6:15–18
77. Brinsden PR, Wada I, Tan SL et al (1995) Diagnosis, prevention and management of ovarian hyperstimulation syndrome. Br J Obstet Gynaecol 102:767–772
78. Tan SL, Maconochie N, Doyle P et al (1994) Cumulative conception and live-birth rates after in vitro fertilization with and without the use of long, short, and ultrashort regimens of the gonadotropin-releasing hormone agonist Buserelin. Am J Obstet Gynecol 171:513–520
79. Venn A, Jones P, Quinn M et al (2001) Characteristics of ovarian and uterine cancers in a cohort of in vitro fertilization patients. Gynecol Oncol 82:64–68
80. Venn A, Watson L, Bruinsma F et al (1999) Risk of cancer after use of fertility drugs with in-vitro fertilization. Lancet 354:1586–1590
81. Rao GD, Chian RC, Son WS et al (2004) Fertility preservation in women undergoing cancer treatment. Lancet 363:1829–1830

82. Bines J, Oleske DM, Cobleigh MA (1996) Ovarian function in premenopausal women treated with adjuvant chemotherapy for breast cancer. J Clin Oncol 14:1718–1719
83. Partridge AH, Gelber S, Peppercorn J et al (2004) Web-based survey of fertility issues in young women with breast cancer. J Clin Oncol 22:4174–4183
84. Chian RC, Uzelac PS, Nargund G (2013) In vitro maturation of human immature oocytes for fertility preservation. Fertil Steril 99:1173–1181
85. Fahy G (2007) Theoretical considerations for oocyte cryopreservation by freezing. Reprod Biomed Online 14:709–714
86. De Santis L, Cino I, Coticchio G et al (2007) Objective evaluation of the viability of cryopreserved oocytes. Reprod Biomed Online 15:338–345
87. Chian RC, Wang Y, Li YR (2014) Oocyte vitrification: advances, progress and future goals. J Assist Reprod Genet 31:411–420

# Chapter 4

# Role of Antioxidants and Antifreeze Proteins in Cryopreservation/Vitrification

**Seul Ki Kim, Hye Won Youm, Jung Ryeol Lee, and Chang Suk Suh**

## Abstract

In recent years, supplementation of antioxidants and antifreeze proteins during cryopreservation/vitrification has significantly improved the survival and function of oocytes and ovarian tissues (OT) in animal models. In this chapter, the experimental protocols for the use of antioxidants and antifreeze proteins in cryopreservation/vitrification are described.

**Key words** Oocyte, Ovarian tissue, Cryopreservation, Vitrification, Antioxidants, Antifreeze proteins, Fertility preservation

## 1 Introduction

Among the various fertility preservation options, embryo cryopreservation is the most widely used, resulting in thousands of births each year from frozen embryos. In 2013, the American Society for Reproductive Medicine (ASRM) and the Society for Assisted Reproductive Technology (SART) announced that modern procedures for oocyte cryopreservation should no longer be considered experimental [1]. Following the earlier use of slow freezing for tissue preservation, rapid advances in the vitrification approach have led to successful cryopreservation of embryo and mature oocyte. While clinical application of cryopreservation in assisted reproduction programs with mature oocytes has resulted in reasonable success rates, the use of immature oocytes in cryopreservation remains in its infancy.

Ovarian tissue (OT) cryopreservation may be the best option for single women and prepubertal girls on the verge of starting cancer treatment, but this method is still considered experimental. To date, a relatively small number (~40) of live births have resulted from using OT cryopreservation followed by reimplantation, worldwide. In addition, unlike embryo and oocyte

Zsolt Peter Nagy et al. (eds.), *Cryopreservation of Mammalian Gametes and Embryos: Methods and Protocols*, Methods in Molecular Biology, vol. 1568, DOI 10.1007/978-1-4939-6828-2_4, © Springer Science+Business Media LLC 2017

cryopreservation, several studies comparing slow freezing and vitrification of ovarian tissue have presented conflicting results [2]. An optimal protocol for vitrification of the ovarian tissue has not yet been established, and there are only a few reports of live births using vitrified human ovarian tissues.

The main problems concerning cryopreservation are cellular injuries caused due to generation of reactive oxygen species (ROS), alteration in the antioxidant defense capability, and meiotic or cytoskeletal spindle disruption [3–5]. One of the effective strategies for overcoming these obstacles is supplementation with protective agents such as antioxidants or antifreeze proteins (AFPs).

Several studies have demonstrated the benefits of antioxidant addition during various phases of the culture, cryopreservation, and transplantation protocols. These improvements in developmental outcomes are in turn interpreted as due to the mitigation of cryopreservation-induced oxidative stress by exogenous antioxidants such as L-carnitine [6, 7], L-glutamine and taurine [8], *N*-acetylcysteine [9], alpha-tocopherol [10], ascorbic acid [11], beta-mercaptoethanol (BME) [12], and SOD and catalase [13].

AFPs (or ice-binding proteins [IBPs]) lower the freezing point of a solution in a non-colligative manner, leading to an increase in the difference between the melting point and the freezing point. This phenomenon known as thermal hysteresis enables to keep ice crystals very small or prevent their formation. Different types of AFPs including FfIBP, fish type III AFP, and LeIBP have been recently studied in the context of oocyte and OT cryopreservation [14–18].

Here, we organize and describe the protocols for the use of antioxidants and AFPs in oocyte and OT cryopreservation/vitrification in previously published studies.

## 2 Materials

### 2.1 Antioxidants in Mouse Oocyte Cryopreservation

#### 2.1.1 L-Carnitine in Immature Mouse Oocyte Cryopreservation [7, 19]

A. Reagent

1. L-carnitine (Sigma-Aldrich, St. Louis, USA): 3.72 mM (0.6 mg/mL).
2. Modified potassium simplex optimized medium with amino acids (mKSOM$^{AA}$; Caisson, North Logan, USA).
3. MEM-alpha medium (Gibco, Carlsbad, USA).
4. Ethylene glycol (EG; Sigma-Aldrich, St. Louis, USA).
5. Dimethyl sulfoxide (DMSO; Sigma-Aldrich, St. Louis, USA).
6. Trehalose (Sigma-Aldrich, St. Louis, USA).
7. Bovine serum albumin (BSA; Sigma-Aldrich, St. Louis, USA).
8. Fetal bovine serum (FBS; Gibco, Burlington, USA).
9. Follicle-stimulating hormone (FSH; Serono, Aubonne, Switzerland).

10. Equine chorionic gonadotrophin (eCG; Sigma-Aldrich, St. Louis, USA).
11. Pyruvate.
12. Streptomycin.
13. Penicillin G potassium salt.
14. Mineral oil.

B. Vitrification Device and Solutions

1. Cryoloop (20 μm wide; 0.5–0.7 mm diameter; Hampton Research, Aliso Viejo, USA; *see* **Note 1**).
2. Basic medium (BM): 10 % FBS in MEM-alpha.
3. Equilibration solution (ES): 7.5 % EG (v:v) + 7.5 % DMSO (v:v) in BM.
4. Vitrification solution (VS): 15 % EG (v:v) + 15 % DMSO (v:v) in BM.
5. Warming solution (WS): 1 M (WS1), 0.5 M (WS2), and 0 M (WS3) trehalose in BM.
6. In vitro maturation (IVM) medium: 5 % FBS + 25 μg/mL pyruvate + 300 ng/mL FSH + 50 μg/mL streptomycin + 75 μg/mL penicillin G potassium salt in MEM-alpha.
7. Oocyte collection medium: 5 % FBS + 50 μg/mL streptomycin + 75 μg/mL penicillin G potassium salt in MEM-alpha.
8. Sperm capacitation medium: 0.9 % BSA in MEM-alpha.
9. Washing medium: 0.4 % BSA in MEM-alpha.
10. IVF medium: 0.4 % BSA in MEM-alpha.
11. Culture medium for cleavage and blastocyst stage embryos: modified potassium simplex optimized medium with amino acids (mKSOM$^{AA}$).

C. Equipment

1. Plastic culture dishes (several types).
2. Plastic disposable tubes (15 mL).
3. Forceps.
4. Curved fine scissors.
5. Fine syringe needle (26 G).
6. Pasteur glass pipette.
7. Timer.
8. Cane.
9. Sterile hood.
10. Liquid nitrogen ($LN_2$).
11. $LN_2$ storage tank.
12. Stereomicroscope.
13. Inverted microscope.

*2.1.2 SOD and Catalase in Mouse Oocyte Cryopreservation [13]*

A. Reagent

1. Superoxide dismutase (SOD; Sigma-Aldrich, St. Louis, USA): 50 IU/mL.
2. Catalase (Sigma-Aldrich, St. Louis, USA): 10 IU/mL.
3. Phosphate-buffered saline (PBS; Gibco, Paisley, UK).
4. Krebs-Ringer bicarbonate medium (Sigma-Aldrich, St. Louis, USA).
5. Propanediol (PrOH; Nakarai Chemical, Tokyo, Japan).
6. Sucrose (Sigma-Aldrich, St. Louis, USA).
7. Bovine serum albumin (BSA; Sigma-Aldrich, St. Louis, USA).
8. Human tubal fluid (HTF; Irvine Scientific, Santa Ana, USA).
9. Pregnant mare serum gonadotrophin (PMSG; Sigma-Aldrich, St. Louis, USA).
10. Human chorionic gonadotrophin (hCG; Sigma-Aldrich, St. Louis, USA).
11. Mineral oil.

B. Slow Freezing Device and Solutions

1. Plastic straw (*see* **Note 1**).
2. Basic medium (BM): 10 mg/mL BSA in PBS.
3. Slow freezing solution:
   (a) 0.5 M, 1.0 M, and 1.5 M PrOH in BM.
   (b) 1.5 M PrOH + 0.1 M Sucrose in BM.
4. Warming solution [20]
   (a) 1.5 M, 1.0 M, and 0.5 M PrOH in BM.
   (b) 0.1 M sucrose in BM.
5. Oocyte handling and dilution medium: 10 mg/mL BSA in PBS.
6. Sperm capacitation medium: 4 mg/mL BSA in Krebs-Ringer bicarbonate medium.
7. Washing medium: 4 mg/mL BSA in HTF.
8. IVF medium: 4 mg/mL BSA in HTF.
9. Culture medium for cleavage and blastocyst stage embryos: 4 mg/mL BSA in HTF.

C. Equipment

1. Plastic culture dishes (several types).
2. Plastic disposable tubes (15 mL).
3. Forceps.
4. Curved fine scissors.
5. Fine syringe needle (26G).
6. Pasteur glass pipette.
7. Timer.

8. Cane.
9. Sterile hood.
10. Controlled programmed freezer (Gryoembryo HP; Daido-Hoxan, Tokyo, Japan; *see* **Note 2**).
11. Liquid nitrogen ($LN_2$).
12. $LN_2$ storage tank.
13. Stereomicroscope.
14. Inverted microscope.

### 2.2 Antioxidants in Human Ovarian Tissue Cryopreservation

#### 2.2.1 L-Glutamine and Taurine in Human Ovarian Tissue Cryopreservation [8]

A. Reagent

1. L-glutamine (Sigma-Aldrich, St. Louis, USA): 25.0 mmol/L.
2. Taurine (Sigma-Aldrich, St. Louis, USA): 0.5 mmol/L.
3. Medium A (*see* **Note 3** for the compositions).
4. Propanediol (PrOH; Sigma-Aldrich, St. Louis, USA).
5. Glycine (Sigma-Aldrich, St. Louis, USA).
6. Raffinose (Sigma-Aldrich, St. Louis, USA).
7. HEPES.
8. Human serum albumin (HSA; Vitrolife Sweden AB, Sweden).

B. Slow Freezing Device and Solutions

1. Cryovial (*see* **Note 1**).
2. Medium A (basic medium).
3. Medium B (freezing medium): consists of Medium A without $CaCl_2$ and supplemented with 3 M PrOH, 50 M glycine, 0.05 M raffinose, and 21.8 M HEPES [21].
4. Thawing medium: Medium A.

C. Equipment

1. Plastic culture dishes (several types).
2. Plastic disposable tubes (15 mL).
3. Forceps.
4. Curved fine scissors.
5. Timer.
6. Cane.
7. Sterile hood.
8. Controlled programmed freezer (Minicool 40 PC, Air Liquide, France; *see* **Note 2**).
9. Liquid nitrogen ($LN_2$).
10. $LN_2$ storage tank.
11. Stereomicroscope.
12. Inverted microscope.

*2.2.2 N-Acetylcysteine in Human Ovarian Tissue Cryopreservation [9]*

A. Reagent

1. *N*-Acetylcysteine (Sigma-Aldrich, St. Louis, USA): 25 mmol/L.
2. Dulbecco's phosphate-buffered saline (D-PBS; Gibco, Paisley, UK).
3. Propanediol (PrOH; Sigma-Aldrich, St. Louis, USA).
4. Human serum (HS).

B. Slow Freezing Device and Solutions

1. Cryovial (Nunc, Roskilde, Denmark; *see* **Note 1**).
2. Basic medium (BM): 30 % HS in D-PBS.
3. Freezing solution: 1.26 M PrOH + 0.175 M in BM.
4. Thawing solution:
   (a) Solution 1: 0.76 M PrOH + 0.175 M sucrose in BM.
   (b) Solution 2: 0.26 M PrOH + 0.175 M sucrose in BM.
   (c) Solution 3: 0.175 M sucrose in BM.

C. Equipment

1. Plastic culture dishes (several types).
2. Plastic disposable tubes (15 mL).
3. Forceps.
4. Curved fine scissors.
5. Timer.
6. Cane.
7. Sterile hood.
8. Controlled programmed freezer (*see* **Note 2**).
9. Liquid nitrogen ($LN_2$).
10. $LN_2$ storage tank.
11. Stereomicroscope.
12. Inverted microscope.

### 2.3 Antifreeze Proteins in Mouse Oocyte Cryopreservation/Vitrification

*2.3.1 Antifreeze Proteins in Immature Mouse Oocyte Cryopreservation [14]*

A. Reagent

1. Type III antifreeze protein (AFP): 500 ng/mL.
2. Dulbecco's phosphate-buffered saline (D-PBS; Gibco, Paisley, UK).
3. Ethylene glycol (EG; Sigma-Aldrich, St. Louis, USA).
4. Dimethyl sulfoxide (DMSO; Sigma-Aldrich, St. Louis, USA).
5. Sucrose (Sigma-Aldrich, St. Louis, USA).
6. Hyaluronidase (Sigma-Aldrich, St. Louis, USA).
7. Mouse tubal fluid (MTF; *see* **Note 4** for the compositions).
8. Medium-199 (Life technology, Burlington, USA).
9. Global® medium (LifeGlobal, Brussels, Belgium).
10. Bovine serum albumin (BSA; Sigma-Aldrich, St. Louis, USA).
11. Fetal bovine serum (FBS; Gibco, Burlington, USA).
12. Follicle-stimulating hormone (FSH; Serono, Aubonne, Switzerland).

13. Human chorionic gonadotrophin (hCG; Serono, Aubonne, Switzerland).
14. Epidermal growth factor (EGF; Sigma-Aldrich, St. Louis, USA).
15. Pregnant mare serum gonadotrophin (PMSG: Sigma-Aldrich, St. Louis, USA).
16. Mineral oil.

B. Vitrification Device and Solutions

1. CryoTop (Kitazato, Fujinomiya, Japan; *see* **Note 1**).
2. Basic medium (BM): 20 % FBS in D-PBS.
3. Equilibration solution (ES): 7.5 % EG + 7.5 % DMSO in BM.
4. Vitrification solution (VS): 15 % EG + 15 % DMSO + 0.5 M sucrose in BM.
5. Warming solution (WS): 1 M (WS1), 0.5 M (WS2), 0.25 M (WS3), and 0 M (WS4) sucrose in BM.
6. In vitro maturation (IVM) medium: 20 % FBS + 75 mIU FSH + 0.5 IU/mL hCG + 10 ng/mL EGF in Medium-199.
7. Sperm capacitation medium: 0.4 % BSA in MTF.
8. Washing medium: 0.4 % BSA in MTF.
9. Fertilization medium: 0.4 % BSA in MTF.
10. Culture medium for cleavage and blastocyst stage embryos: 0.4 % BSA in Global® medium.

C. Equipment

1. Four-well dish (Nunc, Roskilde, Denmark) for vitrification and warming procedures.
2. Organ culture dish (Falcon, Franklin Lakes, USA) for sperm insemination and washing.
3. Tissue culture dish (30 or 60 mm; Falcon, Franklin Lakes, USA) for oil drop culture.
4. Plastic disposable tubes (15 mL).
5. Forceps.
6. Curved fine scissors.
7. Fine syringe needle (26G).
8. Pasteur glass pipette.
9. Timer.
10. Cane.
11. Sterile hood.
12. Liquid nitrogen ($LN_2$).
13. $LN_2$ storage tank.
14. Stereomicroscope.
15. Inverted microscope.

### 2.3.2 *Antifreeze Proteins in Mature Mouse Oocyte Cryopreservation [15]*

A. Reagent

1. Type III antifreeze protein (AFP): 500 ng/mL.
2. Dulbecco's phosphate-buffered saline (D-PBS; Gibco, Paisley, UK).
3. Ethylene glycol (EG; Sigma-Aldrich, St. Louis, USA).
4. Dimethyl sulfoxide (DMSO; Sigma-Aldrich, St. Louis, USA).
5. Sucrose (Sigma-Aldrich, St. Louis, USA).
6. Hyaluronidase (Sigma-Aldrich, St. Louis, USA).
7. Mouse tubal fluid (MTF; *see* **Note 1** for the compositions).
8. Global® medium (LifeGlobal, Brussels, Belgium).
9. Bovine serum albumin (BSA; Sigma-Aldrich, St. Louis, USA).
10. Fetal bovine serum (FBS; Gibco, Burlington, USA).
11. PMSG (Sigma-Aldrich, St. Louis, USA).
12. hCG (Sigma-Aldrich, St. Louis, USA).

B. Vitrification Device and Solutions

1. CryoTop (Kitazato, Fujinomiya, Japan; *see* **Note 1**).
2. Basic medium (BM): 20 % FBS in D-PBS.
3. Equilibration solution (ES): 7.5 % EG + 7.5 % DMSO in BM.
4. Vitrification solution (VS): 15 % EG + 15 % DMSO + 0.5 M sucrose in BM.
5. Warming solution (WS): 1 M (WS1), 0.5 M (WS2), 0.25 M (WS3), and 0 M (WS4) sucrose in BM.
6. Sperm capacitation medium: 0.4 % BSA in MTF.
7. Washing medium: 0.4 % BSA in MTF.
8. Fertilization medium: 0.4 % BSA in MTF.
9. Culture medium for cleavage and blastocyst stage embryos: 0.4 % BSA in Global® medium.

C. Equipment

1. Four-well dish (Nunc, Roskilde, Denmark) for vitrification and warming procedures.
2. Organ culture dish (Falcon, Franklin Lakes, USA) for sperm insemination and washing.
3. Tissue culture dish (30 or 60 mm; Falcon, Franklin Lakes, USA) for oil drop culture.
4. Plastic disposable tubes (15 mL).
5. Forceps.
6. Curved fine scissors.
7. Pasteur glass pipette.
8. Timer.
9. Cane.

10. Sterile hood.
11. Liquid nitrogen ($LN_2$).
12. $LN_2$ storage tank.
13. Stereomicroscope.
14. Inverted microscope.

### 2.4 Antifreeze Proteins in Mouse Ovarian Tissue Cryopreservation [16]

#### 2.4.1 Three Types of Antifreeze Proteins in Mouse Ovarian Tissue Cryopreservation [16]

A. Reagent

1. Antifreeze protein (AFP) types: type III AFP, LeIBP, and FfIBP.
2. Dulbecco's phosphate-buffered saline (D-PBS; Gibco, Paisley, UK).
3. Ethylene glycol (EG; Sigma-Aldrich, St. Louis, USA).
4. Dimethyl sulfoxide (DMSO; Sigma-Aldrich, St. Louis, USA).
5. Sucrose (Sigma-Aldrich, St. Louis, USA).
6. Fetal bovine serum (FBS; Gibco, Burlington, USA).
7. Zoletil (Virbac, Carros, France).
8. Rompun (Bayer, Leverkusen, Germany).

B. Vitrification Device and Solutions

1. Electron microscope (EM) copper grids (IGC 400; Jeol, Tokyo Japan; *see* **Note 1**).
2. Basic medium (BM): 20 % FBS in D-PBS.
3. Equilibration solution (ES): 7.5 % EG + 7.5 % DMSO in BM.
4. Vitrification solution (VS): 20 % EG + 20 % DMSO + 0.5 M sucrose in BM.
5. Warming solution (WS): 1 M (WS1), 0.5 M (WS2), 0.25 M (WS3), and 0 M (WS4) sucrose in BM.

C. Equipment

1. Four-well dish (Nunc, Roskilde, Denmark) for vitrification and warming procedures.
2. Cryovial (1.5 or 1.8 mL; Nunc, Roskilde, Denmark).
3. Forceps.
4. Curved fine scissors.
5. Sterilized filter paper or gauze.
6. 4-0 silk suture.
7. Timer.
8. Cane.
9. Sterile hood.
10. Liquid nitrogen ($LN_2$).
11. $LN_2$ storage tank.

## 3 Methods

Carry out all procedures at room temperature unless otherwise specified and use the sterile hood when processing.

### 3.1 Antioxidants in Mouse Oocyte Cryopreservation

#### 3.1.1 L-Carnitine in Immature Mouse Oocyte Cryopreservation [7, 19]

##### Animals

Eight- to 10-week-old (B6.DBA)F1 female mice were housed under controlled conditions as follows: lighting (12-h light/dark cycle), temperature (21–24 °C), humidity (30–60 %), and fed ad libitum.

##### L-Carnitine

Dissolve the L-carnitine in the vitrification, warming, and IVM medium at a final concentration 0.6 mg/mL and sterilized by filtration.

##### Retrieval of Immature Oocytes and In Vitro Maturation

1. Treat the mice with i.p. injections of 5 IU eCG.
2. After 45–47 h, sacrifice them by cervical dislocation.
3. Dissect both ovaries and puncture the antral follicles using a fine syringe needle (26G).
4. Collect the cumulus-oocyte complexes (COCs) covered with compact cumulus.

##### Vitrification Procedure

Vitrify the COCs with ES and VS in the presence of L-carnitine:

1. Maintain the COCs in ES for 3 min.
2. Transfer the COCs to 50 uL droplet of VS.
3. Dip the cryoloop into the VS so that a thin film is formed by surface tension.
4. Place 3–5 COCs gently on the film of a cryoloop using a fine glass pipette.
5. Directly plunge the cryoloop into a cryovial.
6. Submerge it into $LN_2$ and tightened the cap.
7. Plunge into $LN_2$ for storage.
8. Perform the whole process from VS treatment to plunging into $LN_2$ within less than 1 min.

##### Warming Procedure

All the thawing procedures are conducted at 37 °C:

1. Unscrew and open the cap of the cryovial carefully in $LN_2$.
2. Transfer the cryoloop containing COCs into 1 mL of WS1 quickly.
3. After the COCs fall into the solution, keep it warm for 3 min.

4. Transfer the COCs into 500 μL of WS2 and WS3 for 3 min, respectively.
5. Subject the viable COCs to IVM medium.

In Vitro Maturation and Fertilization

1. After warming, transfer and culture the COCs in 50 μl droplet of IVM medium for 16 h.
2. After IVM, assess the oocyte maturity by checking extrusion of a first polar body under an inverted microscope (200×) (*see* **Note 5**).
3. Collect the mature MII oocyte in 50 μL droplet of IVF medium.
4. Add the mouse epididymal sperm (10 μL) to each microdrop of IVF medium containing COCs (*see* **Note 6**).
5. Incubate them at 37 °C in humidified 5 % $CO_2$ in air.
6. After 5 h, wash the inseminated oocytes thoroughly with the IVF medium by repeating pipetting.
7. Transfer the fertilized oocytes to the oil drop dish of culture medium (Day 0).
8. Twenty-four hours after insemination, assess the fertilization status by checking 2-cell embryo development.
9. Transfer the 2-cell embryos to the oil drop of culture medium.
10. Assess the blastocyst development ratio at Day 4.

*3.1.2 SOD and Catalase in Mouse Oocyte Cryopreservation [13]*

Animals

Six- to 8-week-old ICR female mice are maintained on a constant light/dark cycle.

SOD and Catalase

Dissolve the 50 IU/mL SOD and 10 IU/mL catalase in the freezing and thawing medium.

Collection of In Vivo Matured Oocytes

1. Treat the mice with i.p. injection of 5 IU PMSG followed by 5 IU hCG 48 h later.
2. Fourteen to fifteen hours after hCG administration, sacrifice the mice by cervical dislocation.
3. Excise both of the oviducts.
4. Place the dissected oviducts in PBS.
5. Release the COCs by punching the ampulla of each oviduct with fine needles and wash them with PBS.

Slow Freezing Procedure

1. Expose the COCs to slow freezing solution (0.5 M, 1.0 M, and 1.5 M PrOH in BM) for 5 min at RT, respectively.
2. Then transfer the COCs to the next slow freezing solution (1.5 M PrOH + 0.1 M sucrose in BM) and leave them for 15 min at RT.

3. Load the COCs to a plastic straw.
4. Place the plastic straw in a controlled programmed freezer.
5. Cooling and seeding programs of this procedure are as below:
   (a) From ambient (20 °C) to −7 °C: −2 °C/min of cooling rate.
   (b) Manual seeding at −7 °C: hold for 5 min (*see* **Note 7**).
   (c) From −7 to −30 °C: −0.3 °C/min of cooling rate.
   (d) From −30 to −150 °C: −50 °C/min of cooling rate.
6. Put the plastic straws into a cane.
7. Immediately plunge the cane into $LN_2$ for storage.

Thawing Procedure

1. Place the plastic straw in a 37 °C water bath rapidly.
2. Release the COCs directly into three steps of warming solution (1.5 M, 1.0 M, and 0.5 M PrOH) and finally 0.1 M sucrose in BM for each 5 min.
3. Wash the COCs with BM.
4. Assess the survival of COCs by the morphologic criteria under an inverted microscope (200×) (*see* **Note 5**).

In Vitro Fertilization

1. Inseminate with cauda epididymal sperms (concentration: $2 \times 10^6$/mL) to the COCs and incubate them at 37 °C in humidified 5 % $CO_2$ in air (*see* **Note 8**).
2. After 8 h, wash the inseminated oocytes thoroughly with the fertilization medium by pipetting.
3. Check the two pronuclei under the stereomicroscope.
4. Transfer the fertilized oocyte to the culture medium.
5. Twenty-four hours after insemination, assess the fertilization status by checking 2-cell embryo development.
6. Culture them further at 37 °C in humidified 5 % $CO_2$ in air.

### 3.2 Antioxidants in Human Ovarian Tissue Cryopreservation

#### 3.2.1 L-Glutamine and Taurine in Human Ovarian Tissue Cryopreservation [8]

Ovarian Tissue Samples

Ovarian tissue was collected from 25 women aged 19 to 36 years (mean ± SEM: 27.7 ± 1.0) during the operation treating their ovarian cyst, after giving informed consent to participate in the study.

L-Glutamine and Taurine

Dissolve L-glutamine (25.0 mmol/L) and taurine (0.5 mmol/L) in the freezing medium.

Slow Freezing Procedure

1. Immediately immerse the ovarian tissue in “Medium A” and transport it on ice to keep 4 °C.
2. Remove the medulla and remain the ovarian cortex.
3. Slice the ovarian cortex into an adequate size.
4. Place the ovarian tissue slice into the medium mixture of “Medium A and B” of 1:1 (v/v) under gentle agitation.
5. After 15 min of equilibration at 4 °C, transfer the slices to a cryovial containing 1.5 mL of cryoprotectant.
6. Load the cryovial into a programmable freezer.
7. Cooling and seeding programs of this procedure are as below:
   (a) From 4 to −11 °C: −2 °C/min of cooling rate.
   (b) From −11 to −40 °C: −2 °C/min of cooling rate.
   (c) From −40 to −150 °C: −10 °C/min of cooling rate.
8. Put the cryovial into a cane.
9. Immediately plunge the cane into $LN_2$ for storage.

Thawing Procedure

1. Expose the cryovial containing ovaries into 37 °C water bath for 2 min.
2. Release the ovarian tissue directly into Medium A with gentle shaking and wash it twice for 5 min at 37 °C.

*3.2.2 N-Acetylcysteine in Human Ovarian Tissue Cryopreservation [9]*

Ovarian Tissue Samples

Human ovarian tissue samples were obtained from 10 informed and consenting patients, aged 17–25 years (mean age ± SD: 26.2 ± 6.5) for cryopreservation of ovarian tissue before chemotherapy and radiotherapy.

*N*-Acetylcysteine

Dissolve *N*-acetylcysteine (25 mmol/L) in the freezing and thawing solutions.

Slow Freezing Procedure

1. Wash the ovarian tissue with BM.
2. Remove the medulla and remain the ovarian cortex.
3. Slice the ovarian cortex into an adequate size.
4. Place each slice into the cryovial containing a slow freezing solution.
5. After 60 min exposure to the cryoprotectant at 4 °C for equilibration, transfer the cryovial into a programmable freezer.
6. Cooling and seeding programs of the freezer are as below:
   (a) From 0 to −9 °C: −2 °C/min of cooling rate.
   (b) Manual seeding at −9 °C: held for 10 min (*see* **Note** 7).
   (c) From −9 to −40 °C: −0.3 °C/min of cooling rate.
   (d) From −40 to −140 °C: −10 °C/min of cooling rate.
7. Put the cryovial into a cane.
8. Immediately plunge the cane into $LN_2$ for storage.

Thawing Procedure

1. Expose the cryovial containing ovaries to air (RT) for 30 s.
2. Then immerse into 37 °C water bath at for 2 min.
3. Remove the cryoprotectant by three steps of serial dilutions (thawing solutions 1–3) of the cryoprotectants at 4 °C.
4. Finally, maintain the ovarian tissues for 20 min in BM at 4 °C.

### 3.3 Antifreeze Proteins in Mouse Oocyte Cryopreservation

#### 3.3.1 Antifreeze Proteins (AFPs) in Immature Mouse Oocyte Cryopreservation [14]

Animals

Four- to 6-week-old female BDF-1 mice used were housed under a 12-h light/dark cycle at 22 °C and provided food ad libitum. All procedures were processed at RT except in the case of special comments.

Antifreeze Protein

Type III AFP: The concentration of AFP was determined by the previous researches [14], which showed higher survival and blastocyst formation rates at 500 ng/mL type III AFP. Vitrify the COCs with ES and VS in the presence of type III AFP.

Retrieval of Immature Oocytes

1. Treat the mice with i.p. injections of 5 IU PMSG.
2. After 48 h, kill them by cervical dislocation.
3. Excise both ovaries and place them in 1 mL of washing medium.
4. Collect the cumulus-oocyte complexes (COCs) covered with compact cumulus cells by puncturing the antral follicles using fine syringe needle.

Vitrification Procedure

1. Expose the COCs to ES for 5 min.
2. Transfer the COCs to VS and leave them for 45–60 s.
3. Load the five to six COCs onto a CryoTop.
4. Remove almost of the entire loading solution by aspiration.
5. Immediately plunge the CryoTop into $LN_2$ for storage.

Warming Procedure

1. Plunge the CryoTop into the WS1 quickly for 1 min (37 °C).
2. After the COCs come off from the CryoTop, transfer them to WS2, WS3, and WS4 for 3 min, respectively.
3. Wash the COCs twice with washing medium.
4. Examine the COCs morphologically to assess their viability (*see* **Note 5**).

In Vitro Maturation and Fertilization

1. After warming, transfer and culture the COCs in IVM medium for 17–18 h.

2. After IVM, denude their cumulus cells completely by treatment with 85 IU/mL hyaluronidase.
3. Then assess the oocyte maturity by checking extrusion of a first polar body under an inverted microscope (200×).
4. Collect the mature MII oocyte in the organ dish of fertilization medium.
5. Inseminate the mouse epididymal sperm (concentration: $2 \times 10^6$/mL) to the mature oocytes, and incubate them at 37 °C in humidified 5 % $CO_2$ in air (*see* **Note 9**).
6. After 4 h, wash the inseminated oocytes thoroughly with the fertilization medium by pipetting.
7. Transfer the fertilized oocytes to the oil drop dish of the fertilization medium.
8. Twenty-four hours after insemination, assess the fertilization status by checking 2-cell embryo development.
9. Transfer the 2-cell embryos to the oil drop of culture medium.
10. Assess the blastocyst development ratio 72 h after insemination.

#### *3.3.2 Antifreeze Proteins in Mature Mouse Oocyte Cryopreservation [15]*

##### Animals

Four- to 6-week-old female BDF-1 mice used were housed under a 12-h light/dark cycle at 22 °C and provided food ad libitum. All procedures were processed at RT except in the case of special comments.

##### Antifreeze Protein

Type III AFP: The concentration of AFP was determined by the previous research [15], which showed higher survival and blastocyst formation rates at 500 ng/mL type III AFP. Vitrify the mature oocytes with ES and VS in the presence of type III AFP.

##### Collection of In Vivo Matured Oocytes

1. Treat the mice with i.p. injection of 5 IU PMSG followed by 5 IU hCG 48 h later.
2. Thirteen to fourteen hours after hCG injection, kill the mice by cervical dislocation.
3. Collect both of the oviducts.
4. Place the dissected oviducts in an organ dish containing washing medium.
5. Release the COCs by tearing the ampulla of the oviducts.
6. Remove the cumulus cells enzymatically using 85 IU/mL hyaluronidase and mechanically using a glass pipette.
7. Assess the morphologically normal mature oocytes by checking the presence of a first polar body.

##### Vitrification Procedure

1. Expose the oocytes to ES for 5 min.
2. Transfer the oocytes to VS and leave them for 45–60 s.
3. Load five to six oocytes onto a CryoTop.

4. Remove almost of the entire loading solution by aspiration.
5. Immediately plunge the CryoTop into $LN_2$ for storage.

Warming Procedure

1. Plunge the CryoTop into the WS1 quickly for 1 minute (37 °C).
2. After the oocytes come off from the CryoTop, transfer them to WS2, WS3, and WS4 for 3 min, respectively.
3. Wash the oocytes twice with washing medium and transfer them to the fertilization medium.
4. Assess the survival rate 1 h after incubation by identifying the morphologic appearance of the membrane and ooplasm integrity under an inverted microscope (200×).

In Vitro Fertilization

1. Inseminate the mouse epididymal sperm (concentration: $2 \times 10^6$/mL) to the mature oocytes, and incubate them at 37 °C in humidified 5 % $CO_2$ in air (*see* **Note 9**).
2. After 4 h, wash the inseminated oocytes thoroughly with the fertilization medium by pipetting.
3. Transfer the fertilized oocyte to the oil drop dish of the fertilization medium.
4. Twenty-four hours after insemination, assess the fertilization status by checking 2-cell embryo development.
5. Transfer the 2-cell embryos to the oil drop of culture medium.
6. Assess the blastocyst development ratio 72 h after insemination.

### 3.4 Antifreeze Proteins in Mouse Ovarian Tissue Cryopreservation

#### 3.4.1 Three Types of Antifreeze Proteins in Mouse Ovarian Tissue Cryopreservation

Antifreeze Proteins: Types and Characteristics [16]

| | *Type III AFP* | *LeIBP* | *FfIBP* |
|---|---|---|---|
| Mass (kDa) | 6.5 | ~27 | ~25.3 |
| Structure | Globular | β-helix | β-helix |
| Natural source | Ocean pout, wolfish, and eelpout | *Glaciozyma* sp. | *Flavobacterium frigoris* |
| Thermal hysteresis (TH) | ~1.5 °C at 3 mM | 0.42 °C at 0.4 mM | 2.5 °C at 50 μM |
| Reference | Structure-function relationship in the globular type III AFP: identification of a cluster of surface residues required for binding to ice | Characterization of the ice-binding protein from Arctic yeast *Leucosporidium* sp. AY30 | Structure-based characterization and antifreeze properties of a hyperactive ice-binding protein from the Antarctic bacterium *Flavobacterium frigoris* PS1 |

#### Antifreeze Protein Application

AFPs were added in the ES, VS, and all series of warming solutions (WS1–WS4) according to each type and characteristic of AFPs [16].

#### Animals

Five-week-old female BDF-1 mice used were housed under a 12-h light/dark cycle at 22 °C and provided food ad libitum. All procedures were processed at RT except in the case of special comments.

#### Vitrification Procedure

Vitrify the ovaries with ES and VS supplemented with AFPs (type III AFP, LeIBP, and FfIBP, respectively):

1. After 1 week of mice adaptation, kill the mice by cervical dislocation or in case of autotransplantation, anesthetize the mice by i.p. injection of 30 mg/kg of Zoletil and 10 mg/kg of Rompun.
2. Excise both ovaries and place them in 1 mL of the BM.
3. In case of autotransplantation, suture the skin incisions.
4. Suspend the ovaries in ES for 10 min.
5. Expose them to VS for 5 min.
6. Place each ovary on top of an EM copper grid.
7. Remove the VS droplet through the underlying sterilized filter paper or gauze.
8. Put the grids into a 1.5 or 1.8 mL cryovial (*see* **Note 10**).
9. Put the cryovial into a cane.
10. Immediately plunge the cane into $LN_2$ for storage.

#### Warming Procedure

1. Expose the cryovial containing ovaries to air (RT) for 10 s.
2. Pour the EM copper grids to the sterilized filter paper or gauze.
3. Pick the grids and transfer them to WS1 quickly.
4. Expose the ovaries that came off from the EM grids to WS1, WS2, WS3, and WS4 for 5 min, respectively.

#### Transplantation of the Vitrified-Thawed Ovaries

1. After analgesia with i.p. injection of 30 mg/kg of Zoletil and 10 mg/kg of Rompun, exteriorize the kidney through a dorsal-horizontal incision.
2. Insert the ovaries beneath the kidney capsule through a small hole made by a fine forceps.
3. Suture the skin incisions.

## 4 Notes

1. Various kinds of carriers can be used for this procedure.
2. Various kinds of freezers can be used for this procedure.

3. The composition of "Medium A": L-glutamine (25.0 mmol/L), taurine (0.5 mmol/L), 0.5 % HAS, NaCl (94.7 mmol/L), KCl (4.8 mmol/L), MgSO4 (0.8 mmol/L), NaH2PO4 (1.0 mmol/L), NaHCO3 (25.0 mmol/L), CaCl2 (1.8 mmol/L), sodium lactate (21.3 mmol/L), sodium pyruvate (0.3 mmol/L), and D-glucose (5.5 mmol/L).
4. The composition of MTF (for making 1 L of medium): 0.2514 g $CaCl_2.2H_2O$, 0.0372 g EDTA, 6.6733 g NaCl, 0.3563 g KCL, 0.2933 g $MgSO_4.7H_2O$, 0.162 g $KH_2PO_4$, 0.0407 g Na-pyruvate, 0.6127 g D-glucose, 0.537 g Na-lactate, 0.678 g Na-lactate (60 %), 2.1003 g NaHCO3, 0.0375 g penicillin G, 0.0251 g streptomycin, 10.0 mL MEM-NEA (Gibco), 10.0 mL RPMI (Gibco), 5.0 mL L-glutamine (Gibco), and 2.5 mL MEM Vitamin (Gibco).
5. Oocytes with a spherical and symmetrical shape, containing evenly granulated cytoplasm, were regarded as viable, whereas oocytes apparently with membrane damage, degenerated cytoplasm, or loss of cumulus cells were regarded as nonviable.
6. Sperms were collected from 10–12-week-old (B6.DBA)F1 males. Spermatozoa were allowed to swim out of the epididymis into a droplet of 300 μL pre-warmed sperm capacitation medium and then incubated for 1 h at 37 °C with 5 % $CO_2$ in a humidified incubator for sperm capacitation.
7. Induce an ice formation (seeding) manually by touching the cryovial at the level of the meniscus with precooled scissors/forceps.
8. The epididymal spermatozoa were retrieved from the cauda epididymis of 4–6-month-old ICR male mice, and the sperm suspensions were preincubated for 2 h in capacitation medium.
9. The protocols for mouse sperm preparation and insemination were same as those described previously by Jo et al. [15]. The epididymal spermatozoa were retrieved from the cauda epididymis of 8–10-week-old BDF-1 mice, and the sperm suspensions were preincubated for 1.5 h in capacitation medium.
10. The ovary can be inserted into the cryovial directly without loading on the EM copper grids.

## References

1. Practice Committees of American Society for Reproductive, M. and T. Society for Assisted Reproductive (2013) Mature oocyte cryopreservation: a guideline. Fertil Steril 99(1):37–43
2. Amorim CA et al (2011) Vitrification as an alternative means of cryopreserving ovarian tissue. Reprod Biomed Online 23 (2):160–186
3. Varghese AC, Nagy ZP, Agarwal A (2009) Current trends, biological foundations and future prospects of oocyte and embryo cryopreservation. Reprod Biomed Online 19 (1):126–140
4. Tatone C et al (2010) Cryopreservation and oxidative stress in reproductive cells. Gynecol Endocrinol 26(8):563–567

5. Eroglu A, Toth TL, Toner M (1998) Alterations of the cytoskeleton and polyploidy induced by cryopreservation of metaphase II mouse oocytes. Fertil Steril 69(5):944–957
6. Chankitisakul V et al (2013) Supplementation of maturation medium with L-carnitine improves cryo-tolerance of bovine in vitro matured oocytes. Theriogenology 79 (4):590–598
7. Moawad AR et al (2014) l-carnitine supplementation during vitrification of mouse germinal vesicle stage-oocytes and their subsequent in vitro maturation improves meiotic spindle configuration and mitochondrial distribution in metaphase II oocytes. Hum Reprod 29 (10):2256–2268
8. Sanfilippo S et al (2013) Quality and functionality of human ovarian tissue after cryopreservation using an original slow freezing procedure. J Assist Reprod Genet 30:25–34
9. Fabbri R et al (2015) Effects of N-acetylcysteine on human ovarian tissue preservation undergoing cryopreservation procedure. Histol Histopathol 30(6):725–735
10. Gutnisky C et al (2013) Glycolytic pathway activity: effect on IVM and oxidative metabolism of bovine oocytes. Reprod Fertil Dev 25 (7):1026–1035
11. Castillo-Martin M et al (2014) Addition of l-ascorbic acid to culture and vitrification media of IVF porcine blastocysts improves survival and reduces HSPA1A levels of vitrified embryos. Reprod Fertil Dev 27:1115–1123
12. Hosseini SM et al (2009) Antioxidant supplementation of culture medium during embryo development and/or after vitrification-warming; which is the most important? J Assist Reprod Genet 26(6):355–364
13. Dinara S et al (2001) Effects of supplementation with free radical scavengers on the survival and fertilization rates of mouse cryopreserved oocytes. Hum Reprod 16(9):1976–1981
14. Jo JW et al (2012) The beneficial effects of antifreeze proteins in the vitrification of immature mouse oocytes. PLoS One 7(5):e37043
15. Jo JW et al (2011) Effect of antifreeze protein supplementation in vitrification medium on mouse oocyte developmental competence. Fertil Steril 96(5):1239–1245
16. Lee J et al (2015) Effects of three different types of antifreeze proteins on mouse ovarian tissue cryopreservation and transplantation. PLoS One 10(5):e0126252
17. Wen Y et al (2014) The protective role of antifreeze protein 3 on the structure and function of mature mouse oocytes in vitrification. Cryobiology 69(3):394–401
18. Lee JR et al (2015) Effect of antifreeze protein on mouse ovarian tissue cryopreservation and transplantation. Yonsei Med J 56(3):778–784
19. Moawad AR et al (2013) L-carnitine supplementation during vitrification of mouse oocytes at the germinal vesicle stage improves preimplantation development following maturation and fertilization in vitro. Biol Reprod 88 (104):1–8
20. Murayama S et al (1994) Successful freezing of unfertilized mouse oocytes and effect of cocultures in oviducts on development of in vitro fertilized embryos after thawing. J Assist Reprod Genet 11(3):156–161
21. Schubert B et al (2005) Human ovarian tissue from cortex surrounding benign cysts: a model to study ovarian tissue cryopreservation. Hum Reprod 20:1786–1792

# Part III

## Methodology

# Chapter 5

# Slow Freezing of Human Sperm


## Ashok Agarwal and Eva Tvrda



## Abstract

Cryopreservation of human spermatozoa is a highly efficient procedure for managing male fertility, and much of its successful application seems to have a crucial impact on the reproductive outcome of assisted reproduction technologies. Here, we present, explain, and describe the slow freezing method for preserving human spermatozoa, which is currently the most commonly used freezing technique in most clinical andrology laboratories.



**Key words** Human spermatozoa, Slow freezing, Cryodamage, TEST yolk buffer, Cryopreservation


## 1 Introduction to Sperm Cryopreservation: Why and Who?

Cryopreservation is the collection, freezing, and long-term storage of structurally intact living cells and tissues [1]. Historically, spermatozoa were the first mammalian cells to be successfully cryopreserved [2], due to the discovery of the cryoprotectant glycerol. Since then, many advances in cryopreservation technology have led to the evolution of different methods that allow low-temperature maintenance of a variety of cells, tissues, and organs [3]. Much progress in this field has originated from empirical work and fundamental cryobiology. Furthermore, increased awareness of the causes and consequences of cryo-injury has continuously helped to improve distinct cryopreservation methods [4].

Human sperm cryopreservation has become a widely accepted procedure in the contemporary world [5]. The importance of sperm banking following cryopreservation lies in its potential to preserve the future fertility of men who temporarily or permanently lose their fertility due to a variety of reasons, including cancer [6]. Malignant diseases are associated with fever, malnutrition, altered immune response and hormone imbalance, stress, and inflammation leading to destructive consequences on testicular and sperm function [7, 8]. In addition, cancer treatments are highly toxic to the male reproductive system. Chemotherapy is particularly


Zsolt Peter Nagy et al. (eds.), *Cryopreservation of Mammalian Gametes and Embryos: Methods and Protocols*, Methods in Molecular Biology, vol. 1568, DOI 10.1007/978-1-4939-6828-2_5, © Springer Science+Business Media LLC 2017

harmful during the early stages of spermatogenesis, especially to the proliferating and differentiating spermatogonia, and could lead to temporary or permanent azoospermia or severe oligozoospermia [9, 10]. Radiotherapy is similarly deleterious to spermatogenesis, generally causing sperm DNA fragmentation, damage to Sertoli and Leydig cells, and obstruction of the ejaculatory ducts, leading to reduced semen volume [11, 12]. Patients with other diseases, such as lupus, multiple sclerosis, or ulcerative colitis, can also benefit from sperm banking before treatment [13, 14].

Sperm cryopreservation is often recommended to patients with urological diseases and pathologies, including varicocele and testicular torsion [15]. Men awaiting urogenital surgical procedures should consider sperm banking as a precaution as well as those men opting for surgical contraception (bilateral vasectomy) [14].

Healthy men may choose sperm cryopreservation to prevent health problems from occupational exposure to toxic chemicals, ionizing radiation, or biological contaminants [15, 16]. Males undergoing gender reassignment through hormonal or surgical therapy may also be a potential target population [17].

Cryopreservation is also widely used to store spermatozoa retrieved from oligozoospermic and azoospermic patients as well as from men with ejaculatory dysfunction or spinal cord injury. In the latter cases, testicular sperm extraction or percutaneous epididymal sperm aspiration is used, both of which avoid the need for repeated biopsies and aspiration [5].

Finally, cryopreservation of spermatozoa is mandatory in donor insemination programs. All frozen samples are screened for infectious diseases including HIV, herpes, and hepatitis before being released [5, 14, 16].

Whatever the reason for sperm cryopreservation, when pregnancy is desired, the semen sample is thawed and used in a number of assisted reproductive techniques including intracytoplasmic sperm injection, intrauterine insemination, and in vitro fertilization [6].

This chapter will discuss slow freezing of human spermatozoa, which is currently regarded as a safe, cost-effective, and useful strategy to preserve male fertility.

## 2 Slow Freezing: Principles and Overview

Slow freezing is a cryopreservation technique based on dehydration. A small amount of cryoprotectant is added to a sperm sample, and the mixture is slowly cooled to −196 °C. Thus, the actual dehydration occurs during cooling [6]. At a certain point, ice masses containing pure crystalline water are created. An unfrozen fraction remains between the growing ice masses, where all cells and solutes are retained. The concentrations of sugars, salts, and cryoprotectant increase while the volume of the unfrozen fraction declines. Increasing osmotic pressure causes an efflux of water from the cells, which minimizes intracellular ice formation. As cooling continues, the viscosity of the unfrozen fraction becomes too high for further crystallization. The remaining unfrozen fraction ultimately becomes an amorphous solid with no ice crystals [4].

Slow freezing is the preferred method for cryopreservation of human spermatozoa. Raw or washed fresh semen samples are generally used. At the same time, different techniques are available to preserve surgically retrieved spermatozoa [6]. For example, isolated sperm can be injected into an empty zona pellucida of a hamster oocyte and placed between two air bubbles inside a straw [19, 20]. Other studies have proposed freezing under a layer of paraffin oil with glycerol [21]. A frozen "testicular pill" combining a mixture of sperm and testicular tissue may be possible as well [22].

## 3 Materials and Equipment, Samples, and Reagents

### 3.1 Equipment

1. Phase contrast microscope.
2. MicroCell counting chamber.
3. Automatic pipettor.
4. Vortex.
5. −20 °C freezer.
6. LN2 dewar with 11″ canisters.
7. Sterile specimen container.
8. Sterile 15 mL centrifuge tubes with screw caps.
9. Sterile serological pipettes (2 mL capacity).
10. Sterile cryovials (1.8 mL capacity).
11. Colored cryomarkers.
12. Test tube racks (for 15 mL test tubes).
13. Cryovial racks.
14. Cryocanes.
15. Plastic cryosleeves.
16. Cryogloves.
17. Protective goggles.
18. Vinyl/nitrile gloves.
19. 37 °C incubator.
20. LN2 from supplier.
21. Viscosity treatment system (VTS) (REF 15524, Vitrolife, San Diego, CA).

### 3.2 Reagents

Freezing medium (TEST yolk buffer; TYB). Each bottle is sterile and for onetime use. Stored frozen at −20 °C until time of use.

### 3.3 Specimen

A fresh semen sample is collected in a sterile plastic container by masturbation in the collection room. The client should abstain from ejaculation between 48 and 72 h before collection (*see* **Note 1**).

Even semen samples with poor sperm quality can be cryopreserved. Specimens with poor semen quality can be successfully used by in vitro fertilization or intracytoplasmic sperm injection technique.

## 4 Methods and Procedures

1. The collected sample is labeled and placed in a 37 °C incubator for at least 20 min (*see* **Note 2**). Simultaneously, a bottle of TEST yolk buffer is thawed in the 37 °C incubator for at least 20 min (*see* Fig. 1).
2. After liquefaction, the sample is drawn into a 2 mL sterile pipette, leaving at least 50 μL in the specimen container. The volume is measured and recorded, and any unusual consistency is noted (*see* **Note 3**).

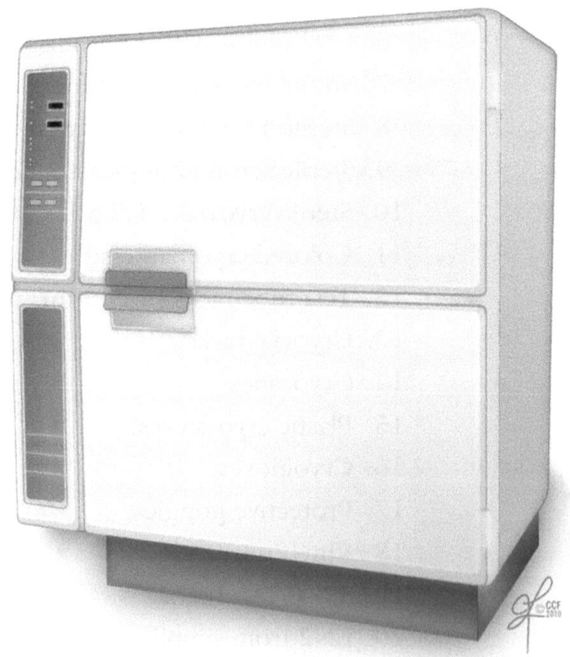

**Fig. 1** 37 °C incubator used for sample incubation and thawing of TEST yolk buffer. Reprinted with permission, Cleveland Clinic Center for Medical Art & Photography © 2010–2013. All Rights Reserved

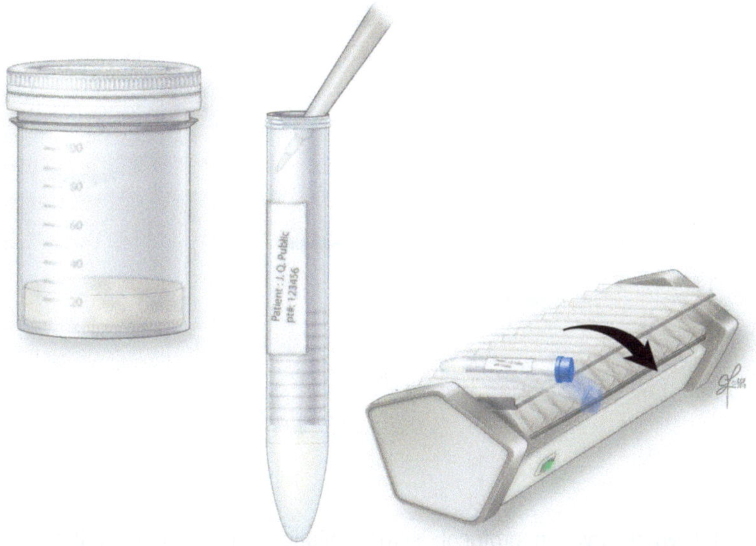


**Fig. 2** Cryopreservation protocol: shaker for the mixing of the cryoprotectants. Reprinted with permission, Cleveland Clinic Center for Medical Art & Photography


3. The specimen is placed into a labeled, sterile 15 mL conical tube previously checked for cracks or defects.
4. The remaining aliquot in the specimen container is analyzed to assess sperm motility (*see* **Note 4**).
5. Within 1 h of collection, an aliquot of freezing medium equal to 25 % of the original specimen volume is added to the centrifuge tube with a sterile pipette.
6. The specimen is gently rocked with the freezing media for 5 min on an aliquot mixer (*see* Fig. 2).
7. **Steps 5** and **6** are repeated three times or until the volume of freezing medium added is equal to the original specimen volume (*see* Fig. 3).
8. During the mixing steps above, cryomarkers are used to label the cryocanes. An additional 1.8 mL cryovial is labeled—this will contain a leftover aliquot of the cryodiluted specimen to be assessed for cryosurvival 24 h after freezing.
9. The cryodiluted specimen is visually inspected for motility. Percent motility is assessed using a MicroCell chamber and phase microscope. The percent motility is documented as "pre-cryo motility %."
10. The well-mixed, cryodiluted sample is evenly distributed into the pre-labeled vials using a 2 mL sterile serological pipette. At least 0.2 mL is added to the post-thaw test cryovial (*see* Fig. 4).

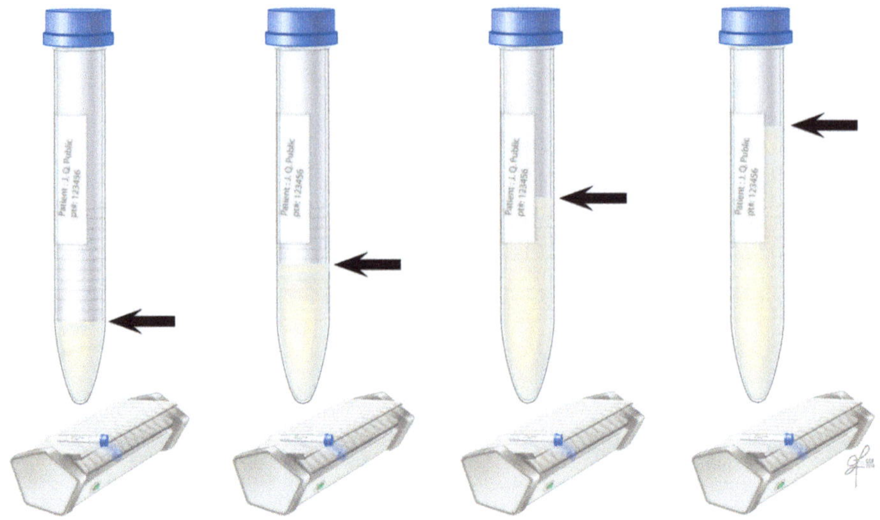

**Fig. 3** Cryopreservation protocol: semen dilution procedure with the TEST yolk buffer. Reprinted with permission, Cleveland Clinic Center for Medical Art & Photography © 2010–2013. All Rights Reserved

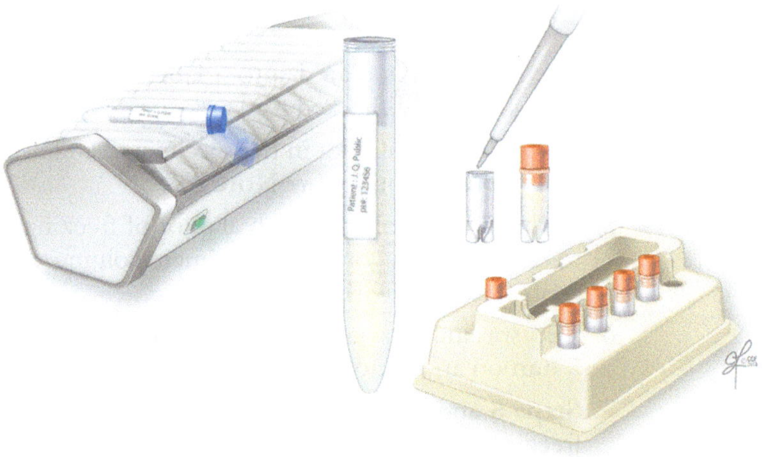

**Fig. 4** Distribution of the cryodiluted sample into pre-labeled vials. Reprinted with permission, Cleveland Clinic Center for Medical Art & Photography © 2010–2013. All Rights Reserved

11. Labeled vials are placed into the labeled cryocane(s) (*see* Fig. 5). A maximum of two cryovials are placed into bottom slots of canes right-side up, while the canes are held upside down (i.e., labeled portion of cryocanes will be facing the ground while placing the cryovials upright).
12. The cryocane(s) is covered with a cryosleeve(s) and placed in the −20 °C freezer for 8 min (*see* **Note 5**).
13. The cryocane(s) is then removed from the −20 °C freezer and placed into a $LN_2$ vapor tank for at least 2 h (*see* **Note 6**). Thereafter, the vials are frozen by the $LN_2$ vapors (−80 °C).

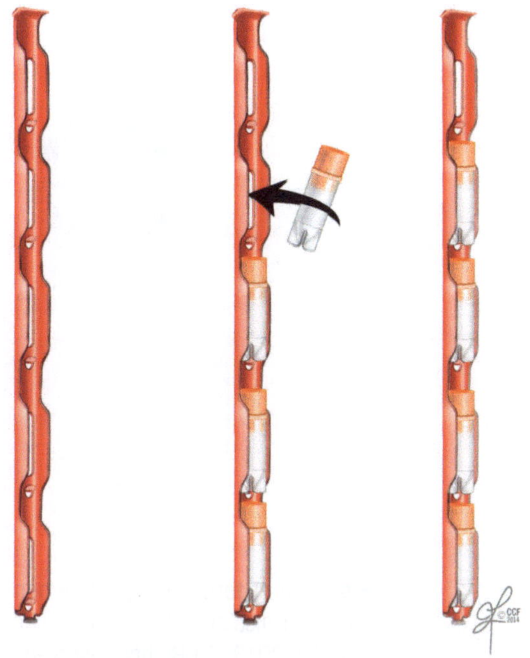

**Fig. 5** Placement of the labeled vials into the cryocane. Reprinted with permission, Cleveland Clinic Center for Medical Art & Photography© 2010–2013. All Rights Reserved

14. After at least 2 h of incubation in the $LN_2$ vapors (−80 °C), the canes are turned upside down and immersed into $LN_2$ (−196 °C) (*see* Fig. 6).
15. At least 24 h in $LN_2$ thereafter, the aliquot is thawed in the 1.8 mL post-thaw cryovial. The cane containing the cryovial is removed and snapped out. The cap is loosened and placed in the 37 °C incubator for 20 min.
16. The vial is mixed well and analyzed using a computer-assisted semen analyzer for spermatozoa motility parameters.
17. The results are recorded and the cryosurvival is evaluated using the following formula (*see* **Note 7**):

$$\frac{\%\text{ motility of post-thaw specimen}}{\%\text{ motility of pre-freeze specimen}}$$

18. For IUI, the thawed specimen is pipetted into a sterile non-spermatotoxic centrifuge tube and rinsed with an equal volume of sperm washing media containing human albumin.
19. After centrifugation at 300 × *g* for 7 min, the supernatant is discarded and the pellet is resuspended to a volume of 0.5 mL using sperm wash medium. Insemination should be performed as soon as possible

**Fig. 6** Short-term storage of semen samples in liquid nitrogen tanks. Reprinted with permission, Cleveland Clinic Center for Medical Art & Photography 

## 5 Notes

1. Sterile technique should be used throughout the sample processing. Gloves are mandatory. At the same time, latex can be toxic to sperm. Therefore, it is important to prevent contamination of the specimen with latex or talc or, preferably, not use them at all. Vinyl or nitrile gloves are recommended.
2. Only one sample is processed at a time to avoid mix-up or possible contamination. If more than one patient is scheduled for a specific time slot, then a second technologist should process the other sample.
3. Specimens that have a moderate or high viscosity that do not liquefy after incubation can be manipulated by pipetting them up and down with a sterile pipette or adding viscosity treatment system (VTS) whenever necessary.
4. The presence of round cells in the semen sample is counted, and a peroxidase staining test for white blood cells is done if $\geq$0.20 M/mL round cells are seen under wet preparation. At this point, any bacteria or foreign cells should be noted [23].
5. Exposure to freezing conditions should occur within 1.5 h following specimen collection [24].
6. Wear cryogloves and protective goggles when exposed to $LN_2$.

7. Quality control of the protocol consists of a pre-freeze and post-thaw evaluation of a new lot number of TEST yolk buffer using a donor sample that meets the following criteria:

   50 % survival, calculated by the following formula:

$$\frac{\%\ \text{post-thaw motility}}{\%\ \text{pre-freeze motility}}$$

## 6 Slow Freezing Versus Alternative Techniques: Pros and Cons

Slow, controlled sperm freezing is the cryopreservation method of choice for human spermatozoa (*see* Figs. 7, 8). The protocol [25, 26] results in thawed spermatozoa with excellent motion parameters including curvilinear velocity, straight-line velocity, and average path velocity [25].

However, in spite of reports showing successful sperm freezing with manual techniques, the reproducibility of this procedure could pose a variety of problems, especially to unexperienced technologists. For this reason, automated, computerized methods are

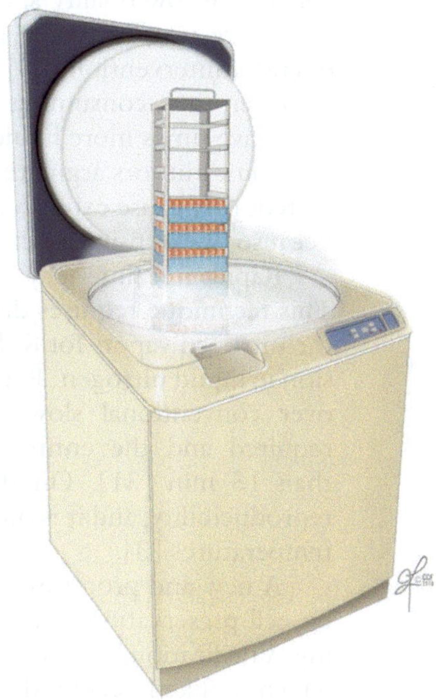

**Fig. 7** Long-term storage of semen sample in liquid nitrogen tank. The figure shows a metal canister containing boxes filled with cryovials. Reprinted with permission, Cleveland Clinic Center for Medical Art & Photography © 2010–2013. All Rights Reserved

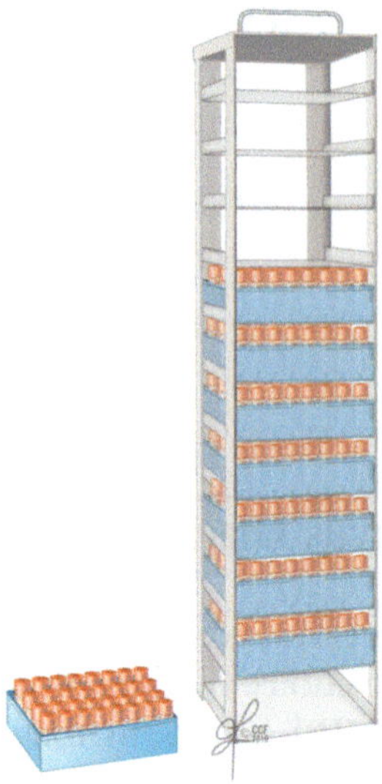

**Fig. 8** Freeze racks containing boxes filled with cryovials. Reprinted with permission, Cleveland Clinic Center for Medical Art & Photography © 2010–2013. All Rights Reserved

thought to be more exact and have been reported to limit cryodamage of low-quality spermatozoa [27]. Furthermore, the programs are usually simple to use and do not require continuous operator intervention [28]. On the other hand, automated freezers are often time-consuming and expensive to purchase and also use up to five times more liquid nitrogen [29].

Some authors argue that slow freezing, either manual or automated, may cause extensive chemico-physical damage to the sperm because of extensive ice crystallization [30].

Rapid freezing may be used as an alternative option [3]. This technique requires direct contact between the cryotubes and the nitrogen vapors for 8–10 min only, followed by a rapid immersion in liquid nitrogen. Rapid sperm freezing has several advantages over conventional slow freezing: no programmable freezer is required and the entire freezing/thawing procedure takes less than 15 min [31]. On the other hand, the technique has low reproducibility, and it is difficult to control the drop in the freezing temperatures [3].

A new and promising freezing technique is vitrification, based on cell preservation at extremely low temperatures without freezing. Vitrification involves the formation of an amorphous solid state which, unlike freezing, does not involve the formation of ice crystals, which can damage delicate cellular structures. Modern vitrification techniques are ice- and cryoprotectant-free, in which a sperm suspension is plunged directly into liquid nitrogen [32, 33]. It is a relatively simple and straightforward approach to preserve sperm motility and fertilizing ability, as it omits the need for permeable cryoprotectants, preventing the lethal effects of osmotic shock [34]. Sperm vitrification is, however, a relatively new technique and has not yet been standardized for use in clinical andrology laboratory.

## 7 Concluding Remarks

The impact of disease, environment, and medical treatment may impair sperm quality and lead to male infertility. Therefore, effective promotion of sperm banking is crucial, based on adequate counseling focused on the severity and estimated risk for infertility, assessment of the patient's plans to have children in the future, and emphasis on the benefits of banking as well as on possible difficulties such as cost or cultural factors.

Continuing research on developing fast, simple, and cost-effective protocols for semen cryopreservation is needed. The potential of cryopreservation continues to be explored currently, focusing on cryoprotectant-free sperm vitrification and refining the freezing and thawing protocols to obtain optimal outcomes.

### References

1. Gupta S, Sekhon LH, Agarwal A (2011) Sperm banking: when, why and how? In: Sabanegh E (ed) Current clinical urology: male infertility: problems and solutions, 1st edn. Springer Science+Business Media, LLC, New York
2. Polge C, Smith AU, Parkes AS (1949) Revival of spermatozoa after vitrification and dehydration at low temperatures. Nature 164:666
3. Di Santo M, Tarozzi N, Nadalini M, Borini A (2012) Human sperm cryopreservation: update on techniques, effect on DNA integrity, and implications for ART. Adv Urol 2012:1–12
4. FAO (2012) Basic principles of cryopreservation. In: FAO (ed) Cryoconservation of animal genetic resources , RomeFAO Animal Production and Health Guidelines No. 12
5. Varghese AC, Nandi P, Mahfouz R, Athayde KS, Agarwal A (2010) Human sperm cryopreservation. In: Rao AR, Agarwal A, Srinivas MS (eds) Andrology laboratory manual. Jaypee Brothers Medical Publishers Ltd., New Delhi
6. Gupta S, Agarwal A, Sharma R, Ahmady A (2014) Sperm banking via cryopreservation. In: Rizk Botros RMB, Aziz N, Agarwal A, Sabanegh E (eds) Medical and surgical management of male infertility. Jaypee Brothers Medical Publishers Ltd., New Delhi
7. Agarwal A, Allamaneni SS (2005) Disruption of spermatogenesis by the cancer disease process. J Natl Cancer Inst Monogr 34:9–12
8. Perdichizzi A, Nicoletti F, La Vignera S, Barone N, D'Agata R, Vicari E, Calogero AE (2007) Effects of tumour necrosis factor-alpha on human sperm motility and apoptosis. J Clin Immunol 27:152–162
9. Ragheb AM, Sabanegh ES (2010) Male fertility-implications of anticancer treatment and strategies to mitigate gonadotoxicity. Anticancer Agents Med Chem 10:92–102
10. Howell SJ, Shalet SM (2005) Spermatogenesis after cancer treatment: damage and recovery. J Natl Cancer Inst Monogr 34:12–17
11. Shalet SM, Tsatsoulis A, Whitehead E, Read G (1989) Vulnerability of the human Leydig cell to radiation damage is dependent upon age. J Endocrinol 120:161–165
12. Colpi GM, Contalbi GF, Nerva F, Sagone P, Piediferro G (2004) Testicular function following chemo-radiotherapy. Eur J Obstet Gynecol Reprod Biol 113(Suppl 1):S2–S6
13. Ranganathan P, Mahran AM, Hallak J, Agarwal A (2002) Sperm cryopreservation for men with nonmalignant, systemic diseases: a descriptive study. J Androl 23:71–75

14. Hamada A, Wasik M, Gupta S, Agarwal A (2012) Sperm banking: indications and regulations. In: AK D (ed) Infertility: diagnosis, management and IVF. Jaypee Brothers Medical Publishers Ltd., New Delhi
15. Anger JT, Gilbert BR, Goldstein M (2003) Cryopreservation of sperm: indications, methods and results. J Urol 170(4 Pt 1):1079–1084
16. Gupta S, Agarwal A, Sharma R, Ahmady A (2011) Recovery, preparation, storage and utilization of spermatozoa for fertility preservation in cancer patients and subfertile men. J Reprod Stem Cell Biotech 1:150–168
17. Cherry N, Moore H, McNamee R, Tulandi T (2008) Occupation and male infertility: glycol ethers and other exposures. Occup Environ Med 65(10):708–714
18. Mishra DP, Shaha C (2005) Estrogen-induced spermatogenic cell apoptosis occurs via the mitochondrial pathway: role of superoxide and nitric oxide. J Biol Chem 280 (7):6181–6196
19. Cohen J, Garrisi GJ, Congedo-Ferrara TA, Kieck KA, Schimmel TW, Scott RT (1997) Cryopreservation of single human spermatozoa. Hum Reprod 12(5):994–1001
20. Hsieh Y, Tsai H, Chang C, Lo H (2000) Cryopreservation of human spermatozoa within human or mouse empty zona pellucidae. Fertil Steril 73(4):694–698
21. Craft I, Tsirigotis M (1995) Simplified recovery, preparation and cryopreservation of testicular spermatozoa. Hum Reprod 10 (7):1623–1626
22. Romero J, Remohi J, Minguez Y, Rubio C, Pellicer A, Gil-Salom M (1996) Fertilization after intracytoplasmic sperm injection with cryopreserved testicular spermatozoa. Fertil Steril 65(4):877–879
23. Saleh RA, Agarwal A, Kandirali E, Sharma RK, Thomas AJ, Nada EA, Evenson DP, Alvarez JG (2002) Leukocytospermia is associated with increased reactive oxygen species production by human spermatozoa. Fertil Steril 78 (6):1215–1224
24. Pasqualotto FF, Pasqualotto EB, Agarwal A (2009) Male germ cell cryobanking. In: Talwar P (ed) A comprehensive manual of cryobiology for assisted reproduction and stem cells. Jaypee Brothers Medical Publishers, New Delhi Chapter 15, pp 211–222
25. Nallella KP, Sharma RK, Allamaneni SS, Aziz N, Agarwal A (2004) Cryopreservation of human spermatozoa: comparison of two cryopreservation methods and three cryoprotectants. Fertil Steril 82(4):913–918
26. Kobayashi H, Ranganathan P, Mahran AM, Sharma RK, Thomas AJ, Agarwal A (2001) Comparison of two cryopreservation protocols for freezing human spermatozoa. Fertil Steril 76(3 Suppl 1):S229–S230
27. Ragni G, Caccamo AM, Dalla Serra AD, Guercilena S (1990) Computerized slow-staged freezing of semen from men with testicular tumors or Hodgkin's disease preserves sperm better than standard vapor freezing. Fertil Steril 53(6):1072–1075
28. Holt WV (2000) Basic aspects of frozen storage of semen. Anim Reprod Sci 62(1–3):3–22
29. Paras L, Freisinger J, Esterbauer B, Schmeller N, Szlauer R, Jungwirth A (2008) Cryopreservation technique: comparison of Test yolk buffer versus SpermCryo and vapour versus computerised freezing. Andrologia 40 (1):18–22
30. Mazur P, Rall WF, Rigopoulos N (1981) Relative contributions of the fraction of unfrozen water and of salt concentration to the survival of slowly frozen human erythrocytes. Biophys J 36(3):653–675
31. Vutyavanich T, Piromlertamorn W, Nunta S (2010) Rapid freezing versus slow programmable freezing of human spermatozoa. Fertil Steril 93(6):1921–1928
32. Nawroth F, Isachenko V, Dessole S, Rahimi G, Farina M, Vargiu N, Mallmann P, Dattena M, Capobianco G, Peters D, Orth I, Isachenko E (2002) Vitrification of human spermatozoa without cryoprotectants. Cryo Letters 23 (2):93–102
33. Isachenko E, Isachenko V, Katkov II, Dessole S, Nawroth F (2003) Vitrification of mammalian spermatozoa in the absence of cryoprotectants: from past practical difficulties to present success. Reprod Biomed Online 6(2):191–200
34. Isachenko V, Isachenko E, Katkov II, Montag M, Dessole S, Nawroth F, van der Ven H (2004) Cryoprotectant-free cryopreservation of human spermatozoa by vitrification and freezing in vapor: effect on motility, DNA integrity, and fertilization ability. Biol Reprod 71(4):1167–1173

# Chapter 6

# Technology of Aseptic Cryoprotectant-Free Vitrification of Human ICSI Spermatozoa

**Vladimir Isachenko, Raul Sanchez, Peter Mallmann, Gohar Rahimi, and Evgenia Isachenko**

## Abstract

The aim of this chapter was to describe the standardized aseptic technology of permeable cryoprotectant-free vitrification of human spermatozoa in capillaries (for ICSI or IVF in microvolume). Spermatozoa, vitrified by this technology, are free from seminal plasma owing to swim-up procedure preceding vitrification and are free from permeable cryoprotectants. They are ready for further use immediately after warming without any additional treatment.

**Key words** Human, Spermatozoa, ICSI, Vitrification, Technique

## 1 Introduction

Cryobiology is a rapidly evolving field which only relatively recently has found broad applications in reproductive medicine. However, as any emerging technology, it has both a great potential and a need for further developments [1]. The use of programmable or non-programmable "slow" (conventional) freezing [2–4] allows to preserve relatively large volumes of diluted ejaculate or sperm preparations from 0.25 to 1.0 mL with acceptable rates of motility of spermatozoa after thawing and adequate integrity of acrosomal and cytoplasmic membranes [5–7]. It is also instrumental in minimizing the cryopreservation-induced cell deterioration resulting in membrane destabilization processes, such as membrane phosphatidylserine translocation [8–10].

One of the relatively recently emerged technologies within the field of the reproductive cryobiology is the spermatozoa vitrification (cryopreservation by direct plunging into liquid nitrogen). This method is based on the rapid cooling of the cells by immersion into liquid nitrogen, and, thereby, is the key factor reducing the chance of the formation of big ice crystals. In contrast to the

Zsolt Peter Nagy et al. (eds.), *Cryopreservation of Mammalian Gametes and Embryos: Methods and Protocols*, Methods in Molecular Biology, vol. 1568, DOI 10.1007/978-1-4939-6828-2_6, © Springer Science+Business Media LLC 2017

programmable ("slow") conventional freezing, vitrification has series of technological advantages useful for the practice: it renders the use of permeable cryoprotectants superfluous and, in addition, is much faster, simpler in application and more cost-effective than conventional freezing. At the same time vitrification can effectively protect spermatozoa from cryo-injuries associated with cooling-warming [11–17]. In particular, the mutagenic effects of permeable cryoprotectants [18] can be avoided.

At the present, most reported applications of this technology describe vitrification of only very small volume of spermatozoa suspensions: from 1 to 10 μL with isolation from liquid nitrogen [15] and from 20 to 30 μL with direct plunging/dropping of suspension with spermatozoa into liquid nitrogen [11, 15, 16]. Vitrification of up to 10 μL spermatozoa suspension, as it is being used in daily practice [15], presumes that open pulled straws [19] are used. Current industrial technology does not yet enable the manufacturing of such open pulled straws with uniform (standard) diameter of pulled part of straw. This shortcoming is reflected in nonuniformity of the vitrification regime. On the other hand, industrial suppliers offer plastic capillaries of regular cylinder shape with stable (fixed) diameter for medical applications. Vitrification process can be standardized using these standard capillaries, which would unfold a great potential for assisted reproduction using vitrified spermatozoa.

The aim of this chapter is to describe the standardized aseptic technology of permeable cryoprotectant-free vitrification of human spermatozoa in capillaries (for ICSI or IVF).

## 2 Materials and Methods

Except where otherwise stated, obtain all chemicals from Sigma (Sigma Chemical Co., St. Louis, MO, USA).

### 2.1 Spermatozoa Preparation

1. After informed consent, obtain ejaculates from patients with oligo-astheno-terato-azoospermia in anamneses who were either patients of an infertility clinic (*see* **Notes 1** and **2**).
2. Carry out vitrification on swim-up prepared spermatozoa (*see* **Note 3**).
3. Prepare solution 0.5 M of sucrose with bi-distillate water (*see* **Note 4**).
4. Prior to cooling (plunging into liquid nitrogen), dilute spermatozoa 1:1 with 0.5 M sucrose at room temperature (*see* **Note 5**).
5. Fill the capillary with 10 μL of spermatozoa suspension by aspiration (Fig. 1a) (*see* **Notes 6** and 7).

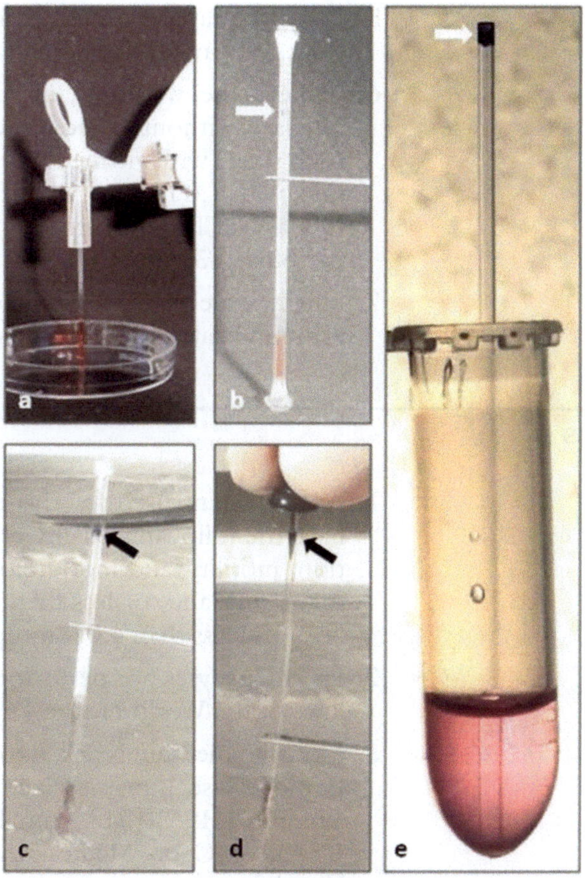


**Fig. 1** Schematic illustration of spermatozoa vitrification using 50-μL capillaries. (*arrows*) Marked end of 50-mL capillary; (**a**) aspiration of spermatozoa suspension in capillary, (**b**) 50-mL capillary sealed in a 0.25-mL straw, (**c**) cutting of 0.25-mL straw, (**d**) expelling of 50-mL capillary from 0.25-mL straw, (**e**) warming of spermatozoa

6. After the aspiration was completed, insert the capillary into 0.25 mL straw (*see* **Note 8**).
7. Seal the end of this straw in advance using heat sealer (Cryo Bio System, Paris, France) (Fig. 1b) (*see* **Note 9**).

### 2.2 Spermatozoa Cooling and Warming

1. Plunge the straw into liquid nitrogen (*see* **Note 10**).
2. For warming, fix tightly the end of the capillary, while the part of straw containing spermatozoa is still submerging in liquid nitrogen (Fig. 1c, d).
3. Disinfect the straw with 70 % ethanol in the area where the marked end of the capillary is located (Fig. 1, arrows).
4. Cut the upper part of the straw with sterile scissors as close as possible to the marked end of the capillary, just above the mark without touching the marked end of the capillary.

5. Expel the capillary with a conical bolt (Fig. 1d) (*see* **Note 11**) from isolating 0.25 mL straw.
6. Achieve the warming up of spermatozoa by immersing of capillary without conical apex (capillary must be open from both sides) with vitrified spermatozoa into 1.8 mL centrifuged tube with 1.0 mL pre-warmed to 43 °C physiological solution (0.9 % NaCl) (*see* **Note 12**) (for approximately 20 s (Fig. 1e) (*see* **Note 13**).
7. Finally, expel the suspension of spermatozoa from the capillary for immediate evaluation of spermatozoa quality and ICSI.

## 3 Notes

1. All specimens used for this methodology of cryopreservation have the following quality criteria for spermatozoa concentration, motility, and morphology: less than five million spermatozoa/mL and/or less than 32 % progressive motile ("a" and "b") and less than 4.0 % morphologically normal spermatozoa.
2. Sperm analyses are performed according to published guidelines of the World Health Organization (WHO) [20].
3. Swim-up medium is an "automatically" basic medium for vitrification. Use human tubal fluid (HTF) with human serum albumin (HSA) [21] from Irvine Scientific (Santa Ana, CA, USA), or Cook Medical Inc. (Bloomington, IN, USA), or Vitrolife (Göteborg, Sweden). Prepared (ready for use) vitrification medium includes swim-up medium and sucrose.
4. Prepare solution of 0.5 M sucrose with bi-distillate water, pass through 0.22 μm filter, and then store frozen until use. Before preparation of vitrification medium, thaw 0.5 M sucrose and maintain at room temperature.
5. The final concentration of sucrose 0.25 M.
6. Specifically for our purposes, 50 μL plastic capillaries were manufactured from hydrophobic material as vehicles for cooing of spermatozoa (Gynétics Medical Products, Lommel, Belgium) (Fig. 1). The end of the straw is labeled on the top to mark the cutting-off position (Fig. 1, arrows).
7. It was absolutely crucial to avoid that the inner surface of the capillary becomes moist during packaging procedure. Aspirating the volume of sperm cell suspension above the mark and correcting it by lowering the fluid level inside the capillary after aspiration are technologically wrong and would result in excess of portion's volume after thawing.
8. Produced by Gynétics Medical Products, Lommel, Belgium.

9. Hermetical heat sealing of 0.25 mL straw can be achieved using flame of alcohol burner and forceps or any commercial equipment (including ultrasound equipment because of large distance between spermatozoa suspension and focus of sonographic appliance).
10. Cooling speed 600 °C/min.
11. For this purpose, instead the conical bolt forceps can be used in place.
12. The type and quality of warming medium place the secondary role because at warming there was no contact between spermatozoa and this warming medium due to the hydrophobic material of the capillary.
13. Volume of vitrified suspension after warming is not increased (Fig. 1e). The above meniscus of spermatozoa suspension is located under the level of warming medium because of hydrophobic material of the capillary.

## References


1. Petrunkina AM, Volker G, Weitze KF et al (2005) Detection of cooling-induced membrane changes in the response of boar sperm to capacitating conditions. Theriogenology 63:2278–2299
2. McLaughlin EA, Ford WC, Hul MG (1990) A comparison of the freezing of human semen in the uncirculated vapour above liquid nitrogen and in a commercial semi-programmable freezer. Hum Reprod 5:724–728
3. Yin HZ, Seibel MM (1999) Human sperm cryobanking. Use of modified liquid nitrogen vapor. J Reprod Med 44:87–90
4. Stanic P, Tandara M, Sonicki Z, Simunic V, Radakovic B, Suchanek E (2000) Comparison of protective media and freezing techniques for cryopreservation of human semen. Eur J Obstet Gyn Reprod Biol 91:65–70
5. Sawetawan C, Bruns ES, Prins GS (1993) Improvement of post-thaw sperm motility in poor quality human semen. Fertil Steril 60:706–710
6. Larson JM, McKinney KA, Mixon BA (1997) An intrauterine insemination-ready cryopreservation method compared with sperm recovery after conventional freezing and post-thaw processing. Fertil Steril 68:143–148
7. Hammadeh ME, Askari AS, Georg T (1999) Effect of freeze-thawing procedure on chromatin stability, morphological alteration and membrane integrity of human spermatozoa in fertile and subfertile men. Int J Androl 22:155–162
8. Glander HJ, Schaller J (2000) Hidden effects of cryopreservation on quality of human spermatozoa. Cell Tissue Bank 1:133–142
9. Schuffner A, Morshedi M, Oehninger S (2001) Cryopreservation of fractionated, highly motile human spermatozoa: effect on membrane phosphatidylserine externalization and lipid peroxidation. Hum Reprod 16:2148–2153
10. Duru NK, Morshedi M, Schuffner A et al (2001) Cryopreservation-thawing of fractionated human spermatozoa and plasma membrane translocation of phosphatidylserine. Fertil Steril 75:263–268
11. Nawroth F, Isachenko V, Dessole S et al (2002) Vitrification of human spermatozoa without cryoprotectants. Cryo Letters 23:93–102
12. Isachenko E, Isachenko V, Katkov II et al (2003) Vitrification of human spermatozoa without cryoprotectants: review of problem and practical success. Reprod Biomed Online 6:191–200
13. Isachenko E, Isachenko V, Katkov II et al (2004) DNA integrity and motility of human spermatozoa after standard slow freezing versus cryoprotectant-free vitrification. Hum Reprod 19:932–939
14. Isachenko V, Isachenko E, Katkov II et al (2004) Cryoprotectant-free cryopreservation of human spermatozoa by vitrification and freezing in vapor: effect on motility, DNA integrity, and fertilization ability. Biol Reprod 71:1167–1173

15. Isachenko V, Isachenko E, Montag M et al (2005) Clean technique for cryoprotectant—free vitrification of human spermatozoa. Reprod Biomed Online 10:350–354
16. Isachenko E, Isachenko V, Weiss JM et al (2008) Acrosomal status and mitochondrial activity of human spermatozoa vitrified with sucrose. Reproduction 136:167–173
17. Isachenko V, Isachenko E, Petrunkina AM et al (2012) Human spermatozoa vitrified in the absence of permeable cryoprotectants: birth of two healthy babies. Reprod Fertil Dev 24:323–326
18. Fraga CG, Motchnik PA, Shigenaga MK et al (1991) Ascorbic acid protects against endogenous oxidative DNA damage in human sperm. Proc Nat Acad Sci USA 88:11003–11006
19. Vajta G, Kuwayama M, Holm P et al (1998) Open pulled straw (OPS) vitrification: a new way to reduce cryoinjuries of bovine ova and embryos. Mol Reprod Dev 51:53–58
20. WHO (2010) World Health Organization (WHO) laboratory manual for the examination and processing of human semen, 5th edn. Geneva, Switzerland
21. Quinn P, Warnes GM, Kerin JF et al (1985) Culture factors affecting the success rate of in vitro fertilization and embryo transfer. Ann N Y Acad Sci 442:195–204

# Chapter 7

# Human Epididymal and Testicular Sperm Cryopreservation

**Pankaj Talwar and Sarabpreet Singh**


## Abstract

Since the advent of intracytoplasmic sperm injection (ICSI) in the early 1990s, surgical techniques to recover samples from the epididymis and testis directly have been used to benefit patients suffering from obstructive and nonobstructive azoospermia. Various studies have demonstrated comparable fertilization, ongoing pregnancy, and implantation rates when fresh and frozen-thawed epididymal sperms were used for ICSI [1]. Injection of fresh and frozen testicular sperms into mature oocytes resulted in similar fertilization rates in cases of obstructive azoospermia. However, in cases of nonobstructive azoospermia, the outcome depends upon the degree of impairment of spermatogenesis, criteria for sperm freezing, and patient selection [2].


**Key words** Azoospermia, Surgical sperm retrieval, Obstructive azoospermia (OA), Nonobstructive azoospermia (NOA), Testicular sperm aspiration (TESA), Testicular sperm extraction (TESE), Sperm cryopreservation, Intracytoplasmic sperm injection (ICSI)

## 1 Introduction

Until the mid-1990s, all patients with obstructive or nonobstructive azoospermia were described as untreatable due to lack of understanding of male reproduction and cryopreservation techniques. The scenario completely changed after the advent of ICSI, which was combined with surgical techniques to retrieve sperm from the epididymis and testis directly.

The technique of testicular sperm extraction was first introduced in 1993. The first birth after using frozen epididymal and testicular sperm for ICSI was achieved in 1995 and 1996, respectively [3].

Zsolt Peter Nagy et al. (eds.), *Cryopreservation of Mammalian Gametes and Embryos: Methods and Protocols*, Methods in Molecular Biology, vol. 1568, DOI 10.1007/978-1-4939-6828-2_7, © Springer Science+Business Media LLC 2017

## 2 Materials

### 2.1 Epididymal/ Testicular Sperm Aspiration and Mechanical Disaggregation

1. Sperm washing media-gamete buffer media, Code K-SIGB, William A Cook, Australia Pty. Ltd.
2. BD Falcon 357575 pipette.
3. BD Falcon 357575 dishes.
4. BD Tuberculin syringe 1 mL.
5. Polystyrene conical tubes.
6. Syringes (3 mL) with 21-gauge needles.
7. Sterile pair of curved iris scissors.
8. Stereo zoom microscope.
9. Inverted microscope.

### 2.2 Enzymatic Procedures

#### 2.2.1 Collagenase Media

1. GM501 Collagenase (GM501 CollagenaseTM, Gynemed, Lensahn, Germany).
2. GM501 SpermAir (GM501 SpermAirTM, Gynemed, Lensahn, Germany).
3. Benchtop centrifuge.

#### 2.2.2 Erythrocyte-Lysing Buffer (RBC Lysis Buffer): Composition

1. Ammonium chloride (NH4Cl) 0.829 g.
2. Potassium bicarbonate (NaHCO3) 0.100 g.
3. Ethylenediaminetetraacetic acid (EDTA) 0.074 g.
4. Tissue culture-grade water 100 mL.

### 2.3 Cryopreservation of Epididymal/ Testicular Sperm

1. Sperm washing medium, gamete buffer media (Code K-SIGB, William A Cook, Australia Pty. Ltd.).
2. The freezing medium (SpermFreeze-FertiPro NV, Beernem, Belgium).
3. RBC lysis buffer.
4. HSV straws (Irvine Scientific; Irvine, CA).
5. Computer-controlled biological freezer (Nicool LM10, France).
6. Cryovials (Code CE 0543, Nunc, Denmark).
7. Aluminum canes.
8. Plastic cryosleeves.
9. Liquid nitrogen dewar.
10. Liquid nitrogen.
11. Personal protective gear for handling liquid nitrogen.

## 3 Methods

### 3.1 How to Obtain Epididymal and Testicular Sperm

#### 3.1.1 Epididymal Sperms Are Obtained by

1. Open microscopic surgery (**MESA**).
2. Percutaneus Epididymal Sperm Aspiration (**PESA**): a 21-gauge "butterfly" or equivalent needle is used to aspirate fluid (Fig. 1).

Epididymal aspirate hence obtained is suspended in the sperm handling media (Fig. 2). Usually, high sperm numbers with pure fraction of spermatozoa are visualized under the microscope. The sample should be prepared by density gradient centrifugation or single wash at low RPM.

#### 3.1.2 Testicular Sperm Is Obtained by

1. Open biopsy (testicular sperm extraction, **TESE** (Fig. 3), or testicular sperm aspiration, **TESA** (Fig. 4))
2. Percutaneous needle biopsy (testicular fine needle aspiration, TEFNA)

In most cases, seminiferous tubules are obtained (Fig. 5).

### 3.2 Processing

The samples obtained after surgical removal should be processed to facilitate release of the sperms from seminiferous tubules. Depending upon their personal experience and choice, biologist worldwide uses various methods.

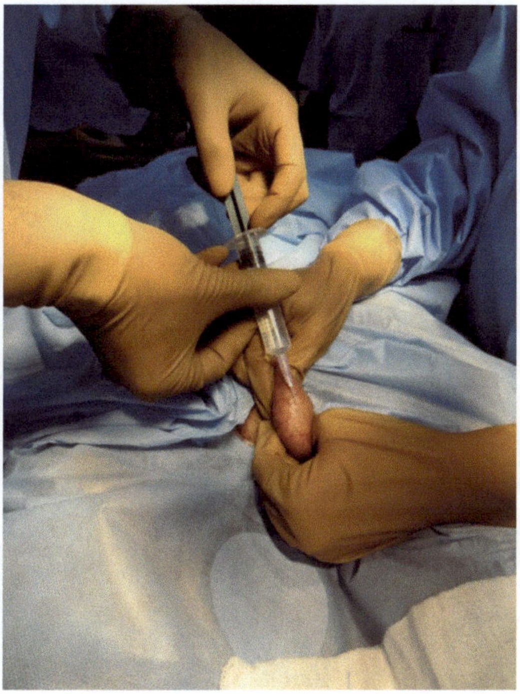

**Fig. 1** Percutaneous epididymal sperm aspiration (PESA) being performed

#### 3.2.1 Mechanical Procedures

Seminiferous tubules are ruptured to release sperms. This is achieved by either of the following methods:

1. Shredding the tissue with sterile glass slides (especially in cases with high tubule elasticity)
2. Needle dissection and teasing (Fig. 6)

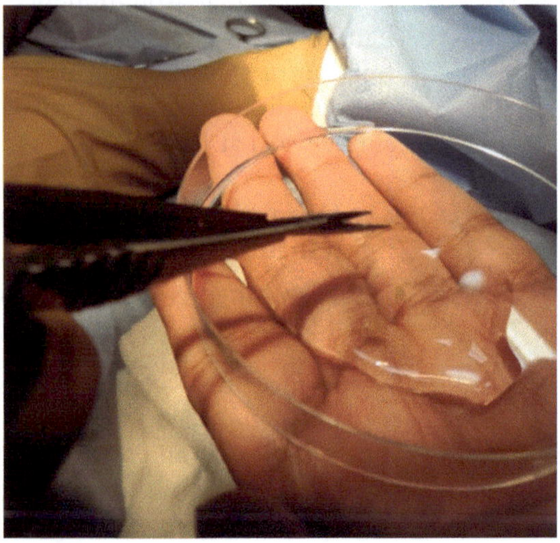

**Fig. 2** Testicular/epididymal aspirates suspended in buffered media

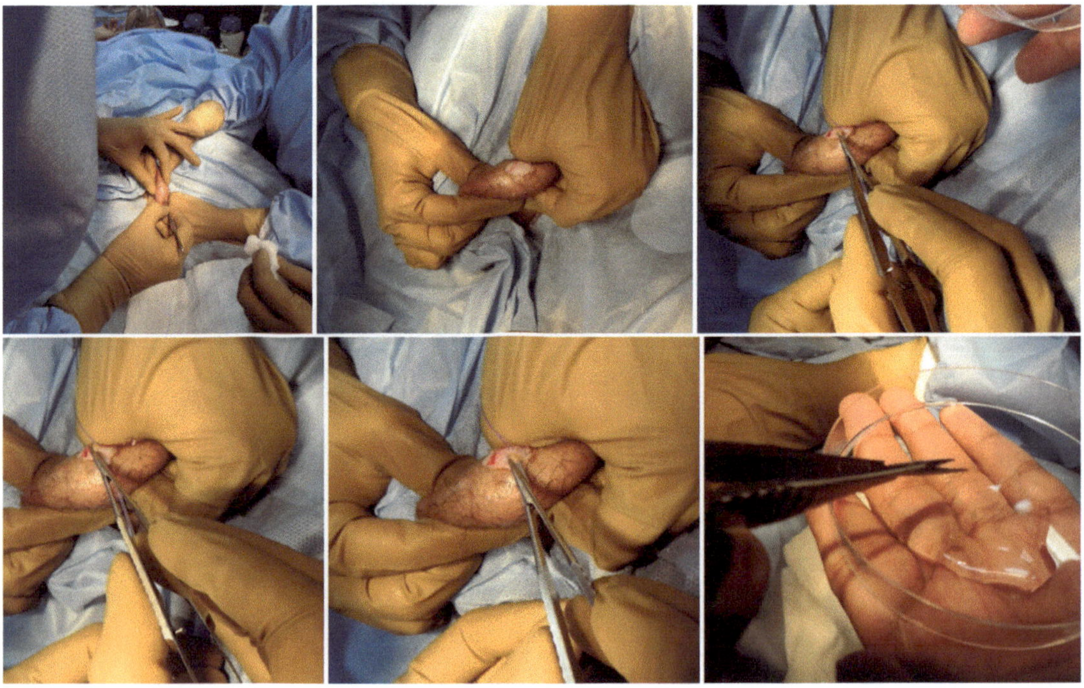

**Fig. 3** Showing testicular sperm extraction (TESE)

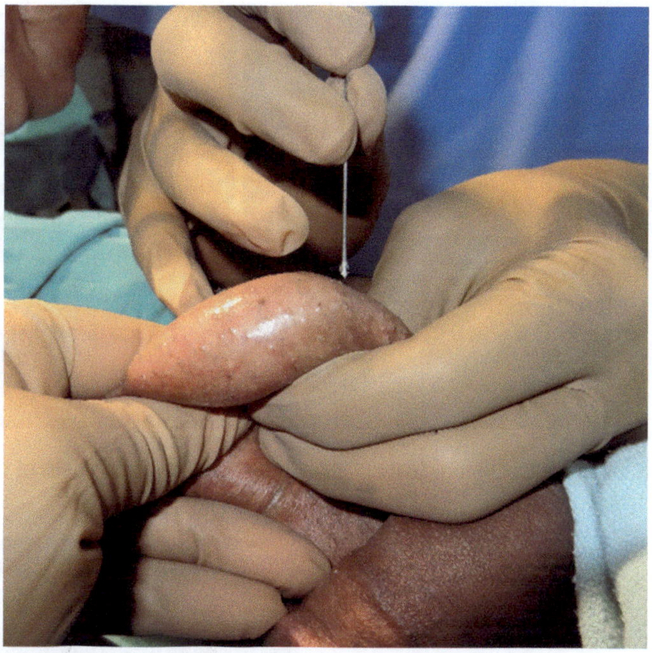

**Fig. 4** Showing testicular sperm aspiration (TESA)

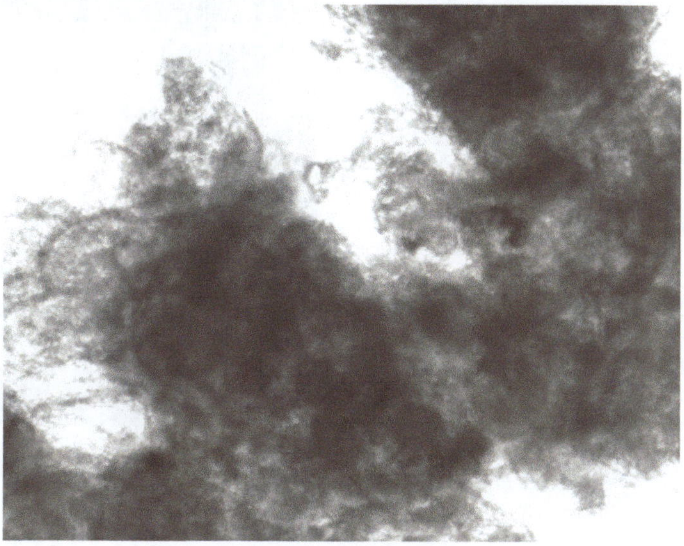

**Fig. 5** Showing convoluted seminiferous tubules after TESA/TESE

3. Dissection of the tissue using micro-scissors (Fig. 7)
4. Macerating the tubules with a micro-grinder

Method and Processing

1. Testicular tissue is obtained by TESA/TESE and can be processed at room temperature or at 37 °C.
2. With the help of a sterile 3 mL pipette (BD Falcon 357575), rinse the tissue in HEPES-based sperm washing medium

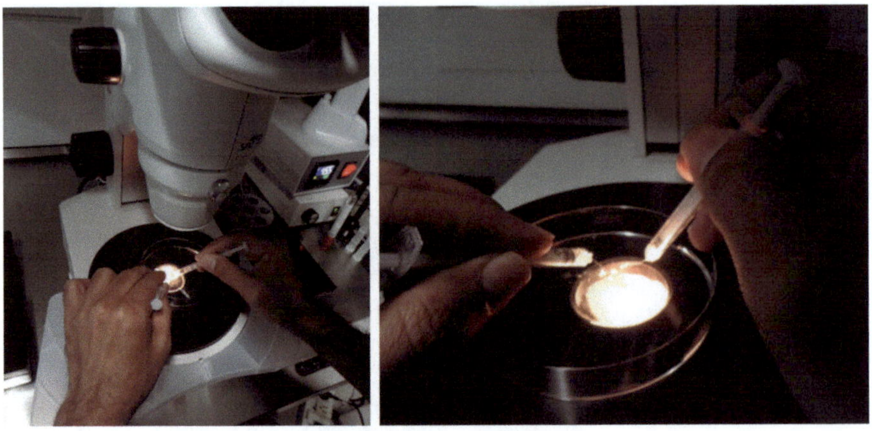

**Fig. 6** Showing dissection of testicular aspirate/tissue with the help of needles

**Fig. 7** Showing dissection of testicular aspirate/tissue with the help of microscissors

(gamete buffer media, Code K-SIGB, William A Cook, Australia Pty. Ltd.) to remove the blood. Transfer it to a sterile petri dish (BD Falcon 353037) containing fresh sperm washing media.

3. Seminiferous tubules are visualized in the biopsy sample under the dissecting microscope (stereoscope) (Fig. 5).
4. The sample is teased and spread out using fine forceps and dissected into small lengths of 1 cm or less. Now, squeeze them out from the middle to an open end.

5. Alternatively, mince the sample with surgical needles (Fig. 6) or sterile curved iris scissors (Fig. 7) under the stereo zoom microscope [4]. Use one of the syringes to hold the testicular tissue and the other to squeeze out the tubules.
6. Take care that the tissue is kept moist with the sperm washing medium at all times during mincing.
7. Examine the suspension under the inverted microscope at 200X/400X magnification.
8. Testicular sperms are seen along with seminiferous tubules, RBCs, and Sertoli cells. They are usually immotile though motile sperms may be seen.
9. Load the testicular suspension into a conical test tube and allow larger cells and tissue pieces to sediment for 1 min.
10. The supernatant medium is aspirated and processed.
11. 2.0 mL sperm washing medium is added to the larger fragments and aspirated in a 3 mL syringe.
12. Then it is made to pass through a 21-gauge needle repeatedly to dissociate the seminiferous tubules and release the sperms from the lumen (Fig. 4).
13. Depending upon the number of sperms appreciated in the aspirate, they can be prepared by either:
    (a) Large numbers of sperms obtained—Single density gradient.
    (b) Low sperm count—centrifuging at 400 × *g* for 10 min after adding sperm processing media. This concentrates sperms in a small volume. A second wash can be done in the sperm washing media, and the pellet hence obtained is agitated in 0.3 mL of the same medium. This is now added to microdroplets in the ICSI injection dish.

Post-preparation, the result can be either:

1. No sperm retrieved (Fig. 8)
2. Only RBCs seen (Fig. 9)
3. Few motile/immotile spermatozoa retrieved (Fig. 10)

Live testicular sperms can be identified and ICSI attempted (Fig. 11).

### 3.2.2 Enzymatic Procedures

Enzymatic digestion of testicular digestion is used when:

1. No spermatozoa are found.
2. Only immotile spermatozoa are seen.
3. After mechanical preparation, spermatozoa attach to other cell types in the suspension.

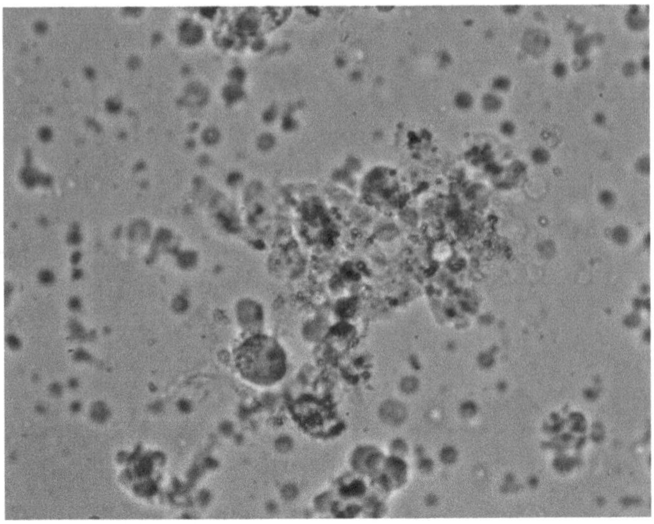

**Fig. 8** Testicular aspirate showing fibrosed tubules. No sperms are appreciated

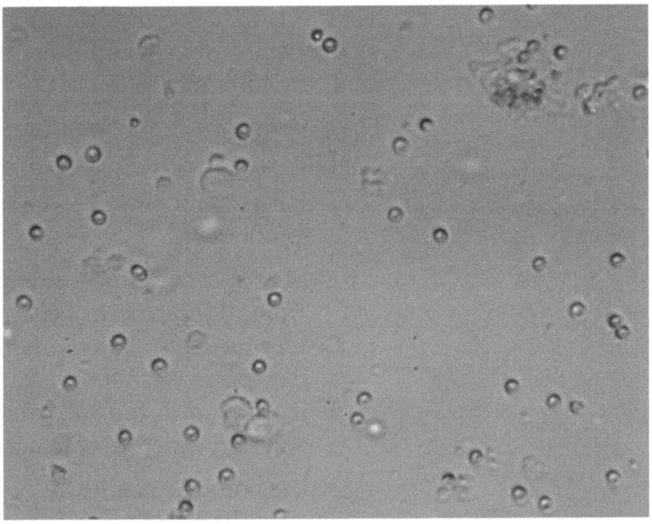

**Fig. 9** Showing testicular aspirate with no sperms retrieved. RBCs are appreciated

The options available are:

(a) *Collagenase type I or IV*: It is used to degrade collagen in basement membrane and matrix especially in NOA patients [5–7]. Collagenase IV has been shown to result in higher number of intact sperm cells compared to collagenase I, and its use is recommended to decrease the number of sperm recovery failures [6].

(b) *Erythrocyte-lysing buffer* is used in aspirates where visualization of sperm is hampered by the presence of numerous

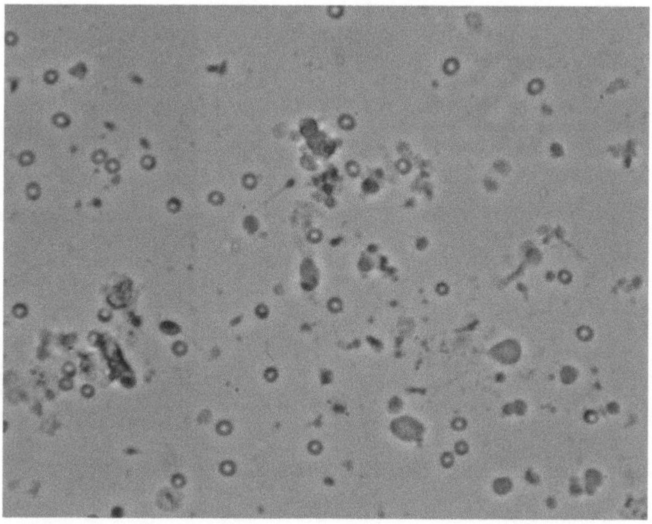

**Fig. 10** Showing testicular aspirate. Few sperms are appreciated

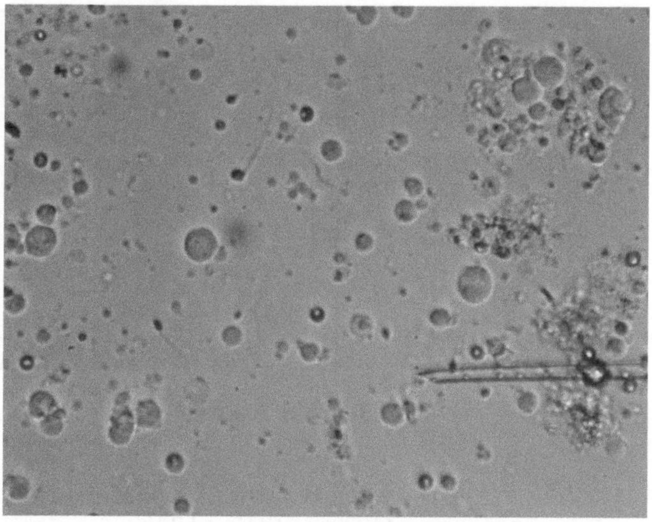

**Fig. 11** Shows attempt being made to aspirate live sperm into the injection pipette

RBCs. In such surgically retrieved samples, treatment with RBC buffer enhances the efficiency of sperm collection. Treatment with RBC buffer doesn't affect fertilization and subsequent embryo development [8].

Collagenase Media

We use the following products from the Gynemed MediaLine [9]:

- GM501 Collagenase (GM501 CollagenaseTM, Gynemed, Lensahn, Germany) (Fig. 12).

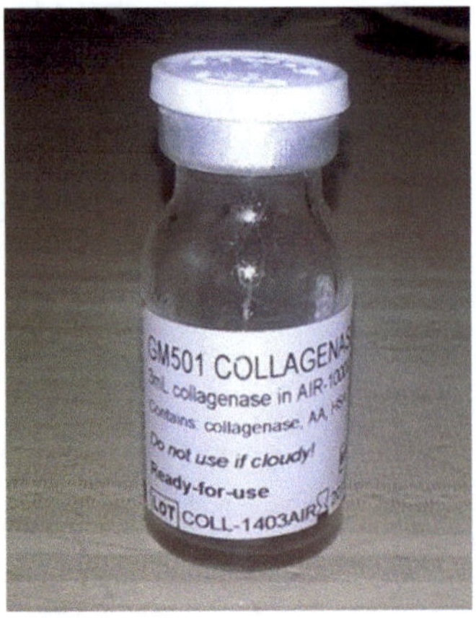


**Fig. 12** Collagenase media

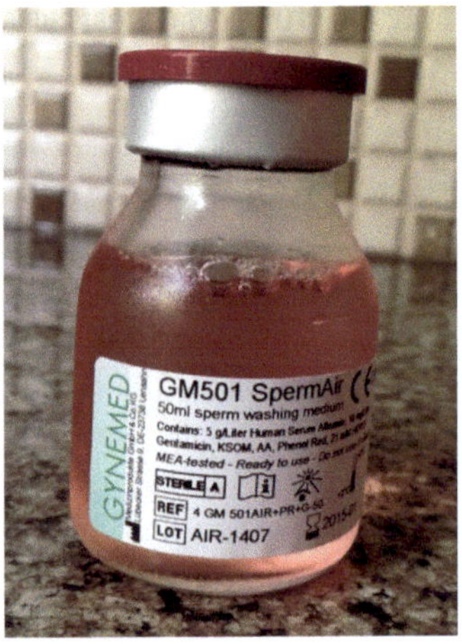


**Fig. 13** GM501 SpermAir

- GM501 SpermAir (GM501 SpermAirTM, Gynemed, Lensahn, Germany) (Fig. 13).

1. Aspirate 1.5 mL GM501 Collagenase out of the bottle.
2. Fill into a 5 mL round bottom tube and allow it to warm to 37 °C.
3. Fill 4.5 mL sperm preparation medium (GM501 SpermAir) into a round bottom tube and warm up to 37 °C.
4. For easier handling, if necessary, fill the tissue suspension into a 60 mm petri dish.
5. Pick pieces of testicular tissue carefully with a fine syringe cannula and transfer into the Collagenase tube.

6. Close the tube tightly and keep it in an incubator for 60 min for digestion of the tissue.
7. Aspirate and release the digested tissue.
8. A suspension of single testicular tissue cells and free sperm has been formed.
9. Now centrifuge the tissue cell suspension and wash twice with each 2 mL sperm preparation medium (GM501 SpermAir).
10. Discard the supernatant and resuspend the pellet in a small volume of 30–80 μL sperm preparation medium (GM501 SpermAir).
11. Examine under microscope after adding it to ICSI microdroplets.
12. If viable spermatozoa are seen, they can be used for ICSI/ cryopreservation.
13. If required, further digest the tissue by repeating the protocol described above.
14. If many RBCs are present, erythrocyte-lysing buffer (ELB) should be used [8].

#### Erythrocyte-Lysing Buffer (RBC Lysis Buffer)

*Composition of RBC lysis buffer*:

Ammonium chloride (NH4Cl) 0.829 g.

Potassium bicarbonate (NaHCO3) 0.100 g.

Ethylenediaminetetraacetic acid (EDTA) 0.074 g.

Tissue culture-grade water 100 mL.

#### *Methodology*

1. Dissolve the abovementioned chemicals and adjust the pH to 7.2.
2. Using a 0.2-mm syringe filter, sterilize the buffer.
3. With the help of a BD Falcon 7575 sterile pipette, add 4 mL of RBC lysis buffer to the pellet and mix it.
4. Centrifuge at 400 × *g* for 5 min.
5. Without disturbing the pellet, discard the supernatant.
6. Add 3 mL of sperm washing media.
7. To remove the RBC lysis buffer, centrifuge the conical tube at 400 × *g* for 10 min.
8. After discarding the supernatant, resuspend the pellet in 0.5–1.0 mL of the sperm washing medium.
9. The suspension can be added to ICSI microdroplets and sperm isolated.

### 3.3 Incubation

The testicular suspension is incubated in the buffered media (gamete buffer media, Code K-SIGB, William A Cook, Australia Pty. Ltd.) for sperms to gain motility. They can also be incubated in fertilization/cleavage media (Cook IVF media group), which helps the spermatozoa to gain motility. This works better in cases of OA, where quite a good number of motile sperms are obtained after 24 h of incubation. Van den Berg recommended incubation at 32 °C in place of 37 °C [10]. Some clinics carry out the biopsy procedure 1 day prior to ovum pickup, if fresh sperms are to be used.

Cryopreserving the biopsy sample after 24 h of incubation post-retrieval gives a higher proportion of motile sperm post-thawing.

### 3.4 Cryopreservation of Surgically Retrieved Sperm

Spermatozoa have low water content and high membrane fluidity. By virtue of these properties, sperm cells are more cold-shock resistant than other cells. During cryopreservation, the cryoprotectant is added gradually allowing equilibration of spermatozoa and cryoprotectant. Initially, the sperm cells shrink, and after equilibration is complete, they return to their original size. The cryoprotectant permeates inside the cell achieving an osmotic equilibrium. This minimizes the intracellular ice crystal formation.

During thawing, the cryoprotectant is washed away and replaced by water again. Usually, about 50 % of the spermatozoa are unable to survive the freezing procedure resulting in loss of motility and vitality.

There are two approaches for cryopreservation.

#### 3.4.1 CRYO-ICSI

Surgical retrieval of sperm is performed prior to the OPU and spermatozoa, if obtained, are cryopreserved. On the day of ovum pickup, sperm are thawed and used for ICSI.

#### 3.4.2 ICSI-CRYO

Spermatozoa are surgically retrieved on the day of ICSI. The spermatozoa are used for ICSI and remaining sample is frozen for use in subsequent cycles.

The aim is to obtain maximum number of sperm and freeze it. The approach is selected depending upon the following factors:

1. *Underlying pathology*: OA or NOA
2. *Sperm retrieval procedure*: TESA/TESE/micro-TESE

#### 3.4.3 Obstructive Azoospermia

The techniques used are:
*PESA*: Sperms are obtained in high numbers in almost 100 % patients after the epididymal aspiration. At times, the urologist will do a TESA or TESE in such cases. Both the approaches are effective in such patients. We prefer to carry out the procedure prior to surgery and freeze the sperms.

*3.4.4 Nonobstructive Azoospermia*

The techniques used are:

*TESE*: Sperms are obtained in 50–60 % cases though in poor numbers and motility. Multiple biopsies may be required. Depending upon degree of impairment of spermatogenesis and the yield of sperm, cryopreservation may/may not be feasible. In cases with severe impairment of spermatogenesis, scheduling fresh TESE on the day of OPU or 1 day prior to OPU is a good option [11].

*TESA*: Usually the yield is low and chances of freezing for subsequent use are poor.
*Micro-TESE*: It is an invasive surgical procedure. It is done by a skilled urologist under an operating microscope, which enables a thorough examination of seminiferous tubules in patients with focal and sporadic spermatogenesis. The seminiferous tubules showing active spermatogenesis are excised. Micro-TESE yields sperms in a higher percentage of cases compared to conventional TESE [12–14]. It avoids complications such as hematoma, fibrosis, and impaired androgen production [12].

*3.4.5 Comparison of CRYO-ICSI and ICSI-CRYO Approaches*

The advantages of one method are the disadvantage of the other such as:

1. CRYO-ICSI approach avoids ovarian stimulation if surgical retrieval fails, whereas in ICSI-CRYO, there is a 50 % risk of pointless stimulation.
2. In ICSI-CRYO approach, simultaneous scheduling of ovum pickup and surgical retrieval has to be planned. CRYO-ICSI approach avoids the pressure of arranging a urologist on the day of the ICSI procedure and allows for independent scheduling of sperm and oocyte retrieval.
3. In CRYO-ICSI approach, patients are relaxed and not worried about a failed retrieval on the day of OPU, as in ICSI-CRYO approach.
4. In CRYO-ICSI approach, there is a loss of sperm quality due to freezing which can affect the prognosis of ICSI. Damage due to cryopreservation is avoided in ICSI-CRYO approach.
5. In CRYO-ICSI approach, due to loss of sperm quality, risk of not finding a motile sperm is high. This problem is avoided in ICSI-CRYO approach.

## 3.5 Cryopreservation of Testicular and Epididymal Sperm

*3.5.1 Indications*

1. Prior to chemotherapy and radiotherapy, which can be gonadotoxic.
2. Testicular tissue cryopreservation prior to the start of cancer treatment offers fertility potential in the future in younger males in whom spermatogenesis has not begun [15].

3. Testicular sperm can be retrieved and cryopreserved during vasectomy, as there is a 20–25 % likelihood of failure in vas-reanastomosis surgeries.
4. Rarely when the partner is unable to produce the sample on day of the OPU and facilities of egg freezing are lacking.
5. In cases where DNA damage to ejaculated sperm is high, using testicular sperm may improve ICSI outcome.

#### *3.5.2 Methods of Cryopreservation*

Freezing can be performed in various ways. Either, the sperm can be frozen rapidly in liquid nitrogen vapor or slowly using computer controlled-rate cooling and freezing. Verheyen et al. in 1993 observed no significant difference in post-thaw sperm quality between the vapor freezing technique and the controlled-rate freezing [16].

#### *3.5.3 Cryopreservation of Epididymal Sperm*

1. The epididymal aspirate is examined under inverted microscope for presence of sperm.
2. If sperms are present, the aspirate is transferred to a conical tube.
3. 2 ml of sperm washing medium at 37 °C is added to the conical tube containing the epididymal fluid.
4. Mix it gently and then centrifuge at 400 × *g* for 10 min.
5. The supernatant is discarded and pellet is examined microscopically.
6. If it is contaminated with erythrocytes, wash with RBC lysis buffer.
7. Resuspend the pellet in 2 mL of RBC lysis buffer. Mix gently.
8. Centrifuge the conical tube at 400 × *g* for 5 min.
9. Gently aspirate the supernatant and discard without disturbing the pellet.
10. Wash the pellet with 1–2 mL of the sperm washing medium by centrifuging the tube at 400 × *g* for 10 min.
11. Resuspend the pellet in 1.0 mL of the sperm washing medium.
12. Add sperm freeze media (in a ratio of 1:1) to epididymal aspirate.
13. Aliquot the mixture into labeled cryovials (Code CE 0543, Nunc, Denmark) for cryostorage.
14. Place the cryovials in a 4 °C refrigerator for 30–45 min and then freeze the cryovials in liquid nitrogen vapor for 1 h.
15. Plunge the vapor-frozen vials into liquid nitrogen for storage.

#### 3.5.4 Cryopreservation of Testicular Sperm

*Procedure*:

1. Bring the reagents to 37 °C by placing in an incubator for 30 min.
2. Aspirate the testicular sample suspension and transfer it into a test tube for 15 min.
3. After the larger fragments settle down, aspirate the supernatant into another conical tube.
4. Aspirate and release the suspension with larger fragments through a 3 mL syringe attached to a 21-gauge needle to further dissociate seminiferous tubules and release sperm into another test tube.
5. Now combine the contents of the two test tubes.
6. Add sperm freeze media (in a ratio of 1:1), drop by drop from the side of the test tube to testicular sperm sample. This should be done over a period of 10 min to minimize hyperosmotic stress, while continuously shaking the tube.
7. The diluted testicular sperm sample is loaded into straws (250 μL volume).
8. Load it into straws using automatic pipette straws.
9. Seal the straw at one end by a cotton plug.
10. Place the straw on the comb and create an air bubble, which will prevent explosion of the straw after thawing.
11. Seal the straws by dipping in polyvinyl alcohol powder and then immerse in water, which polymerizes the powder.
12. The air bubble is moved to the center of the straw.
13. The straws are labeled with details of the patient such as name, id, and date of cryopreservation.
14. Place the straws in the chamber of a computer-controlled biological freezer (Nicool LM10, France) and cool as follows:
    (a) From room temperature to 10 °C at a rate of 1.6 °C/m for 6 min.
    (b) From 10 to −120 °C at a rate of −5.5 °C/min for 20 min.
    Now plunge the straws in the liquid nitrogen at −196 °C [17].
15. *Rapid freezing*: Alternatively, the straws/cryovials (Code CE 0543, Nunc, Denmark) are loaded with diluted testicular sperm sample and placed over the surface of liquid nitrogen (at −80 °C) for 10 min. They are then plunged into liquid nitrogen at −196 °C.

*3.5.5 Thawing Technique for Testicular and Epididymal Sperm*

If the epididymal and testicular sperms are stored in straws:

1. The straw is identified and removed from the liquid nitrogen tank.
2. Thaw at 37 °C (immerse in water at 37 °C) for 1 min.
3. Cut the end of the straw sealed with powder and place it near tip of a conical test tube.
4. Then cut the other end allowing the sample to fall into the tube.
5. 3 mL sperm washing media kept at 37 °C is added to the sample drop by drop while shaking the test tube to avoid osmotic shock.
6. Leave at 37 °C for 10 min.
7. Wash twice by centrifuging at 400 × *g* for 10 m and 300 × *g* for 5 min to get rid of the cryoprotectant.
8. Remove the supernatant and resuspend in a 200 μL of sperm washing media.
9. The preparation of frozen-thawed epididymal sperm is identical to the treatment of frozen-thawed ejaculates.

If the epididymal and testicular sperms are stored in vials:

1. Identify and remove the vial from the liquid nitrogen tank.
2. Thaw by immersing in water kept in a beaker at 37 °C.
3. Put the vial in a heating cabinet at 37 °C for 10 min.
4. Follow **steps 5–8** of the above mentioned protocol.

### 3.6 Cryopreservation of a Single/Few Human Sperm Using a Zona Pellucida

This technique was developed by Cohen and colleagues [18]:

*3.6.1 Freezing Technique*

1. In this method, few sperms or a single spermatozoon is cryopreserved inside an empty zona obtained from mouse, hamster, or human oocytes.
2. The oocyte is denuded enzymatically and mechanically.
3. Hold it with a holding pipette.
4. Two small holes are drilled into the zona with the laser, and the ooplasm is aspirated leaving an empty zona.
5. Motile sperms are identified in the testicular suspension under microscope and transferred to a PVP (10 %) droplet.
6. These sperms are injected into an empty zona using an ICSI needle.
7. Before freezing, zona with sperms is placed in 8 % glycerol solution in phosphate-buffered saline (PBS) containing 3 % HSA.
8. The zona is loaded in a column of cryopreservation medium sandwiched between two air bubbles and then frozen in sterile straws (0.25 mL) and heat sealed.

9. They are then exposed to liquid nitrogen vapor for 2 h, followed by storage in liquid nitrogen.

This is a labor-intensive and difficult technique, but it offers a chance to preserve and retrieve single/few sperm in testicular samples. Various carriers have been used:

1. Alginate beads or agarose microspheres [19].
2. Nylon cryoloop increases the risk of cross-contamination [20].
3. ICSI pipettes [21].

*3.6.2 Thawing*

1. The straw is immersed in a water bath (30 °C) for 30 s.
2. The straw is then cut at one end and contents are released into a center well dish.
3. The zona is identified under a stereoscope and washed with HEPES/MOPS-buffered media and then transferred to a droplet of PVP.
4. The zona is held with a holding pipette, and ICSI injection pipette is used to aspirate the sperms into the PVP drop.
5. The sperms are then immobilized and can be used for ICSI.

*3.6.3 Drawbacks*

1. The presence of zona pellucida 3 protein in human zona induces acrosome reaction in the sperm leading to lower recovery and fertilization rates. The results are better with mouse or hamster eggs [18].
2. It is not only difficult but also impractical to obtain an egg in clinical setting from female partner, as it requires egg retrieval and aspiration of ooplasm to obtain an empty zona.

### 3.7 Advantages of Cryopreservation of Testicular and Epididymal Sperm

1. Freezing of surgically retrieved sperm avoids a repeat surgery in case one treatment cycle fails.
2. Moreover, second or third surgery can increase the chance of complications including hematomas, inflammation, testicular devascularization, fibrosis, and permanent testicular damage [22].

*3.7.1 Challenges with Cryopreservation of Surgically Retrieved Sperm*

Similar principles and methodology apply to the cryopreservation of testicular and epididymal sperm as ejaculated sperm. However, compared to ejaculated sperm, cryopreservation of surgically retrieved testicular sperm is difficult due to:

1. Low number of sperms retrieved.
2. Lack of motility and vitality.
3. Most labs use similar cryopreservation protocols for epididymal and testicular samples. These may not produce optimal results, as there are differences in the membrane permeability between ejaculated and testicular sperm.

4. Contamination with cellular debris and blood cells.
5. Relatively poor survival rates (50 %) obtained after freeze-thawing.
6. Larger pieces of tissue should not be frozen as:
   (a) Different cells of a biopsy respond differently to cryo-protectants and require different concentration of CPAs.
   (b) This results in variation in cooling rates within different parts of the tissue.
   (c) Large pieces of tissue offer increased resistance to heat transfer and penetration of CPAs.
   (d) Seminiferous tubules capture liquid and increase chances of ice formation [15].

#### *3.7.2 Measures to Overcome the Challenges*

1. To avoid these difficulties, cryopreservation of smaller tissue fragments or mincing tissue prior to freezing has been advocated [23]. This has been shown to improve the post-thaw viability (39 vs. 25 %) and motility (9 vs. 4 %) in suspension compared to biopsy [24].
2. Hypoosmotic swelling test helps in identifying viable sperm before injection.

## 4 Notes (For the Beginners)

1. The tricks of making TESE ICSI plate:
   (a) Make multiple dishes with numerous 5-μL drops of sperm washing media, as finding the required number of motile spermatozoa may require examination of many drops.
   (b) Aspirate sperm washing media with the help of a flexipet and add prepared sperm sample into the drops. This allows us to visualize sperms in one plane and aspirate it easily compared to a 5-μL drop where sperms are appreciated in different planes.
2. The tricks of the sperms:
   (a) In our experience, incubating the prepared testicular sperm in equilibrated cleavage media (Cook IVF media group) in Falcon 3037 dishes for 1 h provides some degree of motility to the sperm cells.
   (b) SpermMobil media (GM501 SpermMobilTM, Gynemed, Lensahn, Germany) can also be used. It provides motility to live sperms.
   (c) These two methods make it easier to induce motility in spermatozoa and differentiate between live immotile and dead immotile spermatozoa.

3. The tricks of freezing the sperms harvested after TESE.
   (a) Fill the testicular suspension into a conical tube and allow the large fragments to settle.
   (b) Now, aspirate the supernatant into another conical tube. Ideally, this contains the good number of sperm cells.
   (c) Aspirate and release the suspension through a 3 mL syringe attached to a 21-gauge needle to further dissociate seminiferous tubules and release sperm.
4. Add sperm freeze (in a ratio of 1:1) drop by drop from the side of the test tube to testicular sperm sample over a period of 10 min.
5. We use the rapid freezing technique for freezing the testicular/epididymal sperm. Better sperm survival is obtained with the following modification:
   (a) Identify the surface of liquid nitrogen by lowering an empty cryovial and mark levels of 10 and 5 cm on the thread used to hold the cryovial.
   (b) Cryovials filled with diluted sperm sample are first placed 10 cm above the surface of liquid nitrogen (at −80 °C) for 5 min.
   (c) Then, the cryovials are lowered to a distance of 5 cm above the surface of liquid nitrogen.
   (d) They are then plunged into liquid nitrogen at −196 °C.

## References


1. Nicopoullos JD, Gilling-Smith C, Almeida PA, Norman-Taylor J, Grace I, Ramsay JW (2004) Use of surgical sperm retrieval in azoospermic men: a meta-analysis. Fertil Steril 82(3):691–701
2. Verheyen G, Nagy Z, Joris H, De Croo I, Tournaye H, Van Steirteghem A (1997) Quality of frozen-thawed testicular sperm and its preclinical use for intracytoplasmic sperm injection into in vitro-matured germinal-vesicle stage oocytes. Fertil Steril 67(1):74–80
3. Verheyen G (2010) Outcomes, safety and effectiveness for cryopreservation of ejaculated, testicular and epididymal sperm. Centre for Reproductive Medicine UZ Brussel, Belgium. Paper presented at the ALPHA Conference, Budapest, 30th April–2nd May 2010
4. Verheyen G, De Croo I, Tournaye H, Pletincx I, Devroey P, van Steirteghem AC (1995) Comparison of four mechanical methods to retrieve spermatozoa from testicular tissue. Hum Reprod 10(11):2956–2959
5. Salzbrunn A, Benson DM, Holstein AF, Schulze W (1996) A new concept for the extraction of testicular spermatozoa as a tool for assisted fertilization (ICSI). Hum Reprod 11(4):752–755
6. Crabbé E, Verheyen G, Tournaye H, Van Steirteghem A (1997) The use of enzymatic procedures to recover testicular germ cells. Hum Reprod 12(8):1682–1687
7. Crabbé E, Verheyen G, Silber S, Tournaye H, Van de Velde H, Goossens A, Van Steirteghem A (1998) Enzymatic digestion of testicular tissue may rescue the intracytoplasmic sperm injection cycle in some patients with non-obstructive azoospermia. Hum Reprod 13(10):2791–2796
8. Gangrade BK (2013) Cryopreservation of testicular and epididymal sperm: techniques and clinical outcomes of assisted conception. Clinics 68(S1):131–140
9. Gynemed media catalogue. Edition 03/2012
10. Van den Berg M (1998) The use of epidemiological and testicular sperm in IUF. In: Elder K, Elliott T (eds) World wide conferences on reproductive biology. Ladybrook Publishing, Australia

11. Verheyen G, Vernaeve V, Van Landuyt L, Tournaye H, Devroey P, Van Steirteghem A (2004) Should diagnostic testicular sperm retrieval followed by cryopreservation for later ICSI be the procedure of choice for all patients with non-obstructive azoospermia? Hum Reprod 19 (12):2822–2830
12. Schlegel PN (1999) Testicular sperm extraction: microdissection improves sperm yield with minimal tissue excision. Hum Reprod 14 (1):131–135. doi:10.1093/humrep/14.1.131
13. Okada H, Dobashi M, Yamazaki T, Hara I, Fujisawa M, Arakawa S et al (2002) Conventional versus microdissection testicular sperm extraction for nonobstructive azoospermia. J Urol 168(3):1063–1067
14. Ghalayini IF, Al-Ghazo MA, Hani OB, Al-Azab R, Bani-Hani I, Zayed F et al (2011) Clinical comparison of conventional testicular sperm extraction and microdissection techniques for non-obstructive azoospermia. J Clin Med Res 3(3):124–131
15. Keros V, Hultenby K, Borgstrom B, Fridstrom M, Jahnukainen K, Hovatta O (2007) Methods of cryopreservation of testicular tissue with viable spermatogonia in pre-pubertal boys undergoing gonadotoxic cancer treatment. Hum Reprod 22(5):1384–1395. doi:10.1093/humrep/del508
16. Verheyen G, Pletincx I, Van Steirteghem A (1993) Effect of freezing method, thawing temperature and post-thaw dilution/washing on motility (CASA) and morphology characteristics of high-quality human sperm. Hum Reprod 8(10):1678–1684
17. Yavetz H, Yogev L, Homonnai Z, Paz G (1991) Prerequisites for successful human sperm cryobanking: sperm quality and prefreezing holding time. Fertil Steril 55:812–816
18. Cohen J, Garrisi GJ, Congedo-Ferrara TA, Kieck KA, Sehimmel TW, Scott RT (1997) Cryopreservation of single human spermatozoa. Hum Reprod 12(5):994–1001
19. Isaev DA, Zaletov SY, Zaeva VV, Zakharova EE, Shafei RA, Krivokharchenko IS (2007) Artificial microcontainers for cryopreservation of solitary spermatozoa. Hum Reprod 22: I154–I556
20. Isachenko V, Isachenko V, Katkov II et al (2004) DNA integrity and motility of human spermatozoa after standard slow freezing versus cryoprotectant-free vitrification. Hum Reprod 19(4):932–939
21. Sohn JO, Jun SH, Park LS et al (2003) Comparison of recovery and viability of sperm in ICSI pipette after ultra rapid freezing or slow freezing. Fertil Steril 80:S128
22. Schlegel PN, Su LM (1997) Physiological consequences of testicular sperm extraction. Hum Reprod 12(8):1688–1692
23. Hovatta O (2003) Cryobiology of ovarian and testicular tissue. Best Pract Res Clin Obstet Gynaecol 17(2):331–342 ISSN 521-6934
24. Crabbé E, Verheyen G, Tournaye H, Van Steirteghem A (1999) Freezing of testicular tissue as a minced suspension preserves sperm quality better than whole-biopsy freezing when glycerol is used as cryoprotectant. Int J Androl 22 (1):43–48

# Chapter 8

# Human Oocytes Slow-Rate Freezing: Methodology

C. Zacà and A. Borini

## Abstract

Cryopreservation of human oocytes is an important technique for the treatment of human infertility, as it deals successfully with legal, ethical, and moral issues related to embryo cryopreservation (Coticchio et al., Human Fertil (Camb) 4:152–157, 2001; Tucker et al., Eur J Obstet Gynecol Reprod Biol 113: 524–527, 2004) and maintains reproductive potential in women with diseases or conditions that may compromise reproductive capacity. Here, we describe an oocyte cryopreservation technique involving slow freezing–rapid thawing with differential sucrose concentration (0.2 M in the freezing stage—0.3 M in thawing stage) (Bianchi et al., Reprod Biomed Online. 14:64–71, 2007), describing the technique in detail, from the preparation of the solutions through to the performance of the technique and, finally, to a number of helpful hints for optimum results.

**Key words** Oocytes cryopreservation, Human oocytes, Slow freezing, Differential sucrose concentration, Methodology

## 1 Introduction

Oocyte cryopreservation is a technique that makes it possible to maintain female reproductive potential and it consequently brings several advantages [1]. This technique may be used to rescue cycles complicated by ovarian hyperstimulation syndrome, failure to obtain sperm, in oocyte donation programs, in the treatment of congenital infertility disorders or premature ovarian failure (POF), as an alternative to embryo freezing or, for fertile women, to electively delay childbearing [2, 3].

Cryopreservation makes it possible to freeze cells and tissues at very low temperatures and to keep the biological specimen genetically stable and metabolically inert to preserve it for future use [1]. The oocyte cryopreservation procedure involves initial exposure to cryoprotectants, cooling to subzero temperatures, and final storage in liquid nitrogen, because at this temperature (−196 °C) no known biological reactions take place and thus the storage can be prolonged indefinitely.

Zsolt Peter Nagy et al. (eds.), *Cryopreservation of Mammalian Gametes and Embryos: Methods and Protocols*, Methods in Molecular Biology, vol. 1568, DOI 10.1007/978-1-4939-6828-2_8, © Springer Science+Business Media LLC 2017

The structural features (ϕ120 μm, very sensitive to low temperature, high water content, low surface-to-volume ratio, suboptimal plasma membrane permeability) of mature (metaphase II) human oocytes mean that their cryopreservation is fraught with complex difficulties.

The greatest challenge of oocyte cryopreservation is, therefore, surviving the freezing/thawing process while maintaining the ability to fertilize and develop in vitro to the appropriate embryonic stage without undergoing any structural alterations.

To achieve this, it is necessary to eliminate the two main causes of cell death associated with cryopreservation: ice formation [4] and lethal solute concentrations [5], while maintaining the functional capacity of the intracellular organelles. Devising procedures that take biological samples from a physiological temperature (37 °C) to a cryogenic temperature (−196 °C) is an extremely complex task. Over the years, changes have been made to cryopreservation methods to improve results in terms of biological and clinical outcomes.

The first human births from frozen oocytes were reported in the 1980s by Chen [6] and Al-Hasani [7].

Changes to slow-cooling cryopreservation methods consisted in altering the combination and composition of cryoprotectants [8 –11], to avoid damage to chromosomes and the meiotic spindle, the cytoskeleton, and cortical granules [12, 13].

The first protocol used was devised by Lasalle, based on 1.5 M PROH and 0.1 M sucrose and was applied to embryo freezing [14]; however, this protocol did not ensure adequate water flow out of the cell, which caused the formation of intracellular ice, due to cell death. It was necessary to increase the concentration of sucrose to obtain better dehydration and lesser ice formation. Fabbri et al. [15] therefore increased sucrose concentrations from 0.1 M to 0.3 M, which resulted in a significant increase in the survival and fertilization rate, without, however, achieving good clinical results in terms of implantation and pregnancy rate. Therefore, on the basis of these results, Bianchi et al. [8] amended the oocyte cryopreservation protocol to include a freezing solution containing 1.5 M PROH plus 0.2 M sucrose, in order to obtain lesser shrinkage during cooling and a thawing solution with a higher sucrose concentration (0.3 M). This change made it possible to achieve biological results comparable with those of previous studies and significantly better clinical results.

To stabilize biological materials at cryogenic temperatures, the slow-freezing procedures take into account a number of important aspects:

- **Control of the cooling speed** the rate of cooling has a dramatic effect on these phenomena. Rapid cooling minimizes the solute concentration effects as ice forms uniformly, but it leads to the formation of more intracellular ice, as water has not migrated out of the cell. Slow cooling, on the other hand, results in a greater loss of water from the cell and less ice being formed inside, but also causes in an increase in solution effects. If too much water remains inside the cell, damage can occur due to ice crystal formation and re-crystallization during warming, which is usually lethal.

Controlled slow freezing rates are usually achieved using a computerized freezing system consisting of a computer and freezing chamber connected by a pump to a cylinder from which the liquid nitrogen is extracted in gaseous form. The temperature-reduction curve to be achieved is saved on the computer; according to the temperature detected by the temperature sensor, the computer, opens and closes the solenoid valve and regulates the introduction of gaseous nitrogen into the freezing chamber, thus permitting controlled temperature-lowering.

- **Cryoprotectants** The cryopreservation process involves replacing some of the water inside the cells with other compounds that will not form large ice crystals when frozen; therefore, all freezing methods rely on the presence of one or more cryoprotectants at molar concentrations. One feature that is common to these compounds is their ability to interact with water *via hydrogen bonding* and, prior to slow cooling, they lower the freezing point of the solution and allow greater cell dehydration without excessive shrinkage [16]. Cryoprotectants can be divided into two groups according to their molecular weight and, therefore, their ability to pass through the cell membrane:
  - *Permeating agents*:

    These substances are chemically characterized as having a relatively low molecular weight and are able to penetrate the cell's lipid bilayer. This group of agents includes glycerol, ethylene glycol, DMSO, and 1,2-propanediol (PROH). Permeating cryoprotectants that pass into the cells protect the cell membrane during cooling as they modify the state of the phospholipid bilayer from fluid to partially solid. They also osmotically replace intracellular water, decrease the formation of intracellular ice crystals, lower the freezing point of the solution, and support enhanced cellular dehydration during freezing.
  - *Non-permeating agents*:

    These cryoprotectants remain in the extracellular solution because of their size and they include sugars and macromolecules such as sucrose, raffinose, trehalose, and fructose. Non-permeating cryoprotectants, which are unable to cross the cell membrane due to their molecular size, generate an osmotic gradient that causes cell dehydration and decrease the intracellular water content; consequently, there is a lower likelihood of intracellular ice crystal formation as the temperature drops. The most frequently used non-permeating agents are trehalose and sucrose.

In all freezing techniques, a combination of permeating and non-permeating agents allows a better outcome in terms of oocyte stability.

- **Seeding** when the slow-cooling method is used to cryopreserve oocytes or embryos, the freezing solution is usually manually "seeded" at a temperature of −7 °C to induce ice formation. This is usually achieved by touching the wall of each straw with cold forceps (cooled in liquid nitrogen). The aim of this procedure is to induce the first ice crystal [17] in the freezing solution some distance away from the oocytes and to prevent the whole solution from supercooling. Manual seeding is preferable to spontaneous ice formation because it prevents diffuse supercooling and starts dehydration, both of which reduce the likelihood of ice forming inside the oocytes. Controlled ice formation during freezing is recognized as a key factor in determining the oocyte viability following freezing and thawing.

  During thawing procedures, cryoprotectants are removed by stepwise dilution with final return to a physiological environment and rapid temperature transition is preferred to prevent the recrystallization of water with the potential for ice-crystal damage. Here, caution must be taken to avoid osmotic shock from the permeating cryoprotectant, which now is at a very high concentration in the intracellular space. Therefore, an additional non-permeating cryoprotectant is used. As the permeating cryoprotectant gradually diffuses out of the oocyte, the concentration of the non-permeating cryoprotectant gradually decreases, until the oocyte is returned to standard culture medium [17].

## 2 Materials

### *2.1 Materials for Slow Freezing Procedure*

**Reagents for slow freezing procedure**

- PBS Dulbecco's Phosphate Buffered Saline (Sigma Life Science, Sigma-Aldrich).
- PROH 1,2-propanediol (Sigma-Aldrich, the Netherlands).
- SSS (Serum Substitute Supplement) (Irvine Scientific, Santa Ana, California).
- SUCROSE ≥ 99.5 % MW 342.3 (Sigma Life Science, Sigma-Aldrich).

**Equipment for slow freezing procedure**

- Automated biological vertical cryo-freezer Kryo 360–1.7 (Planer).

- Precision balance.
- Impulse-heat sealer.
- Label printer (Brady).
- Timer.
- Flexipet Adjustable Handle Set (Cook Australia).
- Metal forceps.
- Liquid nitrogen.
- Liquid nitrogen reservoir.
- Liquid nitrogen dewar with ten canister equipped with 65 mm plastic goblets for long-term storage.
- Cryogloves.

**Tools for slow freezing procedure**

- Protettive latex gloves.
- Sterile tube (10 mL, Ref.2001, Falcon).
- Sterile serological pipettes (10 mL, Ref.357551, Falcon, USA).
- Sterile serological pipettes (1 mL, Ref.357521, Falcon, USA).
- 10 mL syringe (BD Emerald, Becton Dickinson, Spain).
- Millex GV Filter Unit 0.22 um, Durapore PVDF Membrane (Merck Millipore LTD, IRL).
- Flexipet-Denuding Pipette tips ID = 170 um (Cook, USA).
- Multidish 4-Well plate Nunclon (NUNC, Denmark).
- White adhesive labels, BMP21 (Brady, USA).
- Adapted syringe: Silicone Tube inner diameter 1.5 mm, outer diameter 3.5 mm connected to an Insulin type syringe 1 mL (U Insulin 1 mL, Terumo Syringe, Terumo, Belgium).
- Sterile straws (CBS High Security embryo straw 0.3 mL, Cryo Bio System,Groupe I.M.V. Technologies, France).
- Weighted ID rod 30 mm for CBS High Security straw (Cryo Bio System, Groupe I.M.V. Technologies, France).
- Multicolor goblet diameter 12 mm.

### 2.2 Slow Freezing Solution Preparation

Freezing solutions for slow freezing of human oocytes are two: 1.5 M PROH in PBS (Equilibration Solution) and 1.5 M PROH in PBS plus 0.2 M sucrose (Loading Solution). Both the solutions need a 20 % protein source supplementation (SSS). Manage all the reagents under laminar flow and reduce artificial light conditions.

1. Prepare the solution into a 10 mL tube.
2. Equilibration Solution: mix 6.9 mL PBS and 1.1 mL PROH.
3. Loading Solution: solubilize 0.2 M sucrose in 6.9 mL PBS plus 1.1 mL PROH.

4. Add 2 mL SSS to each solution and mix.
5. Filter both the solution into a new sterile 10 mL tube.
6. All the solutions are stored in the refrigerator at 4 °C for a maximum of 3 days since preparation.

### 2.3 Materials for Rapid Thawing Procedure

**Reagents for rapid thawing procedure**

- PBS Dulbecco's Phosphate Buffered Saline (Sigma Life Science, Sigma-Aldrich).
- PROH 1,2-propanediol (Sigma-Aldrich, the Netherlands).
- SSS (Serum Substitute Supplement) (Irvine Scientific, Santa Ana, California).
- Sucrose $\geq$ 99.5 % MW 342.3 (Sigma Life Science, Sigma-Aldrich).

**Equipment for rapid thawing procedure**

- Precision Balance.
- Thermometer.
- Timer.
- Small sterile scissors.
- Surgical Gauze.
- Flexipet Adjustable Handle Set (Cook Australia).
- Water bath 30 °C.
- Liquid nitrogen.
- Liquid nitrogen reservoir.
- Cryogloves.

**Tools for rapid thawing procedure**

- Protective latex gloves.
- Sterile Tube 10 mL (Ref.2001, Falcon).
- Flask 50 mL (Falcon, Becton Dickinson, NJ, USA).
- Sterile serological pipettes (1 mL) (Ref.357521, Falcon, NJ, USA).
- Sterile serological pipettes (5 mL) (Ref.357543, Falcon, NJ, USA).
- Sterile serological pipettes (10 mL) (Ref.357551, Falcon, NJ, USA).
- Petri dish 35 × 10 mm (Ref. 353,001, Falcon, Becton Dickinson, NJ, USA).
- Millex-GV 0.22 μm, Durapore PVDF Membrane (Merck Millipore LTD, IRL).
- Syringe 10 mL (BD Emerald, Becton Dickinson, Spain).

- Flexipet-Denuding Pipette tips ID = 170 μm (Cook, USA).
- Multidish 4-well plate Nunclon (NUNC, Denmark).
- Adapted syringe: Silicone Tube inner diameter 1.5 mm, outer diameter 3.5 mm connected to a no needle Insuline type syringe 1 mL (U Insulin 1 mL, Terumo Syringe, Terumo, Belgium).

### 2.4 Rapid Thawing Solution Preparation

Thawing solutions for rapid thawing of human oocytes are four:

Thawing solution 1: 1 M PROH in PBS plus 0.3 M sucrose.

Thawing solution 2: 0.5 M PROH in PBS plus 0.3 M sucrose.

Thawing solution 3: 0.3 M sucrose in PBS.

Thawing solution 4: PBS medium.

All the solutions need a 20 % protein source supplementation (SSS). In order to obtain the four thawing solution it is necessary to prepare a stock solution. Manage all the reagents under laminal flow and reduce artificial light conditions.

1. First mix into a 50 mL flask: 14.5 mL PBS with 1.5 mL PROH (stock solution).
2. Prepare the four thawing solutions into a 10 mL tube.
3. Thawing solution 1 (1 M PROH/0.3 M sucrose): solubilize 0.3 M sucrose in 8 mL of stock solution.
4. Thawing solution 2 (0.5 M PROH/0.3 M sucrose): solubilize 0.3 M sucrose in 4 mL of stock solution and 4 mL PBS.
5. Thawing solution 3 (0.3 M sucrose): solubilize 0.3 M sucrose in 8 mL PBS.
6. Thawing solution 4 (PBS): 8 mL PBS medium only.
7. Add 2 mL SSS to each solution and mix.
8. Filter all the solution into a new sterile 10 mL tube.
9. All the solutions are stored in the refrigerator at 4 °C for a maximum of 3 days since preparation.

## 3 Methods

### 3.1 Slow Freezing Procedure

The oocytes are de-cumulated mechanically and enzymatically around 2 h after transvaginal retrieval to detect the Metaphase II oocytes and undergo slow freezing procedure.

1. Place the solutions to room temperature (22–24 °C) before use. All dehydration steps must be performed at room temperature (RT). Just before starting, the cryo-freezer is turned on.
2. Stamp label with the name, surname, freezing date, couple code, and number of the oocytes and put it on the straw (Fig. 1). One to maximum three oocytes are loaded per straw.

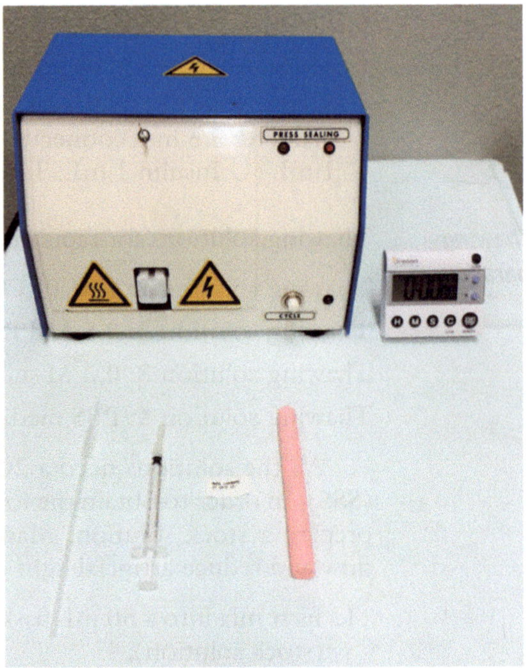

**Fig. 1** Equipment necessary for slow freezing. From the left in the upper of the image: impulse-heatsealer, timer. From the left in the lower of the image: sterile straws, insulin type syringe 1 ml connected to a silicone tube, label (name, surname, sample number, date, couple code) and goblet

Connect an adapted syringe to each straw and prepare a weighted ID rod.

3. Set up a 170 μm denuding pipette.
4. For each oocyte or group of oocytes dispense 0.6 mL of the Equilibration Solution in Well 1 and 0.6 mL of Loading Solution in Well 2 of a 4-well plate labeled with patient's surname, name, and couple code.
5. The denuded mature Metaphase II oocytes are transferred from their culture media to the equilibration solution and incubated for 10 min. The countdown begins when the first oocytes are placed into the Equilibration Solution (Table 1).
6. After this period the oocytes are moved to the Loading Solution beginning with the first oocyte incubated. The passage between one solution and the other is carried out by collecting the oocytes with a 170 μm denuding pipette in a small amount of solution.
7. Within 5 min since the first oocytes have been incubated in the Loading Solution, it is necessary to proceed in loading the oocyte into the straw (Table 1).

**Table 1**
**Slow freezing solution and incubation time**

| Slow cooling (Differential sucrose concentration) | | Volume | Incubation time | Temperature |
|---|---|---|---|---|
| Equilibration Solution | 1.5 M PROH + 20 % SSS | 0.6 mL | 10′ | RT |
| Loading Solution | 1.5 M PROH + 0.2 M Sucrose + 20 % SSS | 0.6 ml | 5′ | RT |

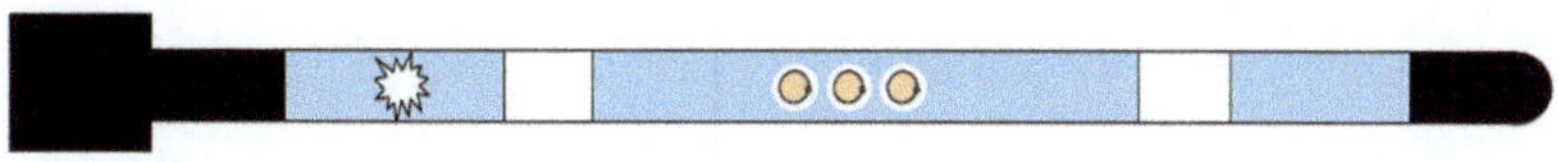

**Fig. 2** Modes of loading the oocytes in the straw and proper position for seeding

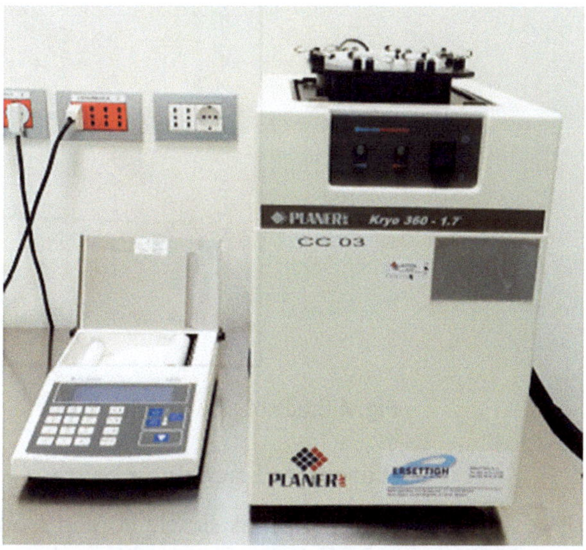


**Fig. 3** Automated biological programmable vertical cryo-freezer Kryo 360–1.7 (Planer)

8. First start loading a small amount of Loading Solution then a bubble air, subsequently Loading Solution containing the oocyte in the middle of the liquid, then another air bubble and at the end another small amount of the same solution (Fig. 2).
9. The straw is then separated from the adapted syringe. Insert each weighted ID rod into the straw and seal it both side.
10. The straws are placed into the automated cryo-freezer (Fig. 3) when the temperature is around 20 °C.
11. All the steps are carried out by two persons (the operator and a second person who assist the operator during the procedure).

| Controlled rate cooling | | |
|---|---|---|
| **Ramp** | **Speed** | **Thermal Interval** |
| 1 | -2.0°C/min | +20°C to -8°C |
| 2 | Hold for 10 min (seeding) | -8°C |
| 3 | -0.3°C/min | -8°C to -30°C |
| 4 | -50°C/min | -30°C to -150°C |
| 5 | Hold for 10 min (plunge into $LN_2$) | -150°C |

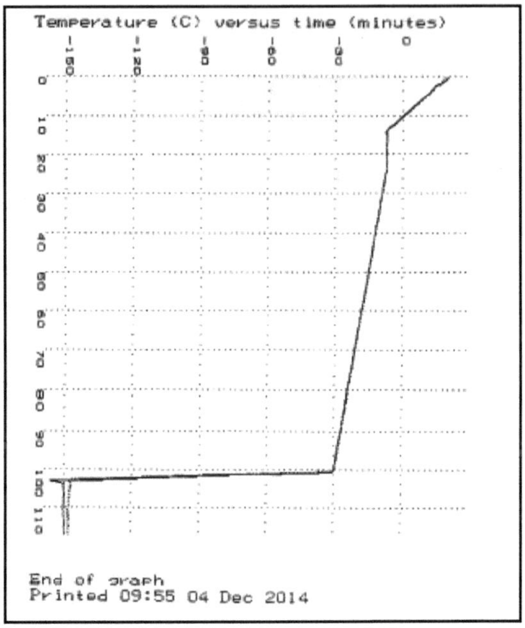


**Fig. 4** Graph of slow freezing curve

### Automated freezing program

The sample is slowly cooled from +20 °C to −7 °C at a rate of −2 °C/min (Fig. 4). Manual seeding of the sample is performed at −7 °C and this temperature is maintained for 10 min. During this time the straws are removed rapidly and not completely to induce ice crystal by touching the straw with very cold metal forceps in proximity of the first small amount of Loading Solution nearby the filter (Fig. 2). Then temperature decreases to −30 °C at a rate of −0.3 °C/min and then rapidly decrease to −150 °C at a rate of −50 °C/min (Fig. 4). An acoustic and visual signal signals the end of the freezing program. Then the straws are directly plunged into identified colored goblet in liquid nitrogen and finally stored into the dewar at −196 °C.

#### *3.2 Rapid Thawing Procedure*

1. Place the thawing solution to room temperature (22–24 °C). All thawing steps must be performed at room temperature (RT).
2. Tag a 4-well plate with patient's surname, name, and couple code. Prepare one 4-well plate for each straw to thaw.

3. Prepare sterile gauze, sterile scissor, adjust syringe, Petri dish, Flexipet Adjustable Handle Set equipped with 170 μm denuding pipette, timer, and 30 °C water bath.
4. With 1 mL serological pipette dispense the thawing solutions into individual wells of the 4-well plate as follows: 0.5 mL of Thawing Solution 1 (1 M PROH/0.3 M sucrose) in Well 1, 0.5 mL of Thawing Solution 2 (0.5 M PROH/0.3 M sucrose) in Well 2, 0.5 mL of Thawing Solution 3 (0.3 M sucrose) in Well 3, and finally 0.5 mL of Thawing Solution 4 (PBS) in Well 4.
5. Remove the samples from the liquid nitrogen dewar and immerse them in the liquid nitrogen reservoir.
6. Start taking out the straw from liquid nitrogen and keep it for 30 s in air (room temperature).
7. Immediately after this put the straw for 40 s in a 30 °C water bath.
8. Subsequently dry the straw gently with a surgical gauze and cut the straw at one end with the small sterile scissors and attach it to the adapted syringe.
9. Then cut the other extremity and empty the solution containing oocytes in a Petri dish under the Stereo zoom microscope to visualize immediately the oocytes.
10. As soon as you see the oocytes move them into the Thawing Solution 1 and incubate for 5 min. Start to calculate the time from this moment.
11. After this time move the oocytes in Thawing Solution 2 and incubate for 5 min, then in Thawing Solution 3 and Thawing solution 4 for 10 min respectively (Table 2).
12. At the end of the thawing procedure evaluate oocyte's integrity with the inverted microscope (20×). The thawed and surviving oocytes are cultured until the insemination time.
13. All the steps are carried out by two persons (the operator and a second person who assist the operator during the procedure).

**Table 2**
**Rapid thawing solution and incubation time**

| Rapid thawing (Differential sucrose concentration) | | Volume | Incubation Time | Temperature |
|---|---|---|---|---|
| Solution 1 | 1.0 M PROH + 0.3 M Sucrose + 20 % SSS | 0.5 mL | 5′ | RT |
| Solution 2 | 0.5 M PROH + 0.3 M Sucrose + 20 % SSS | 0.5 mL | 5′ | RT |
| Solution 3 | 0.3 M Sucrose + 20 % SSS | 0.5 mL | 10′ | RT |
| Solution 4 | PBS + 20 % SSS | 0.5 mL | 10′ | RT |

## 4 Notes

1. We recommend adding 2 mL of Serum Substitute Supplement (SSS) after that the freezing and thawing solutions have been filtered, to prevent a possible protein retaining.
2. Freezing and thawing solutions must be stored at +4 °C, possibly protected from light because of their light-sensitivity.
3. The preparation date can be used as the freezing and thawing solution batch number.
4. An insulin-type syringe without a needle is connected to a narrow silicone tube to connect the syringe to the straw during sample loading.
5. The same pipette can be used to dispense the freezing and thawing solutions if you start with the weaker cryoprotectant solution.
6. Before thawing it is advisable to prepare a small container of liquid nitrogen to remove the straw from goblet.
7. All the steps are carried out by two persons (the operator and a second person who assist the operator during the procedure).
8. On the basis of your experience, you can freeze-thaw more than one straw/oocytes at one time, paying attention to calculate and to consider the different incubation times in the freezing-thawing solutions.
9. During the oocytes loading into the straw it is important to aspire the solution until to wet the filter to obtain vacuum.
10. Slow freezing protocol can be divided into two steps. In the first step permeating cryoprotectants create an osmotic gradient that removes water from the cell by preventing the ice crystal formation. The oocyte dehydrated and shrinked then returns to the original volume after that the cryoprotectant enters the cell replacing water. The extent of shrinkage and swelling cause an osmotic stress that is minimized in the second phase of dehydration when non-permeating cryprotectant is added. In this phase the shrinkage rate is faster with a higher reduction of cell volume before freezing.
11. With the slow-freezing procedure, when moving the oocytes onto next step, collect them in a small amount of the surrounding solution and release them into the next solution. This makes the passage from one molarity to another less shocking.
12. Likewise during the thawing procedure, between steps collect the oocytes in a larger amount of the surrounding solution and release them into the next solution, in order to make rehydration more gradual.

13. At the end of the thawing procedure, rate all thawed oocytes. The survived oocytes have the same morphology as fresh oocytes, whereas degenerated oocytes appear dark and non-three-dimensional.
14. The insemination of surviving oocytes takes place no earlier than 1 maximum 2 h after the end of the thawing procedure and avoid exceeding a total culture time of 7 h and to recover meiotic spindle after thawing.

## References


1. Practice Committees of American Society for Reproductive Medicine, Society for Assisted Reproductive Technology (2013) Mature oocyte cryopreservation: a guideline. Fertil Steril 99:37–43
2. Coticchio G, Garetti S, Bonu MA, Borini A (2001) Cryopreservation of human oocytes. Hum Fertil (Camb) 4:152–157
3. Tucker M, Morton P, Liebermann J (2004) Human oocyte cryopreservation: a valid alternative to embryo cryopreservation? Eur J Obstet Gynecol Reprod Biol 113(Suppl 1):524–527
4. Mazur P (1963) Kinetics of water loss from cells at subzero temperatures and the likelihood of intracellular freezing. J Gen Physiol 47:347–369
5. Kleinhans FW, Mazur P (2007) Comparison of actual vs. synthesized ternary phase diagrams for solutes of cryobiological interest. Cryobiology 54:212–2225
6. Chen C (1986) Pregnancy after human oocyte cryopreservation. Lancet 1:884–886
7. Al-Hasani DK, van der Ven H, Reinecke A et al (1987) Cryopreservation of human oocytes. Hum Reprod 2:695–700
8. Bianchi V, Coticchio G, Distratis V, Di Giusto N, Flamigni C, Borini A (2007) Differential sucrose concentration during dehydration (0.2 mol/L) and rehydration (0.3 mol/L) increases the implantation rate of frozen human oocytes. Reprod Biomed Online 14:64–71
9. Fabbri R, Porcu E, Marsella T, Rocchetta G, Venturoli S, Flamigni C (2001) Human oocytes cryopreservation: new prodpectives regarding oocytes survival. Hum Reprod 16:411–416
10. Stacheki JJ, Willadsen SM (2000) Cryopreservation of mouse oocytes using a medium with low sodium content: effect of plunge temperature. Cryobiology 40:4–12
11. Boldt J, Tidswell N, Sayers A, Kilani R, Cline D (2006) Human oocytes cryopreservation: 5-year experience with a sodium-depleted slow freezing method. Reprod Biomed Online 13:96–100
12. Elliot K, Whelan J (eds) (1977) CIBA foundation symposium 52. The freezing of mammalian embryos. Elsevier, Amsterdam
13. MJ A-S, Farrant J (eds) (1980) Low temperature preservation in medicine and biology. Pitman Press, Bath UK
14. Lassalle B, Testart J, Renard JP (1985) Human embryo features that influence the success of cryopreservation with the use of 1,2 propanediol. Fertil Steril 44:645–651
15. Fabbri R, Porcu E, Marsella T et al (2001) Human oocytes cryopreservation: new perspectives regarding oocytes survival. Hum Reprod 16:411–416
16. Ashwood-Smith MJ (1987) Mechanisms of cryoprotectant action. Symp Soc Exp Biol. 41:395–406
17. Jain JK, Paulson RJ (2006 Oct) Oocyte cryopreservation. Fertil Steril 86(4 Suppl):1037–1046

# Chapter 9

# Slow Freezing and Thawing of Human Cleavage Stage Embryos


## David H. Edgar, Janell Archer, and Debra A. Gook



## Abstract

The ability to store human embryos in a viable state at very low temperatures has been critical to the evolution of responsible practice in clinical Assisted Reproductive Technology (ART). It has encouraged a reduction in the frequency of simultaneous multiple embryo transfer and thereby reduced the risks associated with multiple pregnancy while maintaining high cumulative pregnancy rates from single oocyte collection cycles. In this chapter, we describe a simple slow freezing procedure for human early cleavage stage embryos that results in a high proportion of post-thaw embryos surviving and retaining their implantation potential.



**Key words** Cryopreservation, Human embryos, Cleavage stages, Slow freezing, Dehydration, Rehydration


## 1 Introduction

The almost universal use of ovarian stimulation to produce multiple oocytes/embryos in women undergoing clinical ART, together with the well-recognized risks associated with multiple gestation, has led to the routine application of embryo cryopreservation for embryos that are not transferred fresh in the cycle of oocyte collection. It is then possible for these stored embryos to be thawed and transferred in subsequent menstrual cycles without the need for further ovarian stimulation. For this strategy to be adopted with confidence, the cryopreservation procedure must result in a high proportion of embryos surviving and retaining their developmental/implantation potential.

Water-miscible substances known as cryoprotectants had been shown to play an important role in the removal of intracellular water and thereby minimize the detrimental effects of ice formation during cryopreservation of living cells. Pregnancy and live birth following transfer of human cryopreserved embryos was first achieved [1, 2] with a protocol, involving the use of the permeating


Zsolt Peter Nagy et al. (eds.), *Cryopreservation of Mammalian Gametes and Embryos: Methods and Protocols*, Methods in Molecular Biology, vol. 1568, DOI 10.1007/978-1-4939-6828-2_9, © Springer Science+Business Media LLC 2017

cryoprotectant dimethyl sulfoxide (DMSO), that had been used previously to successfully cryopreserve mouse embryos [3]. It soon became apparent that more reproducible cryopreservation of human cleavage stage embryos could be achieved when 1,2 propanediol (PROH, usually at 1.5 M) was used as the permeating cryoprotectant and intracellular dehydration was further promoted by subsequent exposure to PROH plus sucrose, a non-permeating cryoprotectant at a concentration of 0.1 M [4, 5]. This method was quickly adopted and remained in widespread clinical use for over 20 years. In terms of efficiency, it could result in the majority of post-thaw embryos having at least 50 % of their original blastomeres intact (the conventional criterion for embryo survival) with around half of all frozen embryos remaining fully intact after thawing; the latter being an important correlate of retention of implantation potential [6].

More recently, cryopreservation by vitrification has become a popular methodology in clinical ART and the only published randomized controlled trial comparing this approach to slow freezing with cleavage stage embryos demonstrated improved survival associated with vitrification [7]. Most notably, the proportion of all cryopreserved embryos remaining fully intact after warming/thawing was increased from 45.7 to 73.9 %.

At the same time, our group was developing a modified approach to slow freezing of human cleavage stage embryos. It had been shown that the outcomes from slow cooling of human unfertilized oocytes could be improved by increasing the concentration of the non-permeating cryoprotectant (sucrose) prior to freezing and manipulating the post-thaw rehydration conditions [8–11] and we had reported that this principle could also be applied to cleavage stage embryos that had been biopsied for preimplantation genetic diagnosis [12]. We have employed a similar approach with human non-biopsied cleavage stage (day 2) embryos but also introduced single-step dehydration with simultaneous exposure to 1.5 M PROH and 0.2 M sucrose. This method, which also involves post-thaw rehydration via single-step removal of PROH in high (0.5 M) sucrose, was shown to result in 92.6 % of embryos surviving by the conventional criterion and a high proportion (80 %) of all embryos surviving with all blastomeres intact [13]. Details of our method are described in this chapter.

## 2 Materials

Prepare all solutions in a laminar flow hood into sterile disposable tissue culture plasticware (e.g., Falcon®, Becton Dickinson). Filter sterilize through a 0.2 μm membrane (PALL Life Sciences), discarding the first 1–2 mL to pass through the membrane.

***2.1 Ingredients for Freezing and Thawing Solutions***

1. Basal buffer: Quinn's Advantage® Medium with HEPES, SAGE® Media, Cooper Surgical, a commercially available buffer designed for handling human gametes/embryos in ART under atmospheric conditions (*see* **Note 1**).
2. Sucrose: BioXtra, ≥99.5 % (GC) (Sigma). FW 342.3.
3. 1,2-Propanediol (PROH): ACS reagent, ≥99.5 % (Sigma-Aldrich). FW 76.09; density 1.036 g/mL.
4. Human Serum Albumin (HSA) stock solution (100 mg/mL in isotonic saline), SAGE® Media, Cooper Surgical (*see* **Note 2**).

***2.2 Preparation of Freezing and Thawing Solutions***

1. Sucrose stock solution (1 M): Dissolve 342.3 g of sucrose in basal buffer and make up to a final volume of 1 L (*see* **Note 3**). Filter sterilize and store at 4 °C.
2. Embryo freeze solution (1.5 M PROH/0.2 M sucrose/10 mg HSA/mL): To 59 mL of basal buffer, add 11 mL PROH (*see* **Note 4**), 20 mL sucrose stock solution, and 10 mL HSA stock solution. Mix, filter sterilize, and store at 4 °C.
3. Embryo thaw solution A (0.5 M sucrose/4 mg HSA/mL): To 46 mL of basal buffer, add 50 mL sucrose stock solution and 4 mL HSA stock solution. Mix, filter sterilize, and store at 4 °C.
4. Embryo thaw solution B (0.2 M sucrose/4 mg HSA/mL): To 76 mL of basal buffer, add 20 mL sucrose stock solution and 4 mL HSA stock solution. Mix, filter sterilize, and store at 4 °C.
5. Embryo thaw solution C (4 mg HSA/mL): To 96 mL of basal buffer, add 4 mL HSA stock solution. Mix, filter sterilize, and store at 4 °C.
6. If desired, solutions 1–4 can be checked for osmolality prior to storage (*see* **Note 5**).
7. Storage (*see* **Note 6**).

***2.3 Equipment and Consumables***

1. Programmable freezing machine—KRYO 360–1.7/MRV (Planer Products), attached to pressurized liquid nitrogen source.
2. Liquid nitrogen storage vessel(s), canes, and goblets (Fig. 1).
3. Binocular dissecting microscope with working range approximately x6-60 (for manipulating embryos).
4. Inverted microscope (for embryo assessment).
5. Sterile plastic serological pipettes (e.g., Falcon®, Becton Dickinson) for dispensing solutions.
6. Sterile pipettes for manipulating embryos (*see* **Note 7**).
7. Sterile plastic freezing straws (approximate capacity 0.25 mL) and plugs for sealing straws (Ref: 005592 and 007352 respectively; IMV Technologies, France).
8. Sterile 1 mL syringes.

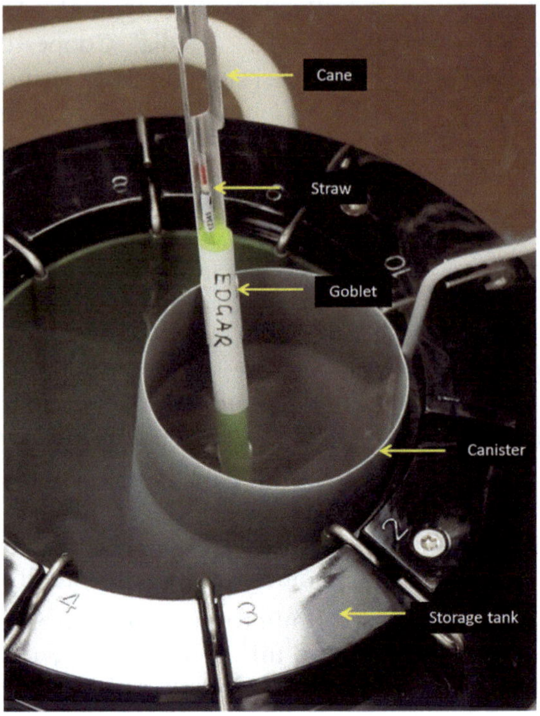


**Fig. 1** Storage of slow frozen embryos in liquid nitrogen

9. Forceps or cotton buds (for seeding straws).
10. Sterile culture dishes (Nunc 4-well, Denmark).
11. Water bath set at 30 °C.
12. Sterile scissors.

## 3 Methods

The freezing machine should be programmed as follows:

Set start temperature at +16 °C.

1st ramp: Cool from +16 °C to −7 °C at 2 °C/min.

2nd ramp: Hold at −7 °C for 10 min (to allow manual seeding).

3rd ramp: Cool from −7 °C to −30 °C at 0.3 °C/min.

4th ramp: Cool from −30 °C to −150 °C at 50 °C/min.

5th ramp: Hold at 150 °C for approximately 30 min (to allow transfer of straws to liquid nitrogen storage vessel).

### *3.1 Dehydration of Cleavage Stage Embryos*

1. Under sterile conditions, add 0.5 mL of embryo freeze solution to a labeled well of a Nunc culture dish for each embryo to be cryopreserved and equilibrate to room temperature (20–22 °C). Prepare a separate dish containing embryo freeze solution to be used for rinsing straws.

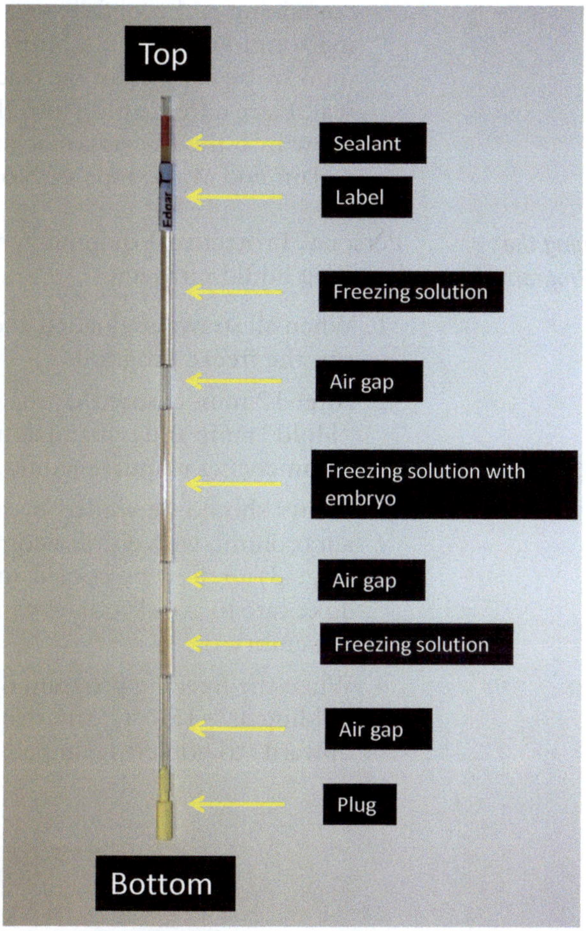


**Fig. 2** Configuration of loaded straw

2. Start the freeze program. Pause/hold the freeze program when it reaches the start temperature.
3. Label one straw for each embryo to be cryopreserved; the label should be at the top (sealant) end of the straw (*see* Fig. 2). Attach a 1 mL syringe (preloaded with air) to the labeled/sealant end of the straw and rinse the straw with freezing solution (*see* **Note 8**).
4. When the freezing medium has reached room temperature, remove the embryo culture dish from the incubator and transfer each individual embryo, selected for cryopreservation, into a separate well containing embryo freeze solution for 10 min (*see* **Note 9**).
5. Load embryos into individual pre-labeled and pre-rinsed straws. The loading configuration should be freezing solution, followed by air bubble, followed by freezing solution

containing embryo, followed by air bubble, followed by freezing solution (Fig. 2). Continue aspiration until the initial column of freezing solution reaches the sealant and activates the seal. Place a freeze plug into the open end of the straw and put straw into the freezing machine in a vertical position with the sealant end at the top (*see* **Note 10**).

### 3.2 Running the Freezing Program

Personal Protective Equipment must be worn at all times when handling liquid nitrogen.

1. When all straws are loaded and placed in the freezing machine, run the freeze program.
2. After 12 min, ensure that the freezing machine has entered the "Hold" ramp and is maintaining a temperature of −7 °C before commencing manual seeding.
3. Straws should be seeded by touching the top (closest to sealant) column with either a cotton bud soaked in liquid nitrogen (Fig. 3) or forceps cooled in liquid nitrogen (*see* **Note 11**). Take care to avoid contact with the middle column (containing the embryo).
4. When the freezing program is complete and the temperature is holding at −150 °C, transfer the straws (still with sealant end upward) to goblets (clamped onto canes—*see* Fig. 1) that have

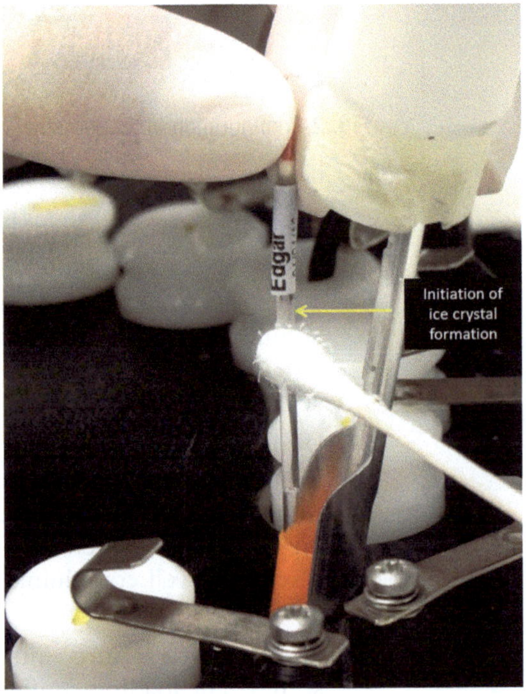


**Fig. 3** Manual seeding of straw

been precooled by immersion in liquid nitrogen, and quickly transfer the canes/goblets to the liquid nitrogen storage vessel (*see* **Note 12**).

### 3.3 Thawing and Rehydrating Slow Frozen Cleavage Stage Embryos

1. At least 3 h prior to thawing embryos, ensure that culture dishes for receiving thawed embryos are set up. These dishes should contain, for each thawed embryo, one well with 0.5 mL culture medium (for washing embryos post-rehydration) and one well with a drop of culture medium under sterile oil. These dishes should be equilibrated in the recommended gas environment (dependent on the culture medium used) at 37 °C.
2. Prior to thawing embryos, Nunc culture dishes for rehydrating thawed embryos should be set up and allowed to equilibrate to room temperature. Each dish should contain one well with 0.5 mL embryo thaw solution A, one well with 0.5 mL embryo thaw solution B, and one well with 0.5 mL embryo thaw solution C. The fourth well should be empty.
3. Remove straw from liquid nitrogen storage location and hold in air at room temperature for 30 s. Wipe straw to remove condensation (*see* **Note 13**).
4. Immerse straw in a tube of sterile water (in a 30 °C water bath) for 40 s (*see* **Note 14**). Remove and wipe straw with a lint-free tissue until dry.
5. Cut the sealed (plugged) end of the straw with sterile scissors and attach a 1 mL syringe, preloaded with air.
6. Cut the labeled (sealant) end of the straw and expel contents into the fourth (empty) well. Quickly locate embryo(s) and transfer, in minimum volume, to well containing embryo thaw solution A for 5 min at room temperature (*see* **Note 15**).
7. Transfer embryo(s), in minimum volume, to well containing embryo thaw solution B for 5 min at room temperature (*see* **Note 16**).
8. Transfer embryo(s), in minimum volume, to well containing embryo thaw solution C for 5 min at room temperature (*see* **Note 16**).
9. Remove culture dish from the incubator, transfer embryo(s) to wash well in minimal volume, rinse thoroughly, and transfer to culture drop before returning to the incubator.

### 3.4 Assessment of Embryos

The criteria for the selection of embryos for cryopreservation may vary between clinics but should be based on implantation data from the individual clinic and a clinical decision on the minimum acceptable predicted implantation potential.

1. Prior to dehydration/cryopreservation, the developmental characteristics of each individual embryo should be recorded.

The details should include number of cells at the time of cryopreservation, degree of fragmentation, evidence of early first cleavage (if recorded), and presence of multinucleation.

2. Post-thaw assessment should, in addition to the pre-freeze information, include details of any blastomere loss, both when the embryo is expelled from the straw and during the rehydration steps. This information is of prognostic significance when assessing the implantation potential of thawed embryos [14–16].
3. If post-thaw culture is continued overnight prior to embryo transfer, any resultant resumption of mitosis/increase in the number of blastomeres should be detailed as this information is also of prognostic significance when assessing the implantation potential of thawed embryos [15–18].

## 4 Notes

1. Alternative basal buffers are available but, for freezing and thawing solutions, it is preferable to use the basal buffer that is consistent with the in-house culture system to minimize stress to the embryo. This buffer should be designed to allow handling in atmospheric conditions and should, therefore, be buffered with, e.g., HEPES or MOPS.
2. As in **Note 1**, alternative products are available but, again, it is preferable to use the albumin source that is present in the embryo culture system if possible.
3. Ideally, the 1 M sucrose solution should be made up in a volumetric flask. From experience, the total volume of buffer required to make 1 L of 1 M sucrose stock solution is between 780 and 790 mL. If a smaller volume of stock solution is required, the amount of sucrose can be reduced pro rata, e.g., 34.23 g in a final volume of 100 mL (requires approximately 78–79 mL of buffer). To ensure that all the sucrose is dissolved, the solution should be shaken vigorously and left overnight before filtration.
4. PROH is a viscous liquid. To ensure accurate delivery, it is important to pay attention to rinsing of the pipette during addition to the solution.
5. To ensure that solutions are made up correctly and to distinguish between solutions, the osmolality of each solution can be determined if an osmometer is accessible. The osmolalities of the freezing solution and the 1 M sucrose stock solution are both significantly higher than the normal standards supplied so appropriate dilutions will be required before using the

osmometer. Osmolality should be measured prior to the addition of the HSA.

6. When solutions are opened for use during storage, sterility of the remaining solution should be maintained by ensuring that aliquots are dispensed in a laminar flow hood. We apply a storage limit of one month for all solutions.
7. We use hand-pulled sterile Pasteur pipettes for transferring embryos between dishes/wells but commercially available pipettes with an internal diameter of 150–190 μm can also be used to minimize carry-over volume.
8. When rinsing straws (to remove any debris and to wet the internal surface of the straw), ensure that the rinsing does not involve contact of the embryo freeze solution with the sealant. After rinsing until loading, the rinsed straws should be stored vertically in sterile tubes with the open ends downward to allow residual rinsing solution to reach the opening. The syringe should still hold some air to facilitate expulsion of this solution prior to loading the straw.
9. When transferring an embryo from culture to the freezing medium, ensure that minimal volume is carried over. We routinely load the transfer pipette with a column of the destination medium (embryo freezing solution) before picking up the embryo from the culture dish and transferring in minimal volume. It is also important to ensure that the embryo is fully submerged in the freeze solution since the viscosity will result in a tendency to float. Submerging the embryo also ensures that it is fully exposed to its new environment.
10. When in the chamber of the freezing machine, we house the straws in plastic goblets clamped onto the canes that are suspended in the freezing chamber. We attempt to restrict capacity to a single straw per goblet/cane but have performed cryopreservation with two straws in goblets on occasion without apparent detriment to the outcome.
11. We prefer to use cotton buds as they will tend to be more efficient at holding the low temperature required for effective seeding. Seeding should always be confirmed by observation of the formation of ice crystals in the straw.
12. In order to minimize the possibility of any rewarming during the transfer of straws into the goblets, it is preferable that this is carried out under liquid nitrogen.
13. As an infection control measure, we use an alcohol (70 % isopropyl alcohol) swab to wipe the straw before immersing in sterile water.

14. To reduce risk of cross contamination, straw(s) from each patient should have a dedicated tube of sterile water used for thawing.
15. When transferring the embryo into thaw solution A in minimum volume, ensure that the embryo is fully submerged in the solution since the embryo will tend to float in the viscous 0.5 M sucrose solution. This will ensure that the embryo is fully exposed to the osmotic conditions. The embryo should undergo significant shrinkage in this solution.
16. At these two steps it is also important to ensure that any minimal carry-over volume is dispersed and that the embryo is fully submerged in the solution. The embryo should undergo re-expansion as the sucrose concentration in the external solution is decreased.

## References


1. Trounson A, Mohr L (1983) Human pregnancy following cryopreservation, thawing and transfer of an eight-cell embryo. Nature 305:707–709
2. Zeilmaker GH, Alberda AT, van Gent I, Rijkmans CM, Drogendijk AC (1984) Two pregnancies following transfer of intact frozen-thawed embryos. Fertil Steril 42:293–296
3. Whittingham DG, Leibo SP, Mazur P (1972) Survival of mouse embryos frozen to -196 degrees and -269 degrees C. Science 178:411–414
4. Lassalle B, Testart J, Renard JP (1985) Human embryo features that influence the success of cryopreservation with the use of 1,2 propanediol. Fertil Steril 44:645–651
5. Testart J, Lassalle B, Belaisch-Allart J, Hazout A, Forman R, Rainhorn JD et al (1986) High pregnancy rate after early human embryo freezing. Fertil Steril 46:268–272
6. Edgar DH, Bourne H, Speirs AL, McBain JC (2000) A quantitative analysis of the impact of cryopreservation on the implantation potential of human early cleavage stage embryos. Hum Reprod 15:175–179
7. Balaban B, Urman B, Ata B, Isiklar A, Larman MG, Hamilton R et al (2008) A randomized controlled study of human Day 3 embryo cryopreservation by slow freezing or vitrification: vitrification is associated with higher survival, metabolism and blastocyst formation. Hum Reprod 23:1976–1982
8. Fabbri R, Porcu E, Marsella T, Rocchetta G, Venturoli S, Flamigni C (2001) Human oocyte cryopreservation: new perspectives regarding oocyte survival. Hum Reprod 16:411–416
9. Bianchi V, Coticchio G, Distratis V, Di Giusto N, Flamigni C, Borini A (2007) Differential sucrose concentration during dehydration (0.2 mol/L) and rehydration (0.3 mol/L) increases the implantation rate of frozen human oocytes. Reprod BioMed Online 14:64–71
10. Borini A, Sciajno R, Bianchi V, Sereni E, Flamigni C, Coticchio G (2006) Cinical outcome of oocyte cryopreservation after slow cooling with a protocol utilizing a high sucrose concentration. Hum Reprod 21:512–517
11. De Santis L, Cino I, Rabellotti E, Papaleo E, Calzi F, Fusi FM et al (2007) Oocyte cryopreservation: clinical outcome of slow-cooling protocols differing in sucrose concentration. Reprod Biomed Online 14:57–63
12. Jericho H, Wilton L, Gook DA, Edgar DH (2003) A modified cryopreservation method increases the survival of human biopsied cleavage stage embryos. Hum Reprod 18:568–571
13. Edgar DH, Karani J, Gook DA (2009) Increasing dehydration of human cleavage-stage embryos prior to slow cooling significantly increases cryosurvival. Reprod Biomed Online 19:521–525
14. Van den Abbeel E, Camus M, Van Waesberghe L, Devroey P, Van Steirteghem AC (1997) Viability of partially damaged human embryos after cryopreservation. Hum Reprod 12:2006–2010
15. Guerif F, Bidault R, Cadoret V, Couet ML, Lansac J, Royere D (2002) Parameters guiding selection of best embryos for transfer after cryopreservation: a reappraisal. Hum Reprod 17:1321–1326

16. Edgar DH, Archer J, McBain J, Bourne H (2007) Embryonic factors affecting outcome from single cryopreserved embryo transfer. Reprod Biomed Online 14:718–723
17. Van der Elst J, Van den Abbeel E, Vitrier S, Camus M, Devroey P, Van Steirteghem AC (1997) Selective transfer of cryopreserved human embryos with further cleavage after thawing increases delivery and implantation rates. Hum Reprod 12:1513–1521
18. Ziebe S, Bech B, Petersen K, Mikkelsen AL, Gabrielsen A, Andersen AN (1998) Resumption of mitosis during post-thaw culture: a key parameter in selecting the right embryos for transfer. Hum Reprod 13:178–181

# Chapter 10

# Human Oocyte Vitrification

**Laura Rienzi, Ana Cobo, and Filippo Maria Ubaldi**


## Abstract

Discovery and widespread application of successful cryopreservation methods for MII-phase oocytes was one of the greatest successes in human reproduction during the past decade. Although considerable improvements in traditional slow-rate freezing were also achieved, the real breakthrough was the result of introduction of vitrification. Here we describe the method that is most commonly applied for this purpose, provides consistent survival and in vitro developmental rates, results in pregnancy and birth rates comparable to those achievable with fresh oocytes, and does not result in higher incidence of gynecological or postnatal complications.


**Key words** Vitrification, Oocyte, Human, Technique, Fertility preservation, Donation

## 1 Introduction

Although the first baby born from a cryopreserved, slow-rate frozen human oocyte was reported in 1986 [1], the low efficiency did not allow widespread application. Vitrification was applied successfully first in 1999 [2], and after further improvements described in 2005 [3], it has become acknowledged worldwide as a technique of choice. Meanwhile, improvements in traditional freezing technique have also been achieved [4–8], and for a few years the two approaches were equally used for the purpose. However, during the past few years vitrification has become the preferred method in most clinics. Most vitrification techniques are relatively simple, no expensive instruments are required, and by using the right method the overall efficiency approaches or reaches that achievable with fresh oocytes [9–12]. Concerns regarding the potential harmful effect of vitrification for further development in vitro and in vivo including pregnancy rates, complications in utero, stillbirths, developmental abnormalities, or other diseases have not been justified by follow-up studies. Oocyte vitrification is not regarded as

Zsolt Peter Nagy et al. (eds.), *Cryopreservation of Mammalian Gametes and Embryos: Methods and Protocols*, Methods in Molecular Biology, vol. 1568, DOI 10.1007/978-1-4939-6828-2_10, © Springer Science+Business Media LLC 2017

experimental any more by ASRM [13] and ESHRE [14] and it is used worldwide in the majority of infertility units.

The need for oocyte cryopreservation was initially underestimated and the technique was used only in special legal and logistic situations. However, with the increasing efficiency the application areas have broadened and now include fertility preservation for medical and nonmedical indications—including malignant diseases and/or their treatments, incipient premature ovarian failure, or wish to postpone maternity—oocyte donation; legal ban of or ethical concerns regarding embryo cryopreservation; accidental unavailability of male gametes. According to some suggestions, cryopreservation of oocytes may also be used as an alternative to that of embryos, to avoid possible consequences of separation/divorce of couples.

During the past decade many different techniques were published for human oocyte vitrification. Most of them were slight variations of the consistent and reliable method of Kuwayama et al. [3] by using slightly modified carrier tools or storage procedures. Other methods attempted to eliminate the supposed toxic effect of DMSO or the potential transmission of diseases via direct contact with contaminated liquid nitrogen. Controversially the embryo toxicity of DMSO was disproven by recent publication [15], and after hundreds of thousands of embryos transferred after application of open vitrification methods not a single infection was reported. On the other hand, vast majority of reports describing high overall efficiency are based on open systems and DMSO-ethylene glycol as cryoprotectants. Accordingly, in this chapter we describe the procedure based on the original publication of Kuwayama [3] with slight practical adjustments.

## 2 Materials

Here we describe two techniques for oocyte vitrification. Both techniques are based on the Cryotop method described by Kuwayama et al. [3].

The first technique (Protocol 1) is the original approach with slight modification, based on commonly used tissue culture dishes.

The second technique (Protocol 2) is based on a set containing dishes developed by Kitazato Biopharma, Shizuoka, Japan, and follows the protocol published in detail on the company's homepage (http://www.kitazato.co.jp/pdf/book/cryotop_protocol_ver8_140709.pdf).

These protocols are the most commonly applied procedures for human MII phase oocyte vitrification worldwide.

According to the authors, results achieved with the original [1] and new [2] protocols are similar.

**2.1 Media for Protocol 1 and 2**

Composition of all solutions used for Cryotop vitrification were published by Kuwayama et al. [3] and also included in the early Kitazato kits. Although recent manuals and homepages of Kitazato do not contain full information, the composition is supposed to be identical with the original described below (*see* **Note 1**) except for a recent change in macromolecule supplementation (*see* **Note 2**). The authors of this chapter did not find significant difference between results achieved with the two solutions.

1. Basic Solution (BS) of all media is Hepes-buffered TCM-199 (*see* **Note 3**), supplemented with Serum Substitute Supplement (SSS) or trehalose) (*see* **Note 2** and **4**).
2. Equilibration Solution (ES) consists of 7.5% ethylene glycol (EG) and 7.5% dimethylsulfoxide (DMSO) (*see* **Note 5**) dissolved in BS.
3. Vitrification Solution (VS) consists of 15% EG, 15% DMSO, and 0.5 M sucrose dissolved in BS.
4. Warming Solution (WS; *see* **Note 6**) consists of 1.0 M sucrose dissolved in BS.
5. Dilution Solution (DS) consists of 0.5 M sucrose dissolved in BS.
6. All media should be pre-warmed to room temperature (24–26 °C) before application except for WS which should be pre-warmed to 37 °C (*see* **Note 7**).

**2.2 Carrier Tools and Other Equipment**

1. Kitazato Cryotop carrier tool equipped with a protective cap (*see* **Note 8**).
2. Styrofoam box for liquid nitrogen.
3. Pipettes, forcepses, fine tip cryo-markers.

**2.3 Dishes for Protocol 1**

1. Four-well dishes and 33 mm diameter petri dishes (Nunc, Roskilde, Denmark).

**2.4 Dishes for Protocol 2**

1. Repro Plate (Kitazato Biopharma, Tokyo, Japan).

## 3 Methods

Principles used for selection of the method shown below are identical to those explained in Subheading 2.

**3.1 Oocyte Recovery and Preparative Steps**

1. Collect oocytes at 35 h post-HCG administration by transvaginal ultrasound-guided aspiration.
2. After incubation in vitro standard laboratory conditions (*see* **Note 9**), expose them to 10–40 IU/mL hyaluronidase solution and remove the corona radiata by gentle pipetting.

3. Evaluate and select MII phase oocytes without any signs of degeneration (*see* **Note 10**). Those with dark cytoplasm, a centrally located granular area, vacuoles, or a large polar body should be excluded from vitrification.
4. Perform vitrification immediately after denudation and selection.

### 3.2 Equilibration and Cooling, Protocol 1

1. Place oocytes (up to three together) in a 20 μL drop of BS on the lid of a 33 mm diameter petri dish (*see* **Note 11**) (Fig. 1).
2. Place a 20 μL drop of ES close to the BS drop. Merge the two drops. Wait for 3 min (*see* **Note 12**).
3. Add another 20 μL ES drop and merge it with the previously merged drops. Wait for 3 min.
4. Move each oocyte individually into a single 20 μL drop of ES and incubate for 6–9 min.
5. Transfer all oocytes simultaneously into 1 mL of VS prepared in a well of a Nunc four-well dish and incubate for 1 min.
6. Transfer each oocyte in a minimum volume to the filmstrip of a pre-labeled Cryotop. Remove excess solution after loading (Fig. 2).
7. With a single uninterrupted rapid movement, immerse the Cryotop into liquid nitrogen (*see* **Note 13**).
8. While submerged in liquid nitrogen, place the protective plastic cap over the filmstrip and fasten it to the plastic handle.

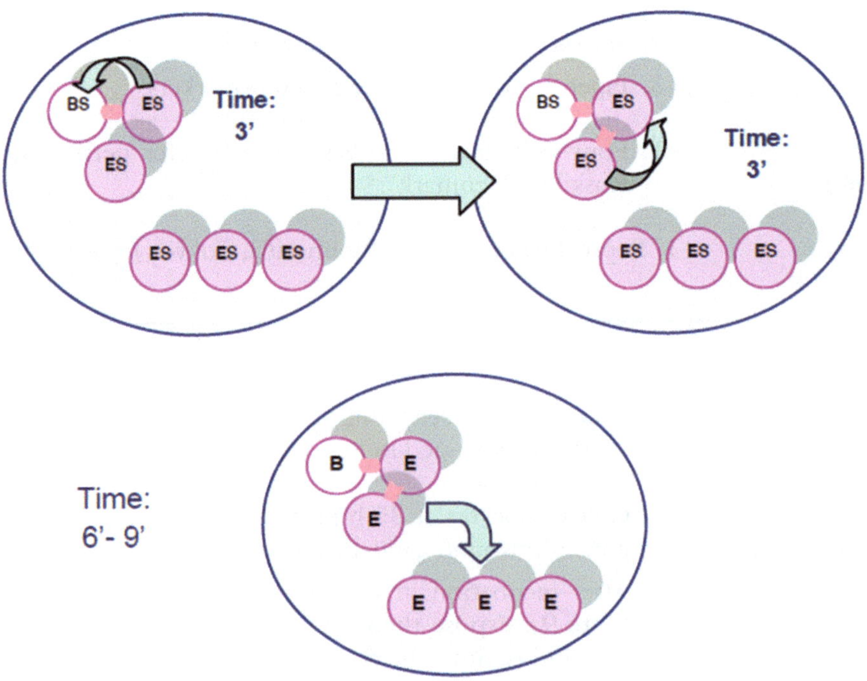


**Fig. 1** Stepwise exposition of oocytes to equilibration solution (ES). *See* explanation in the text

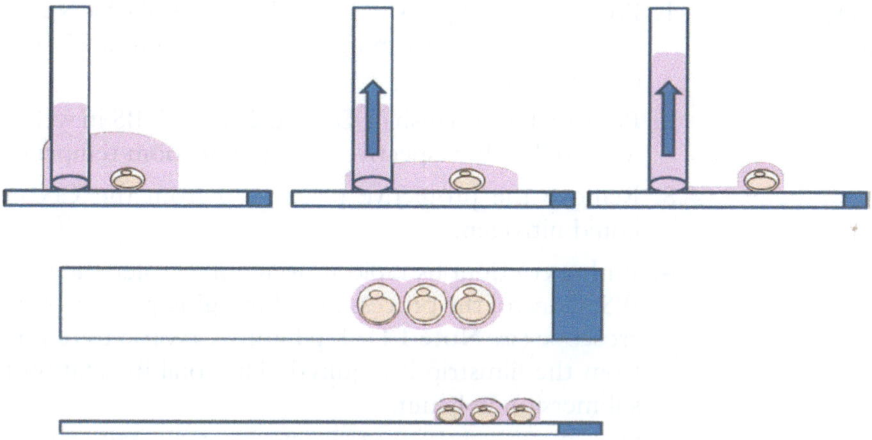

**Fig. 2** Loading of oocytes on the filmstrip of Cryotop. *See* explanation in the text

### *3.3 Equilibration and Cooling, Protocol 2*

1. Drop 20 μL BS into 1st well and 300 μL VS each into 2nd and 3rd well of the Repro Plate.
2. Place the oocytes (up to 16) with minimal volume of medium at the bottom of BS well. Compare the width of perivitelline space with the thickness of zona pellucida and record it.
3. Add 20 μL ES surrounding the previous BS drop. Wait for 3 min.
4. Add another 20 μL ES gently and wait for 3 min.
5. Add 240 μL ES gently and wait for 6–9 min. The volume of the oocytes is required to recover completely.
6. Place the oocytes on the surface of VS. Aspirate the remaining ES surrounding the oocytes and throw it out of the well. Place the oocytes at the bottom of the well and continue removing the remaining ES around them changing their position. All this process must be done in 30 s.
7. Place the oocytes at the bottom of the 2nd VS well. During 30 s, move them throughout the well and remove surrounding solution.
8. Carry the oocytes at the tip of the pipette and place them on the Cryotop strip with minimum volume. Aspirate the excess of solution minimizing the VS drop (Fig. 2). No more than four oocytes per Cryotop must be loaded.
9. Plunge the Cryotop directly into liquid nitrogen and move it quickly (*see* **Note 13**). Keep the Cryotop under liquid nitrogen and cover it with the straw cap.

### *3.4 Storage*

1. Vapor-phase nitrogen freezers (CBS; Custom Biogenic Systems) are used for storage.

### 3.5 Warming and Dilution, Protocol 1

1. Prepare warming dish (1 mL WS in a well of a Nunc four-well dish) 2 h before warming, and place it into a 37 °C incubator (*see* **Note 6**).
2. Prepare dilution dish (1 DS and 2 × 1 mL BS in wells of a Nunc four-well dish, respectively, keep it at room temperature.
3. Remove the protective plastic cap from the Cryotop under liquid nitrogen.
4. With a continuous rapid movement transfer the Cryotop into WS. Control the process by submerging process by a stereomicroscope (*see* **Note 14**). Tap lightly Cryotop to remove oocytes from the filmstrip if required. The total incubation time after submersion is 1 min.
5. Transfer oocytes into the well containing DS and incubate for 3 min.
6. Transfer oocytes into the first well containing BS and incubate for 5 min.
7. Transfer oocytes into the second well containing BS and incubate for 1 min.
8. Transfer oocytes into 1 mL Sage Fertilization Medium (Cooper-Surgical, Trumbull, CT, USA) and incubate at 37 °C in standard laboratory conditions.

### 3.6 Warming and Dilution, Protocol 2

1. Warm WS (TS according to manufacturers' nomenclature) vial (*see* **Note 6**) and a Petri Dish up to 37 °C at least 1.5 h (*see* **Note 7**).
2. Remove the cover straw from the Cryotop under liquid nitrogen. Drop 300 μL DS into 1st well and 300 μL BS each into 2nd and 3rd well of the Repro Plate (*see* **Note 6**).
3. Pour the full content of WS (TS) vial into the petri dish. Immerse the Cryotop into WS situated on the microscope stage with a fast movement. Do not try to recover the oocytes before 40 s (*see* **Note 14**).
4. After 1 min immersion in WS (TS), aspirate the oocytes and take two extra centimeters of WS (TS).
5. Blow WS (TS) from the pipette directly to the bottom of the DS well. Place the oocytes into the WS layer. Wait for 3 min.
6. Aspirate the oocytes in DS and load two extra cm of DS. Blow DS to the bottom of the BS1 (WS according to manufacturers' nomenclature) well and place the oocytes into the DS layer. Wait for 5 min.
7. Aspirate the oocytes with the minimal volume of BS (WS) and transfer them to the top center of BS2. After the oocytes free-fall to the bottom of WS2, repeat the process in WS2.

8. After 1 min in BS2 (WS2), transfer the oocytes to a culture dish containing the appropriate culture medium. Incubate them at 37 °C.

***3.7 Preparation for ICSI***

1. All oocytes should be fertilized with intracytoplasmic sperm injection (ICSI).
2. The in vitro culture period between warming-dilution and ICSI should be done exactly 2 h after warming [11].

## 4 Notes

1. Several laboratories prepare in-house solutions with the same components by using Hepes buffered TCM-199, ethylene glycol, DMSO from Sigma-Aldrich (St. Louis, MO, USA), and SSS from Irvine (Santa Ana, CA, USA)
2. In VT60 series of Kitazato, protein supplementation was replaced with synthetic products and plant derivates including hydroxypropyl cellulose.
3. Although several vitrification methods use other basic solutions and buffers, according to our unpublished experience and that of others, more consistent results can be achieved by using Hepes buffered TCM-199.
4. Other protein sources containing both albumin and globulin for example plasmanate (Bayer, Leverkusen, Germany) were also found suitable.
5. For in-house media preparation: concentrated DMSO exposed to air for a long period of time may become toxic; purchase in 5 mL ampoules and room temperature storage for no more than 3 weeks after opening is suggested.
6. Kitazato kits and protocols use the term "thawing," which is against the acknowledged terms to be applied for vitrification [16]. Accordingly, in this protocol we use the term Warming Solution (WS) instead of Kitazato's Thawing Solution. To avoid further confusion, we replaced Kitazato's abbreviations for washing solutions (WS1, WS2) with Basic Solution (BS) reflecting clearly the composition.
7. For pre-warming WS, do not use incubators with $CO_2$ supply; it may modify the pH of the Hepes-buffered medium.
8. To eliminate the danger of cross-contamination via liquid nitrogen during storage, Kitazato applied the sealed container-straw technique originally described by Vajta et al. [17]. This product (Cryotop SC) was not used in our laboratories.

9. Incubation time between 2 and 4 h does not have any influence on the outcome of vitrification and further development [18]
10. Although a general consensus regarding the morphology and developmental competence is still missing [11], oocytes with dark cytoplasm, a centrally located granular area, vacuoles or a large polar body should be excluded from vitrification.
11. By using the different surface characteristics of the lid vs. the bottom of the petri dish, drops remain higher and handling of oocytes is easier.
12. Immediately after merging the drops, oocytes will shrink drastically, with considerable mishap. However, they usually regain most of their original size after 3 min incubation. The same phenomenon occurs after merging with the second droplet, and after the 6–9 min incubation in the ES droplet.
13. To immerse quickly after loading is important to prevent excessive evaporation of the minuscule amount of solution surrounding oocytes. Rapid passage through the vapor of the liquid nitrogen is also important to increase the cooling rate.
14. Oocytes get extremely transparent immediately after warming, and may float to different levels of WS, accordingly retrieval may become difficult. Before the start of warming, the stereomicroscope should be adjusted to low magnification to see the whole well conveniently. After submersion and detachment of oocytes from the strip, keep optical control continuously during the incubation period in WS.

## References

1. Chen C (1986) Pregnancy after human oocyte cryopreservation. Lancet 1:884–886
2. Kuleshova L, Gianaroli L, Magli C et al (1999) Birth following vitrification of a small number of human oocytes: case report. Hum Reprod 14:3077–3079
3. Kuwayama M, Vajta G, Kato O et al (2005) Highly efficient vitrification method for cryopreservation of human oocytes. Reprod Biomed Online 11:300–308
4. Fabbri R, Porcu E, Marsella T et al (2001) Human oocyte cryopreservation: new perspectives regarding oocyte survival. Hum Reprod 16:411–416
5. Borini A, Lagalla C, Bonu MA et al (2006a) Cumulative pregnancy rates resulting from the use of fresh and frozen oocytes: 7 years' experience. Reprod Biomed Online 12:481–486
6. Borini A, Sciajno R, Bianchi V et al (2006b) Clinical outcome of oocyte cryopreservation after slow cooling with a protocol utilizing a high sucrose concentration. Hum Reprod 21:512–517
7. Boldt J, Tidswell N, Sayers A et al (2006) Human oocyte cryopreservation: 5-year experience with a sodium-depleted slow freezing method. Reprod Biomed Online 13:96–100
8. Borini A, Bianchi V, Bonu MA et al (2007) Evidence-based clinical outcome of oocyte slow cooling. Reprod Biomed Online 15:175–181
9. Cobo A, Kuwayama M, Perez S et al (2008) Comparison of concomitant outcome achieved with fresh and cryopreserved donor oocytes vitrified by the Cryotop method. Fertil Steril 89:1657–1664
10. Nagy ZP, Chang CC, Shapiro DB et al (2009) Clinical evaluation of the efficiency of an oocyte donation program using egg cryo-banking. Fertil Steril 92:520–526
11. Rienzi L, Romano S, Albricci L et al (2010) Embryo development of fresh versus vitrified

metaphase II oocytes after ICSI: a prospective randomized sibling-oocyte study. Hum Reprod 25:66–73

12. Cobo A, Meseguer M, Remohi J et al (2010) Use of cryo-banked oocytes in an ovum donation programme: a prospective, randomized, controlled, clinical trial. Hum Reprod 25:2239–2246
13. Practice Committees of American Society for Reproductive Medicine, Society for Assisted Reproductive Technology (2013) Mature oocyte cryopreservation: a guideline. Fertil Steril 99:37–43
14. ESHRE Task Force on Ethics and Law, Dondorp W, de Wert G, Pennings G et al (2012) Oocyte cryopreservation for age-related fertility loss. Hum Reprod 27:1231–1237
15. Vajta G, Reichart A, Ubaldi F et al (2013) From a backup technology to a strategy-outlining approach. The success story of cryopreservation. Expert Rev Obstet Gynecol 8:181–190
16. Shaw JM, Jones GM (2003) Terminology associated with vitrification and other cryopreservation procedures for oocytes and embryos. Hum Reprod Update 9:583–605
17. Vajta G, Holm P, Kuwayama M et al (1998) Open Pulled Straw (OPS) vitrification: a new way to reduce cryoinjuries of bovine ova and embryos. Mol Reprod Dev 51:53–58
18. Ubaldi F, Anniballo R, Romano S et al (2010) Cumulative ongoing pregnancy rate achieved with oocyte vitrification and cleavage stage transfer without embryo selection in a standard infertility program. Hum Reprod 25:1199–1205

# Chapter 11

# Human Embryo Vitrification

## Juergen Liebermann

## Abstract

Cryopreservation is one of the keystones in clinical infertility treatment. In particular *vitrification* has become a well-established and widely used routine procedure that has allowed important expansion of therapeutic strategies when IVF is used to treat infertility. Vitrification of human blastocysts allows us to maximize the potential for conception from any single in vitro fertilization cycle and prevents wastage of embryos. The technology may even be used to eliminate fresh embryo transfers for reasons of convenience, uterine receptivity, fertility preservation, preimplantation genetic diagnosis, or emergency management. In this chapter, the application of vitrification technology for cryopreserving human blastocysts will be revealed through step-by-step protocols. The results that are presented using the described protocols underscore the robustness of the vitrification technology for embryo cryopreservation.

**Key words** Human blastocyst, Cryopreservation, Embryo transfer, Artificial collapsing, Assisted hatching, Cryostorage, Vitrification, Birth weight

## 1 Introduction

Since the announcement in 1972 of the survival of mouse embryos after cryopreservation at −196 °C [1], the impact of cryopreservation on the growth and improved efficiency of assisted reproduction in humans has become increasingly appreciated. Moreover, cryopreservation has also been shown to increase overall pregnancy rates, while allowing for further selection of embryos. Indeed it is possible to achieve implantation and pregnancy rates with frozen-thawed embryos as high as those achieved with fresh embryos. Routine in vitro blastocyst culture and cryopreservation have been shown to increase pregnancy rates, while allowing for better selection of embryos. With more reliable cryopreservation techniques such as "vitrification," lower numbers of embryos are now being transferred, resulting in less high-order multiple pregnancies, as well as increased healthy implantations. The fundamental objectives for successful cryostorage of cells in liquid nitrogen at −196 °C can be summarized as follows: (1) arresting the metabolism reversibly; (2) maintaining structural and genetic integrity; (3)

Zsolt Peter Nagy et al. (eds.), *Cryopreservation of Mammalian Gametes and Embryos: Methods and Protocols*, Methods in Molecular Biology, vol. 1568, DOI 10.1007/978-1-4939-6828-2_11, 

achieving acceptable survival rates after thawing; (4) maintenance of developmental competency post-thaw; and (5) the technique has to be reliable and repeatable.

The major disadvantage of using cryostorage is that it can lead to crystallization of water and thereby can create new and unwanted physical and chemical events that may injure the cells that are being preserved. Vitrification, however, avoids ice formation altogether during the cooling process by establishing a glassy or vitreous state, wherein molecular translational motions are arrested without structural reorganization of the liquid in which the reproductive cells are suspended. The physical definition of vitrification is the solidification of a solution at low temperature, not by ice crystallization, but by extreme elevation in viscosity during cooling [2, 3]. In general, vitrification solutions are aqueous cryoprotectant solutions that do not freeze when cooled at high rates to very low temperatures. In addition, the rate of cooling/warming and the concentration of the cryoprotectant required to achieve vitrification are inversely related. In addition, recent publications have shown the relatively greater importance of warming rate over cooling rates with regard to the survival of oocytes subjected to a vitrification procedure [4, 5].

The earliest attempts using vitrification as an ice-free cryopreservation method for embryos were first reported in 1985 [6]. In 1993 successful vitrification of mouse embryos was demonstrated [7]. Furthermore, bovine oocytes and cleavage-stage embryos were vitrified and warmed successfully a few years later [8]. In 1999 and 2000 successful pregnancies and deliveries after vitrification and warming of human oocytes were reported [9, 10]. Since that time and because it seems to be that both entities appear to be especially chill-sensitive cells in ART, blastocysts appear to have received a significant boost in survival rates by avoiding ice crystallization through the use of vitrification [11].

Vitrification is very simple, requires no expensive programmable freezing equipment, and relies principally on the placement of the embryo in a very small volume of vitrification medium that must be cooled at extreme rates not obtainable in traditional enclosed cryostorage devices such as straws and vials.

One "drawback" of vitrification considered by embryologists not familiar with the vitrification technique is the use of high concentrations of cryoprotectants, which does potentially mean that vitrification solutions are potentially more toxic than their counterpart solutions used for conventional slow freezing. This is necessitated by the practical limit for the rate of cooling, and the biological limit of tolerance of the cells for the concentration of toxic cryoprotectants being used to achieve the cryopreserved state. It is important to note that recently published papers [12–15] have shown that the use of relatively high concentration of cryoprotectants such as 15 % (v/v) ethylene glycol (EG) used in an equimolar mixture with dimethyl sulfoxide (DMSO) had no negative effect on the perinatal outcomes from blastocyst transfers following vitrification when compared with those from fresh blastocyst transfers.

Cryoprotective agents (CPAs) are essential for the cryopreservation of cells. Basically two groups of cryoprotectants exist: (1) permeating (e.g., glycerol, ethylene glycol and dimethyl sulfoxide) and (2) non-permeating (e.g., disaccharides, proteins, and polymers) agents. The key component of a vitrification solution is the permeating agent. These compounds are hydrophilic nonelectrolytes with a strong dehydrating effect. Furthermore, these CPAs are able to depress the "freezing point" of the solution. Additionally, these highly permeating cryoprotectants are also more likely to diffuse rapidly out of the cells, so that the cells quickly regain their original volume upon warming, thus preventing osmotic injury. The second components of a vitrification solution are the non-permeating CPAs such as disaccharides, for example, sucrose, which does not penetrate the cell membrane, but helps to draw out more water from cells by osmosis, and therefore lessens the exposure time of the cells to the toxic effects of the permeating CPAs. The non-permeating sucrose also acts as an osmotic buffer to reduce the osmotic shock that might otherwise result from the dilution of the cryoprotectant after cryostorage. Furthermore, permeating agents are able to bind with intracellular water, and therefore water is very slowly removed from the cell. During warming using a high extracellular concentration of sucrose (e.g., 1.0 M) counterbalances the high concentration of the cryoprotectant agents in the cell, as it reduces the difference in osmolarity between the intra- and extracellular compartments. The high sucrose concentration cannot totally prevent the cell from swelling, but it can reduce the speed and magnitude of swelling [16–18].

Most important, residual fluid in the blastocoel can therefore reduce post-warming survival of embryos. Vanderzwalmen et al. [19] showed that survival rates in cryopreserved, expanded blastocysts could be improved by artificial reduction of the blastocoel cavity, and others consider that blastocoel collapse is necessary pre-vitrification on whichever day the blastocyst forms [19, 20].

Although one study has suggested that expanded blastocysts show no significant differences in viability, implantation potential, or pregnancy outcome when frozen on day 5 versus day 6 [21], our "body of data" [13–15] refutes the comparable implantation rates for blastocysts cryopreserved on day 5 or 6. And while there are several possibilities why day 6 blastocysts perform less well than day 5 blastocysts, one explanation could be that the older blastocysts are often more expanded with a significantly larger blastocoel impacting water loss and CPA absorption. Being mindful of previously published research on artificial collapse (AC) of blastocysts prior to cryopreservation, we looked for an opportunity that could potentially help us improve the outcome for day 6 blastocysts using AC of the blastocoel prior to vitrification. Because the concept showed improvement for the day 6 blastocysts, we applied to day 5 blastocysts that had well-expanded blastocoels.

AC of the blastocoel can be performed using different techniques, such as microneedles, sucrose solutions, or lasers [22–27]. In 2003 and 2004, two groups independently reported a beneficial effect by applying AC to blastocysts prior to vitrification. Son et al. [22] observed an improved clinical pregnancy rate of 48 % and an implantation rate of 29 % with the use of AC. Hiraoka et al. [23] collapsed day 5 and day 6 blastocysts by manually pipetting embryos until they collapsed and achieved a clinical pregnancy rate of 50 %, with an implantation rate of 33 % after warming. Moreover, Mukaida et al. [24] found that the survival rate of vitrified blastocysts was negatively correlated with the size of the blastocoel. They speculated that a large blastocoel may disturb the efficacy of vitrification. They collapsed the blastocoel by puncturing it with a microneedle or by making a hole between two trophectoderm cells with a laser pulse. After applying AC, the survival improved from 86 to 97.2 %. Moreover, their pregnancy rate went up from 34.1 to 60.2 %, with an implantation rate of 46.7 %. Iwayama et al. [25] used a laser pulse or osmotic shock resulting from exposure of the whole embryo to sucrose, and the implantation rate was significantly higher in both groups compared to the control group without AC (59.7 % and 49.3 % vs. 34.2 %). Furthermore, Hur et al. [26] looked at the effect of AC, achieved using a 29-gauge needle or laser pulse, on clinical outcomes in fresh transfers, and they observed a significant increase in the clinical pregnancy in the study group compared to the control group (58.8 % vs. 39.0 %). All publications mentioned, including Liebermann and Conaghan [27], conclude that AC has a beneficial effect both in frozen blastocyst transfers and for overall cumulative pregnancy and implantation rates.

## 2 Materials and Methods

### 2.1 Materials

1. HSV (High Security Vitrification kit (Catalog # 022137, Cryo Bio System)).
2. Heat sealer (Cryo Bio System).
3. Polycarbonate micropipettes, 300 μm end hole (MidAtlantic Diagnostics# KFPIP-1170-10BS).
4. Brady labeler (TS 2000).
5. Brady labels (PTL-19-427).
6. 90 × 15 mm petri dish (Nunclon # 150362).
7. Center-Well Organ Culture Dish (Falcon 3037).
8. Styrofoam container.
9. Visotubes 10 mm (IMV 5561).
10. Cryo canes aluminum (Thermo Scientific 5015-0001).

### 2.2 Reagents

1. Serum Substitute Supplement (SSS) (Irvine).
2. Modified human tubal fluid (HTF-HEPES) (Irvine).
3. Vit Kit-Freeze (Irvine Scientific #90133DSOC).
4. Vit Kit-Thaw (Irvine Scientific # 90137DSOC).

### 2.3 Equipment

1. Dissecting stereomicroscope (Olympus SZX-12, Bausch Lomb or Leica) with warming stage.
2. Laminar flow hood (Origio).
3. Inverted microscope (Olympus IX-71).
4. Infrared 1.48 μm diode laser (Hamilton Thorne—Zilos laser (Hamilton Thorne Research, Beverly, MA).

### 2.4 Methods

#### 2.4.1 Stepwise Blastocyst Vitrification Procedure

Vitrification of blastocysts is undertaken utilizing a "closed system" (HSV, High Security Vitrification kit; Cryo Bio System, L'Aigle, France; FDA 510(k) clearance for cleavage-stage embryos in blastocysts) (*see* **Notes 1** and **2**) after a two-step loading with cryoprotectant agents at 24 °C (*see* **Notes 3** and **4**). If assisted collapsing (AC) is done prior to vitrification, then the blastocyst is put on an inverted microscope equipped with a laser system (ZILOS-tk, Hamilton Thorne), the junction of two trophectoderm cells in each blastocyst is located, and one shot (100 % power, 500 μs pulse length) is applied to the cell junction. The blastocyst is then moved back in the incubator for 5–10 min. Briefly, blastocysts have to be placed in equilibration solution, which is the base medium (M199 with 20 % Serum Substitute Supplement (SSS) containing 7.5 % (v/v) ethylene glycol (EG) and 7.5 % (v/v) dimethyl sulfoxide (DMSO) (*see* Table 1). After 5–7 min, the blastocysts need to be washed quickly in vitrification solution, which is the base medium containing 15 % (v/v) DMSO, 15 % (v/v) EG, and 0.5 M sucrose, for 45–60 s and transferred onto the HSV using a micropipette.

**Table 1**
**Summary of vitrification and warming solutions (Irvine Vit Kit-"Freeze" and "Warm")**

| | Composition | | |
|---|---|---|---|
| | Ethylene glycol (%) | DMSO (%) | Sucrose (M) |
| Equilibration solution (ES) | 7.5 | 7.5 | 0 |
| Vitrification solution (VS) | 15 | 15 | 0.5 |
| Warming solution (TS) | 0 | 0 | 1.0 |
| Diluent solution (DS) | 0 | 0 | 0.5 |
| Wash solution (WS) | 0 | 0 | 0 |

(M-199 + 20 % SSS) [Wash solution—WS]

Immediately after the loading of not more than two blastocysts in less than a 1 μL drop on the HSV, the straws are heat sealed, then plunged in LN2, and finally stored inside 5 mL liquid nitrogen prefilled canes (Visotube Rond, IMV, France). Each single step is described in detail below:

1. Aseptic techniques are required at all stages. For equilibration and vitrification procedures, ensure the benchwarmer is at room temperature (~24 °C) (*see* **Notes 3** and **4**).
2. Take reagents from the refrigerator and allow them to warm to room temperature.
3. Move blastocysts to freeze into a separate well. Bring this dish to the inverted microscope and with the embryo positioned with the laser objective, use a single pulse to hit the blastocysts between two trophectoderm cells to collapse the embryo. Place the dish back into the incubator for 5–10 min.
4. Label a petri dish with the patient's name under the lid as follows: HTF-HEPES, ES, and VS. Prepare 2 × 50 μL of HTF-HEPES (+20 % SSS—mHTF), 2 × 50 μL of ES, and 4 × 50 μL of VS (*see* Fig. 1).
5. Brady label should include the patients' last name, first name, accession #, medical personal identification #, and date plus number and type of embryos.
6. Before vitrification, use a stripper tip with 300 μm end hole for loading the blastocysts on the top.
7. Fill Styrofoam container with $LN_2$.
8. Each sample that is vitrified will be done in a separate hood and verified by a second embryologist before proceeding. Vitrify good expanded/hatching blastocysts on day 5/6/7.

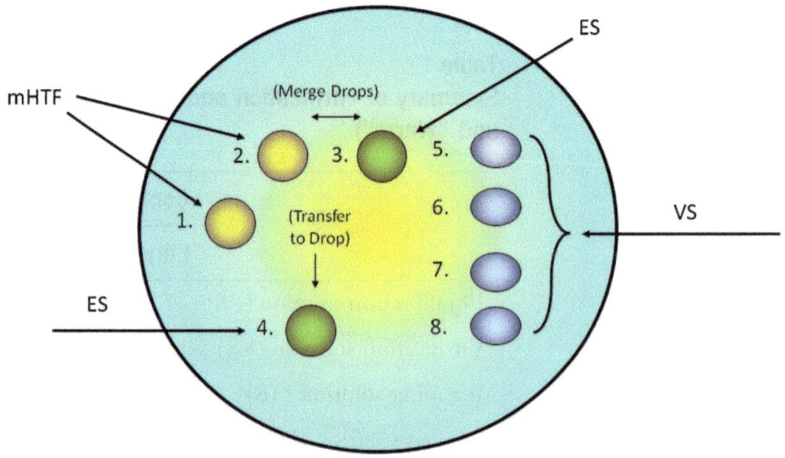


**Fig. 1** Setup for the vitrification procedure on a plain 90 mm dish lid surface. (ES = Equilibration solution; VS = Vitrification solution; mHTF = Wash solution)

9. Remove embryos from culture dishes using a stripper tip into the mHEPES (drop 1), gently aspirating to remove any residual culture medium.
10. Pipette from mHTF (drop 1) to the other drop of mHTF (drop 2) and immediately merge it with the first drop of ES (drop 3). Set the timer for 5 min.
11. After 5 min, transfer the embryos to the remaining drop of ES (drop 4). Set the timer for 3 min. Place the embryos on the top of the drop and let them settle to the bottom.
12. Then load the blastocysts in a VS back-loaded stripper tip, and rinse through the four droplets of VS (drops 5–8), and between each droplet clean the tip.
13. Placement into the VS and loading of the Cryotop should take **less than 1 min**, so that the total incubation time in VS is 30 s. After 30 s, gently transfer them to the tip of the HSV by using a stripper tip to load the blastocyst(s) in as small volume (less than 0.5 μL) as possible onto the edge of the stick (*see* **Note 5**).
14. Visually confirm placement (*see* **Note 6**).
15. Before loading, apply the label to the open end of the empty straw. Load the HSV stick into the empty straw, the side with the embryos first. Use the blue handle to make sure the stick is in as far as it goes. Then, using the heat sealer, seal the open end of the stick and plunge the whole straw into the $LN_2$. Place the straw in a pre-cooled aluminum cane for further storage (*see* **Notes** 7 and **8**).
16. Store at the cane in a nitrogen tank.
17. Make sure to record cane location on the freezing worksheet and cryo inventory log.
18. Complete all paperwork and recheck that all vial locations are logged into the Embryo Inventory.

*2.4.2 Stepwise Blastocyst Warming Procedure*

It is important to mention that regardless of the day of cryopreservation of the embryo (whether day 5, 6, or 7), at thawing, blastocysts should be treated as if they had been cryopreserved on the fifth day of development. To remove the cryoprotectants, blastocysts need to be warmed and diluted in a three-step process. With the HSV submerged in $LN_2$, the inner straw should be removed, and then the carrier with the blastocysts can be removed from the $LN_2$ and placed directly into a pre-warmed (37 °C) organ culture dish containing 1 mL of 1.0 M sucrose (*see* **Note 9**). Blastocysts can be picked up directly from the HSV and placed in a fresh drop of 1.0 M sucrose at 24 °C and immediately connected with a drop of 0.5 M sucrose. After 5 min blastocysts are transferred to a 0.5 M sucrose solution and connected with drops of base medium for additional 5 min. When switching the embryos between different concentrations of warming solutions, be sure to fill up the pipette with the next lower concentration of warming solution before picking up the cells for moving into the next dilution (*see* **Note 10**). Finally the blastocysts are washed in the base medium for 3 min and then returned to culture medium (SAGE + 20 % SSS) (SAGE Blastocyst Medium, Trumbull, CT, USA) in the incubator until transfer. Each single step is described in detail below:

1. Take reagents from the refrigerator and allow them to warm to room temperature. All cryoprotectants are removed at 24 °C.
2. Place a 200 μL drop of TS on a petri dish and place on a warming plate (*see* **Note 9**).
3. Label a petri dish (Nunc) with the patient's name under the lid as follows: TS, DS, and WS. Prepare 1 × 50 μL of TS, 4 × 50 μL of DS, and 6 × 50 μL of WS (*see* Fig. 2).
4. Before warming, use a stripper tip with 200 μm end hole for removing the blastocysts from the HSV tip.
5. Fill Styrofoam container with $LN_2$.
6. Confirm location and identification with a second embryologist before warming any HSV kit. Warm one kit at a time.
7. Each sample that is warmed is done in a separate hood and verified by a second embryologist before proceeding.
8. With the HSV kit under $LN_2$, open the kit by cutting the outer straw. Use the blue handle to remove the inner stick.

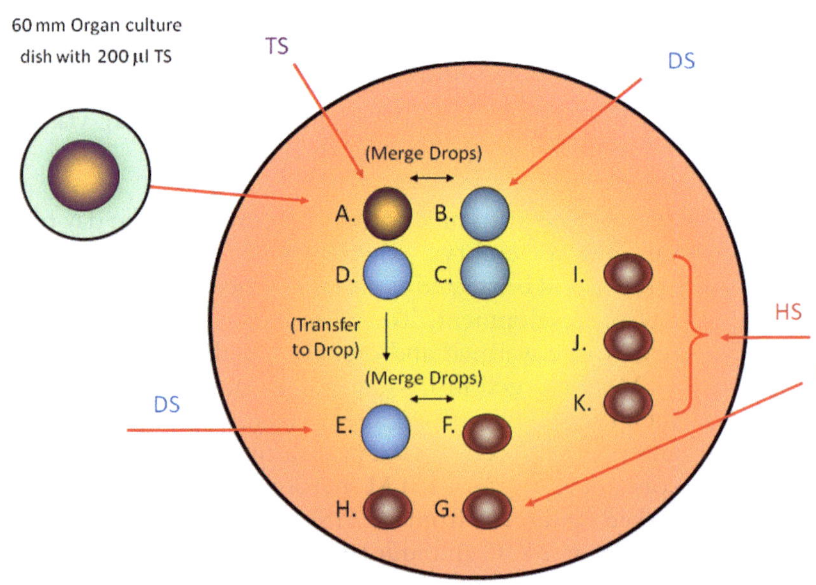


**Fig. 2** Setup for the warming procedure on a plain 90 mm dish lid surface. (TS = Thawing solution; DS = Diluent solution; HS = Holding/Washing solution)

9. Submerge HSV kit directly in the pre-warmed 37 °C drop containing TS, which should be as close as possible to the $LN_2$ Styrofoam container (*see* **Note 11**). As soon as the HSV kit contents liquefy (within 1 s), try to locate the blastocyst(s) before removing it (them) with a stripper tip. After locating all the blastocysts, remove them from the HSV tip and place them in the droplet of TS (drop A) at room temperature and then connect immediately with the first droplet of DS (drop B). Wait for shrinkage and re-expansion.
10. When they start to wrinkle, connect with the second droplet (drop C) and finally with the third droplet of DS (drop D).
11. When they stop reacting and start to reshrink, transfer blastocysts to 0.5 M sucrose (drop E) by placing at the top of this drop, so they float to the bottom. When the reaction is complete, connect with the first droplet of WS (drop F; wait for about 90 % re-expansion).
12. After 100 % expansion, connect with droplet #2 (drop G) and then with droplet #3 (drop H) of WS. Turn on the bench-warmer, and finally dilute through a series of three wash drops of HS (I to K).
13. Place the blastocysts into a culture dish, and put it in the incubator for subsequent culture.
14. Record the survival and appearance of all blastocysts. Update log with warming data, and notify the physician of result (*see* **Note 12**).

## 3 Results on Blastocyst Vitrification

Between 2004 and 2014 the Fertility Centers of Illinois "IVF Laboratory River North"(Chicago) has vitrified 24,390 blastocysts from 6268 patients (Table 2). The majority of blastocysts were vitrified on day 5 (52.4 %) and 45.7 % on day 6, with a minority on day 7 (1.9 %). After 11 years of vitrifying blastocysts using an open (Cryotop) as well as closed (HSV) system and close to 5000 FETs with an average number of 1.7 embryos transferred, the

**Table 2**
**Retrospective data from 6268 patients (average age 34.2 ± 4.9) with blastocyst cryopreservation by vitrification between 2004 and 2014**

| | Day of development | | | |
|---|---|---|---|---|
| | **Day 5** | **Day 6** | **Day 7** | **Total** |
| Number of blastocysts vitrified | 12,781 (52.4%) | 11,148 (45.7%) | 461 (1.9%) | 24,390 (100%) |

**Table 3**
**Perinatal outcome of vitrified blastocysts after more than 4100 transfers between 2004 and 2013 (babies delivered until February 2013). Values are numbers unless otherwise described, and percentages are indicated between brackets**

| | | Day of development | |
|---|---|---|---|
| | Day 5 + Day 6 | Day 5 | Day 6 |
| Deliveries (total) | 1209 | 748 | 461 |
| Babies born (total) | 1495 | 939 | 556 |
| Female | 766 | 488 | 278 |
| Male | 729 | 451 | 278 |
| Singletons | 931 (77.0) | 561 (75.0) | 370 (80.0) |
| Twins | 270 (22.0) | 183 (24.5) | 87 (19.0) |
| Triplets | 8 (1.0) | 4 (0.5) | 4 (1.0) |

**Table 4**
**Live birth weight (LBW) and gestational age (GA) of babies ($n = 1363$) delivered after fresh elective single blastocyst transfers (eSBT; $n = 437$) compared to those delivered from vitrified blastocyst transfers (VBT; $n = 926$). Values are numbers unless otherwise described, and percentages are indicated between brackets**

| | | | Day of development | |
|---|---|---|---|---|
| | eSBT—Day 5 | VBT—Day 5 + Day 6 | VBT—Day 5 | VBT—Day 6 |
| Average age | $31.1 \pm 3.1^{c}$ | $34.7 \pm 4.9^{c}$ | $34.7 \pm 5.0$ | $34.8 \pm 4.8$ |
| Deliveries | 437 | 926 | 561 | 365 |
| Female | $238^{a}$ | $479^{a}$ | 288 | 191 |
| Male | $199^{a}$ | $447^{a}$ | 273 | 174 |
| GA | $38.1 \pm 2.5^{a}$ | $37.9 \pm 2.7^{a}$ | $38.0 \pm 2.6^{a}$ | $37.9 \pm 2.9^{a}$ |
| LBW (g) | $3294.9 \pm 609.4^{b}$ | $3390.1 \pm 710.6^{b}$ | $3419.1 \pm 712.6^{a}$ | $3345.5 \pm 705.8^{a}$ |

T-test (Satterthwaite-Unequal variance): $^{a}P > 0.05$: $^{b}P = 0.013$; $^{c}P < 0.001$

perinatal outcomes are as follows: The number of babies delivered until February 2013 equals 1495 (766 girls and 729 boys) (Table 3). No abnormalities were recorded.

Table 4 summarizes the data on gestational age and live birth weight from 1363 children born. 926 babies were from vitrified day 5 and day 6 transfers, whereas 437 babies were born following fresh elective single embryo transfers (eSET). Furthermore, comparing the 437 newborns from the eSET group to the 926 newborns from frozen-vitrified group (VBT), there was a significant difference in the mean age of the patients ($p < 0.001$). In addition, looking at gestational age (GA) and live birth weight (LBW) of newborns derived from vitrified day 5 blastocysts ($n = 561$) and day 6 blastocysts ($n = 365$), the following data were observed: T-test results indicate no significant difference in gestational age ($p = 0.71$) nor birth weight ($p = 0.124$) for the frozen transfers based on day of development. Obviously, gestational age is highly correlated with birth weight, but there is no significant difference in gestational age between eSET and VBT. However, Table 4 shows that one-way ANOVA of live birth weight by both groups indicates a statistically significant difference between the means of eSET vs. VBT ($p = 0.013$) with heavier babies in the VBT group. This observation is confirmed and published by Pinborg et al. (28). As shown in Table 4, chi-square statistic indicates no difference in the gender distribution between fresh and vitrified transfers ($\chi^2$; $P = 0.3454$, respectively). Running a T-test on gender weight combining the gender from fresh with vitrified transfers indicates a significant difference in average weight by gender ($P < 0.001$; Table 5). Males were on average 157.2 g heavier than females. Table 5 shows the weight distribution for babies born in both groups. Overall, 110 babies weighed less than 2500 g (*considered as low birth weight*), 159 babies born from both groups weighed between 4000 and 4500 g (*considered as large birth weight*), and 33 babies weighed more than 4500 g (*considered as macrosomic birth weight*).

The outcome with regard to day of development and age of the patient, between 2004 and 2014, is summarized in Tables 6 and 7. In good prognosis patients under 35 years old when transferring day

**Table 5**
**Live birth weight (LBW) and gestational age (GA) of babies ($n = 1363$) delivered after fresh elective single blastocyst transfers (eSBT; $n = 437$) compared to those delivered from a vitrified blastocyst transfers (VBT; $n = 926$). Values are numbers unless otherwise described, and percentages are indicated between brackets**

| | eSBT—Day 5 (n) [%] | VBT—Day 5 + Day 6 (n) [%] | Total (n) [%] |
|---|---|---|---|
| LBW (g) < 2500 | 34 *[7.8]* | 76 *[8.2]* | 110 |
| LBW (g) ≥ 4000 | 38 *[8.7]* | 154 *[16.6]* | 192 |
| LBW (g) ≥ 400 and ≤4500 | 32 *[7.3]* | 127 *[13.7]* | 159 |
| LBW (g) > 4500 | 6 *[1.4]* | 27 *[2.9]* | 33 |
| Gender | Girls from eSET and VBT | Boys from eSET and VBT | |
| N | 717 | 646 | |
| GA | $37.9 \pm 2.6^a$ | $38.0 \pm 2.6^a$ | |
| LBW | $3284.1 \pm 663.8^b$ | $3441.3 \pm 691.0^b$ | |

T-test & $\chi^2$ test: $^aP > 0.05$; $^bP < 0.001$

**Table 6**
**Retrospective outcome data (2004–2014) at the *Fertility Centers of Illinois, Chicago*, from vitrified day 5 blastocysts in regard to the patients age. Values are numbers, unless otherwise described**

| | Patient's age (y) | | | | |
|---|---|---|---|---|---|
| | <35 | 35–37 | 38–40 | >40 | Donor |
| Average age | 31.2 ± 2.4 | 35.9 ± 0.8 | 38.9 ± 0.8 | 42.6 ± 2.1 | 43.6 ± 4.7 |
| Cycles | 1404 | 594 | 394 | 195 | 274 |
| Transfers | 1403 | 593 | 392 | 195 | 274 |
| Blastocysts survived | 98.4 | 98.3 | 98.9 | 98.3 | 99.0 |
| Blastocysts transferred (mean) | 1.7 | 1.6 | 1.8 | 1.8 | 1.7 |
| Positive pregnancy/VET (%) | 62.6 | 60.0 | 57.9 | 54.9 | 60.6 |
| Clinical pregnancy/VET (%) | 55.0 | 51.0 | 47.0 | 44.0 | 53.0 |
| Ongoing/delivered pregnancies (%) | 48.0 | 42.0 | 36.0 | 31.0 | 43.0 |
| Implantations | 1033 | 382 | 234 | 105 | 185 |
| Implantation rate (%) | 42.6 | 39.1 | 33.8 | 30.8 | 39.2 |

**Table 7**
**Retrospective outcome data (2004–2014) at the *Fertility Centers of Illinois, Chicago*, from vitrified day 6 blastocysts in regard to the patients age. Values are numbers, unless otherwise described**

| | Patient's age (y) | | | | |
|---|---|---|---|---|---|
| | <35 | 35–37 | 38–40 | >40 | Donor |
| Average age | 31.2 ± 2.3 | 36.0 ± 0.8 | 38.9 ± 0.8 | 42.6 ± 1.8 | 43.2 ± 4.8 |
| Cycles | 868 | 456 | 340 | 222 | 153 |
| Transfers | 861 | 451 | 339 | 219 | 152 |
| Blastocysts survived | 97.5 | 98.5 | 98.5 | 97.0 | 99.6 |
| Blastocysts transferred (mean) | 1.8 | 1.7 | 1.8 | 1.7 | 1.7 |
| Positive pregnancy/VET (%) | 49.2 | 45.0 | 44.5 | 41.6 | 45.4 |
| Clinical pregnancy/VET (%) | 42.0 | 38.0 | 38.0 | 32.0 | 37.0 |
| Ongoing/delivered pregnancies (%) | 34.0 | 30.0 | 30.0 | 21.0 | 27.0 |
| Implantations | 465 | 216 | 167 | 81 | 65 |
| Implantation rate (%) | 30.1 | 27.5 | 27.5 | 21.4 | 24.7 |

5 blastocysts, an ongoing pregnancy and implantation of 48.0 % and 42.6 % were noted (Table 6). In contrast, transferring day 6 blastocysts in patients younger than 35 of age, an ongoing pregnancy and implantation of 34.0 and 30.1 % were recorded (Table 7).

**Table 8**

**A comparison of retrospective data from the cryopreservation program (*Fertility Centers of Illinois, Chicago*) of vitrified day 5, day 6, and day 7 blastocysts using aseptic vitrification technology between October 2007 and 2014. Values are numbers unless otherwise described, and percentages are indicated between brackets**

| | Day 5 + Day 6 + Day 7 | Day 5 | Day 6 | Day 7 |
|---|---|---|---|---|
| Patient's age (y) | 35.6 ± 5.0 | 35.3 ± 2.3 | 36.1 ± 4.8 | 36.0 ± 3.9 |
| Transfers | 3327 | 2068 | 1231 | 28 |
| Blastocysts warmed | 5748 | 3496 | 2205 | 47 |
| Blastocysts survived | 5699 (99.1) | 3475 (99.4) | 2177 (98.7) | 47 (100.0) |
| Blastocysts transferred | 5584 | 3407 | 2130 | 47 |
| Blastocysts transferred (mean) | 1.7 | 1.6 | 1.7 | 1.7 |
| Implantations | 2094 (37.5) | 1447 (42.3)[a] | 640 (30.0)[a] | 7 (14.9)[a] |
| Positive pregnancy/VET | 1935 (58.2) | 1309 (63.3)[b] | 615 (50.0)[b] | 11 (39.3)[b] |
| Clinical pregnancy/VET | 1637 (49.2) | 1123 (54.3)[c] | 508 (41.3)[c] | 6 (21.4)[c] |
| Ongoing/delivered pregnancies | 1349 (40.5) | 948 (45.8)[d] | 396 (32.2)[d] | 5 (17.9)[d] |
| Live births (n) | 905 | 598 | 303 | 4 |

[a, b, c, d] $P < 0.001$; *VET* Vitrified embryo transfer

In October 2007 the Fertility Centers of Illinois "IVF Laboratory River North" moved forward from the use of an open carrier system (Cryotop—embryos are in direct contact with liquid nitrogen) to a closed system (embryos are sealed before contact with liquid nitrogen). Using a closed carrier (High Security Vitrification kit, HSV) for aseptic vitrification, the following data from day 5, day 6, and day 7 blastocysts were observed and are summarized in Table 8: (a) clinical pregnancy rate (cPR), 54.3 % vs. 41.3 % vs. 21.4 %; (b) ongoing pregnancy (oPR), 45.8 % vs. 32.2 % vs. 17.9 %; and (c) implantation rate (IR), 42.3 % vs. 30.0 % vs. 14.9 % (Table 8). As shown in Table 8, oPR, cRP, and IR occurring in the day 5 blastocyst group were significantly higher than transferring day 6 or even day 7 blastocysts.

Between 2007 and July 2014, the Fertility Centers of Illinois "IVF Laboratory River North" (Chicago) performed 1675 vitrified blastocyst transfers (VBT) without collapsing prior to vitrification, with a mean patient age of 35.4 ± 5.0 years (group A), and in 1193 VBT (group B) with a mean patient age of 36.4 ± 4.9 years, where artificial collapse was performed prior to vitrification (Table 9). On average 1.7 embryos were transferred in groups A and B, which means 30 % of all VBT were single embryo transfers. Survival in group A versus group B was not significantly different (98.7 % vs. 99.7 %). However, there was a significant improvement in group B

**Table 9**

**A comparison of retrospective data from the cryopreservation program (*Fertility Centers of Illinois, Chicago*) of vitrified blastocysts without AC (group A) and with AC (group B) using aseptic vitrification technology between 2007 and 2014. VET = vitrified embryo transfer. Values are numbers unless otherwise described, percentages are indicated between brackets**

| | Technique | |
|---|---|---|
| | Group A (no AC) | Group B (with AC) |
| Patient's age (y) | 35.4 ± 5.0 | 36.4 ± 4.9 |
| Transfers | 1675 | 1193 |
| Blastocysts warmed | 3004 | 2075 |
| Blastocysts survived | 2966 (98.7) | 2068 (99.7) |
| Blastocysts transferred | 2911 | 2042 |
| Blastocysts transferred (mean) | 1.7 | 1.7 |
| Implantations | 952 (32.7)[a] | 887 (43.4)[a] |
| Positive pregnancy/VET | 861 (51.4)[b] | 812 (68.1)[b] |
| Clinical pregnancy/VET | 733 (43.8)[c] | 687 (57.6)[c] |
| Ongoing/delivered pregnancies | 603 (36.0)[d] | 575 (48.2)[d] |

[a]$P < 0.01$; [b,c,d]$P < 0.001$

compared with group A for the following: (a) clinical pregnancy rate (cPR), 57.6 % vs. 43.8 %; (b) ongoing pregnancy (oPR), 48.2 % vs. 36.0 %; and (c) implantation rate (IR), 43.4 % vs. 32.7 % (Table 9).

When the vitrified-warmed blastocysts were divided into day 5 and day 6 groups, the following data were observed (Table 10): in 983 VBT transferring day 5 blastocysts from group A ($n = 983$; mean age of 35.3 ± 5.1), the IR, cPR, and oPR were 37.4, 48.7, and 40.9 %, compared to 47.5, 61.6, and 52.6 % of day 5 blastocysts from group B ($n = 799$; mean age of 35.0 ± 4.9). As shown in Table 10, implantation, cPR, and oPR occurring from the day 5 blastocysts in group B were significantly higher than from the day 5 blastocyst in group A ($\chi^2$; P < 0.001, respectively).

If we compare day 6 in group A ($n = 692$; mean age of 35.6 ± 4.9) with day 6 outcome in group B ($n = 394$; mean age of 36.4 ± 4.9), the following data in terms of implantation, cPR, and oPR were observed: 26.2 %, 36.7 %, 29.0 % vs. 34.7 %, 49.5 %, and 39.3 %, respectively (Table 9). As shown in Table 9, implantation, cPR, and oPR occurring in the day 6 blastocysts of group B were significantly higher than transferring day 6 blastocysts from group A ($\chi^2$; P < 0.001 for any comparison, respectively).

In Table 11 the results for patients under 35 years in groups A (no assisted collapsing) and B (assisted collapsing) are summarized.

**Table 10**
**A comparison of retrospective data from the cryopreservation program (*Fertility Centers of Illinois, Chicago*) of vitrified day 5 and day 6 blastocysts without AC (group A) and with AC (group B) using aseptic vitrification technology between 2007 and 2014. *VET* vitrified embryo transfer. Values are numbers unless otherwise described, and percentages are indicated between brackets**

| | Technique | | | |
|---|---|---|---|---|
| | Group A (no AC) | | Group B (with AC) | |
| | Day 5 | Day 6 | Day 5 | Day 6 |
| Patient's age (y) | 35.3 ± 5.1 | 35.6 ± 4.9 | 35.0 ± 4.9 | 36.4 ± 4.9 |
| Transfers | 983 | 692 | 799 | 394 |
| Blastocysts warmed | 1736 | 1268 | 1354 | 721 |
| Blastocysts survived | 1721 (99.1) | 1245 (98.2) | 1350 (99.7) | 718 (99.6) |
| Blastocysts transferred | 1687 | 1224 | 1340 | 702 |
| Blastocysts transferred (mean) | 1.7 | 1.8 | 1.7 | 1.8 |
| Implantations | 631 (37.4)[a] | 321 (26.2)[e] | 637 (47.5)[a] | 250 (34.7)[e] |
| Positive pregnancy/VET | 558 (56.8)[b] | 303 (43.8)[f] | 571 (71.5)[b] | 241 (61.2)[h] |
| Clinical pregnancy/VET | 479 (48.7)[c] | 254 (36.7)[g] | 492 (61.6)[c] | 195 (49.5)[g] |
| Ongoing/delivered pregnancies | 402 (40.9)[d] | 201 (29.0)[h] | 420 (52.6)[d] | 155 (39.3)[h] |

Day 5: [a]$P < 0.01$; [b,c,d]$P < 0.001$; Day 6: [e]$P < 0.01$; [f,g,h]$P < 0.001$

Comparing day 5 from group A ($n = 480$; mean age 31.2 ± 2.3) with day 5 from group B ($n = 418$; mean age 31.3 ± 2.3), we found the following for IR, cPR, and oPR: 41.1 % vs. 50.9 %, 51.9 % vs. 64.8 %, and 45.6 % vs. 56.9 %. Looking at day 6 outcomes for group A versus group B, we observed the following for IR, cPR, and oPR: 31.4 % vs. 40.6 %, 43.0 % vs. 55.9 %, and 36.1 % vs. 45.5 % (*see* Table 11).

## 4 Conclusions and Future Directions

Vitrification is a very promising cryopreservation method with many advantages and an ever increasingly consistent clinical track record. A standardized vitrification protocol applicable to all stages of the preimplantation embryo may not be realistic because of (a) different surface-to-volume ratios; (b) differing cooling rate requirements between oocytes, zygotes, cleavage-stage embryos, and blastocysts; and (c) variable chill sensitivity between these different developmental stages. Currently however, the most widely used protocol applied to any embryo stage is the two-step equilibration in an equimolar combination of the cryoprotectants

**Table 11**
**A comparison of retrospective data from the cryopreservation program (*Fertility Centers of Illinois, Chicago*) of vitrified day 5 and day 6 blastocysts without AC (group A) and with AC (group B) using aseptic vitrification technology in patients younger than 35 years old between 2007 and 2014. Values are numbers unless otherwise described, and percentages are indicated between brackets**

| | Technique | | | |
|---|---|---|---|---|
| | Group A (no AC) | | Group B (with AC) | |
| | Less than 35 years old | | Less than 35 years old | |
| | Day 5 | Day 6 | Day 5 | Day 6 |
| Patient's age (y) | 31.2 ± 2.3 | 31.5 ± 0.6 | 31.3 ± 2.3 | 31.3 ± 2.5 |
| Transfers | 480 | 316 | 418 | 143 |
| Blastocysts warmed | 862 | 591 | 693 | 258 |
| Blastocysts survived | 848 (98.4) | 578 (97.8) | 691 (99.7) | 257 (99.6) |
| Blastocysts transferred | 829 | 564 | 691 | 256 |
| Blastocysts transferred (mean) | 1.7 | 1.8 | 1.7 | 1.8 |
| Implantations | 341 (41.1)[a] | 177 (31.4)[e] | 352 (50.9)[a] | 104 (40.6)[e] |
| Positive pregnancy/VET | 279 (58.1)[b] | 158 (50.0)[f] | 308 (73.7)[b] | 98 (68.5)[f] |
| Clinical pregnancy/VET | 249 (51.9)[c] | 136 (43.0)[g] | 271 (64.8)[c] | 80 (55.9)[g] |
| Ongoing/delivered pregnancies | 219 (45.6)[d] | 100 (36.1)[h] | 238 (56.9)[d] | 65 (45.5)[h] |

Day 5: [a,b,,c,d] $P < 0.001$; Day 6: [e,f,g,h] $P < 0.001$; *VET* vitrified embryo transfer

ethylene glycol and DMSO, at a concentration of 15 % each (v/v) supplemented with 0.5 mol/L sucrose. For the adoption of vitrification in ART, as with all new technologies, there has been initial resistance; but as clinical data has been accrued, this technology is becoming more commonly adopted as standard procedure in many IVF programs worldwide. With this increased use in human-assisted reproduction will come evolution of the vitrification process as it is fine-tuned to clinical needs, so pushing forward its development to higher levels of clinical efficiency, utilization, and universal acceptance.

## 5 Practical Implications Vitrifying at the Blastocyst Stage

Our data have shown that freezing at the blastocyst stage provides excellent survival, implantation, and clinical pregnancy rates. To achieve these outcomes, consider these points: (a) Without a successful blastocyst vitrification storage program, extended culture should never be attempted. (b) The blastocyst is composed of more cells and is therefore better able to compensate for cryoinjury. (c) The cells are smaller, which makes cryoprotectant penetration faster. On average, fewer embryos per patient are cryostored, but each one has a greater potential for implantation when thawed.

## 6 Addendum: Special Notes for the Clinical Embryologist

1. Special care must be given to the selection of the vitrification carrier type. It is necessary to use types of carrier or vessel material with rapid heat transfer that also support the process of uniform heat exchange to achieve higher cooling rates.
2. In addition, although no reports of contamination in human IVF following cryopreservation exist, the user should be encouraged to choose a closed carrier system, which in our experience works for blastocysts without any problems.
3. To minimize the toxicity of the cryoprotectant, a stepwise exposure of cells to pre-cooled concentrated solutions (approximate room temperature of 24 °C) is recommended.
4. Utilizing higher concentrations of cryoprotectant allows shorter exposure times to the cryoprotectant—but be careful—the potential toxicity of the cryoprotectant increases at higher concentrations. As almost all cryoprotectants are toxic to some extent, it is important to carefully monitor the duration of exposure to the final cryoprotectant before plunging into liquid nitrogen.
5. To facilitate vitrification by higher cooling rates, it is also necessary to minimize the volume of the vitrification solution (VS) as much as practical (preferably less than 1 μL). From this point of view, it is very important to use a small pulled pipette. Furthermore, by collecting the blastocysts in one place and loading no more than two blastocysts at the same time in the pipette, it is possible to keep the volume small. However, if the load of media is too large, it can still be reduced before plunging in $LN_2$ by "drawing down" the droplet to flatten the blastocysts slightly while removing all surplus vitrification solution.
6. To make sure that the blastocysts are loaded on the carrier, perform the loading process under a stereomicroscope. Always confirm the number of loaded blastocysts.
7. After sealing the carrier, submerge the carrier loaded with the blastocysts directly in liquid nitrogen by passing rapidly through the vapor phase (nitrogen gas).
8. Store the cryo-cane in a pre-chilled PVC cryo-sleeve sitting in the goblet in the dewar. It is essential to maintain exposure of

the HSV to $LN_2$ at all times to eliminate risk of warming and de-vitrification.

9. Before moving the carrier quickly from the liquid nitrogen into the warming solution, have a stripper tip (micropipette) ready. Fill the pipette with a small amount of the first warming solution (TS). When using the HSV as the vitrification carrier, rinse the open edge of the straw after placing into the pre-warmed (37 °C) 1.0 M sucrose; because the droplet is so small, it warms immediately, and it is essential to pull/stir the blastocysts off the carrier surface as soon as possible to avoid any toxic effect of the VS. A stirring motion is recommended when plunging into the warm TS to agitate the cell off the carrier surface without need to remove it actively.
10. When switching the cells between different concentrations of warming solutions, fill up the pipette with the next lower concentration of warming solution, before picking up the blastocysts to move into the next concentration (dilution effect).
11. In general, during vitrification and warming the $LN_2$ Styrofoam box needs to be as close as possible to the working area to minimize any lag in cooling and warming rates.
12. Be aware of the expiration dates of the vitrification and warming media; once opened, shelf life is 6 weeks (according to recommendation of Irvine Scientific).

## Acknowledgment

Juergen Liebermann wants to thank the Fertility Centers of Illinois (FCI) and the embryologists at the FCI IVF Laboratory River North (Elissa Pelts; BS, Jill Matthews; BS, Sara Sanchez; BS, Rebecca Brohammer; BS, Yuri Wagner; BS, and Ewelina Pawlowska, MS) for their invaluable contributions and support in pushing vitrification to become our standard protocol for cryopreservation of human oocytes and blastocysts within our program since 2004.

### References

1. Whittingham DG, Leibo SP, Mazur P (1972) Survival of mouse embryos, frozen to −196 °C and −289 °C. Science 178:411–414
2. Fahy GM, MacFarlane DR, Angell CA et al (1984) Vitrification as an approach to cryopreservation. Cryobiology 21:407–426
3. Fahy GM (1986) Vitrification: a new approach to organ cryopreservation. In: Merryman HT (ed) Transplantation: approaches to graft rejection. Alan R Liss, New York, pp 305–335
4. Seki S, Mazur P (2009) The dominance of warming rate over cooling rate in the survival of mouse oocytes subjected to a vitrification procedure. Cryobiology 59:75–82
5. Mazur P, Seki S (2011) Survival of mouse oocytes after being cooled in a vitrification solution to −196 °C at 95° to 70,000 °C/min and warmed at 610° to 118,000 °C/min: A new paradigm for cryopreservation by vitrification. Cryobiology 62:1–7

6. Rall WF, Fahy GM (1985) Ice-free cryopreservation of mouse embryos at −196 °C by vitrification. Nature 313:573–575
7. Ali J, Shelton JN (1993) Vitrification of preimplantation stages of mouse embryos. J Reprod Fertil 98:459–465
8. Vajta G, Holm P, Kuwayama M et al (1998) Open pulled straws (OPS) vitrification: a new way to reduce cryoinjuries of bovine ova and embryos. Mol Reprod Dev 51:53–58
9. Kuleshova L, Gianaroli L, Magli C et al (1999) Birth following vitrification of a small number of human oocytes: case report. Hum Reprod 14:3077–3079
10. Yoon TK, Chung HM, Lim JM et al (2000) Pregnancy and delivery of healthy infants developed from vitrified oocytes in a stimulated in vitro fertilization-embryo transfer program. Fertil Steril 74:180–181
11. Walker DL, Tummon IS, Hammitt DG et al (2004) Vitrification versus programmable rate freezing of late stage murine embryos: a randomized comparison prior to application in clinical IVF. Reprod Biomed Online 8:558–568
12. Takahashi K, Mukaida T, Goto T et al (2005) Perinatal outcome of blastocyst transfer with vitrification using cryoloop: a 4-year follow-up study. Fertil Steril 84:88–92
13. Liebermann J, Tucker MJ (2006) Comparison of vitrification versus conventional cryopreservation of day 5 and day 6 blastocysts during clinical application. Fertil Steril 86:20–26
14. Liebermann J (2009) Vitrification of human blastocysts: an update. Reprod Biomed Online 19(Suppl 4):105–114
15. Liebermann J (2011) More than six years of Blastocyst Vitrification—What is the verdict? US Obstetrics and Gynecology 5:14–17
16. Liebermann J, Tucker MJ (2002) Effect of carrier system on the yield of human oocytes and embryos as assessed by survival and developmental potential after vitrification. Reproduction 124:483–489
17. Liebermann J, Nawroth F, Isachenko V et al (2002) Potential importance of vitrification in reproductive medicine. Biol Reprod 67:1671–1680
18. Liebermann J, Dietl J, Vanderzwalmen P et al (2003) Recent developments in human oocyte, embryo and blastocyst vitrification: where are we now? Reprod Biomed Online 7:623–633
19. Vanderzwalmen P, Bertin G, Debauche C et al (2002) Births after vitrification at morula and blastocyst stages: effect of artificial reduction of the blastocoelic cavity before vitrification. Hum Reprod 17(3):744–751
20. Mukaida T, Oka C, Goto T, Takahashi K (2006) Artificial shrinkage of blastocoeles using either a microneedle or a laser pulse prior to the cooling steps of vitrification improves survival rate and pregnancy outcome of vitrified human blastocysts. Hum Reprod 21:3246–3252
21. Richter KS, Shipley SK, McVearry I, Tucker MJ, Widra EA (2006) Cryopreserved embryo transfers suggest that endometrial receptivity may contribute to reduced success rates of later developing embryos. Fertil Steril 86 (4):862–866
22. Son WY, Yoon SH, Yoon HJ, Lee SM, Lim JH (2003) Pregnancy outcome following transfer of human blastocysts vitrified on electron microscopy grids after induced collapse of the blastocoele. Hum Reprod 18:137–139
23. Hiraoka K, Hiraoka K, Kinutani M, Kinutani K (2004) Blastocoele collapse by micropipetting prior to vitrification gives excellent survival and pregnancy outcomes for human day 5 and 6 expanded blastocysts. Hum Reprod 19:2884–2888
24. Mukaida T, Oka C, Goto T, Takahashi K (2006) Artificial shrinkage of blastocoeles using either a micro-needle or a laser pulse prior to the cooling steps of vitrification improves survival rate and pregnancy outcome of vitrified human blastocysts. Hum Reprod 21:3246–3252
25. Iwayama H, Hochi S, Yamashita M (2011) In vitro and in vivo viability of human blastocysts collapsed by laser pulse or osmotic shocks prior to vitrification. J Assist Reprod Genet 28:355–361
26. Hur YS, Park JH, Ryu EK, Yoon HJ, Yoon SH, Hur CY, Lee WD, Lim JH (2011) Effect of artificial shrinkage on clinical outcome in fresh blastocyst transfer cycle. Clin Exp Reprod Med 38:87–92
27. Liebermann J and Conaghan J (2013) Artificial collapse prior blastocyst vitrification: improvement of clinical outcome. J Clin Embryol 16(1):107–118
28. Pinborg A, Henningsen AA, Loft A et al (2014) Large baby syndrome in singleton born after frozen embryo transfer (FET): is it due to maternal factors or the cryotechnique? Hum Reprod 29:618–627

# Chapter 12

# Human Ovarian Tissue Slow Freezing


## Debra A. Gook



### Abstract

Cryopreservation of human ovarian tissue is now being accepted as a mainstream ART laboratory procedure. The procedure described has been validated for use at the histological and functional level, and, recently, unequivocal evidence of preservation of primordial follicles has been demonstrated following twin births from cryopreserved ovarian tissue grafted at a heterotopic site.



**Key words** Cryopreservation, Dehydration, Rehydration, Cortex, Medulla


## 1 Introduction

Although the concept of ovarian tissue cryopreservation was explored in the 1950s [1], it was not until 1994 that Gosden and his team [2] established that a live offspring could result from the transplantation of cryopreserved ovarian tissue in an oophorectomized ewe. At this time, chemotherapy regimens used for many young adult cancer patients were increasing survival rates, but a consequence of this treatment was the onset of premature menopause [3]. The above proof of principle from animal studies suggested that young women about to receive cytotoxic treatments may be able to preserve their fertility by cryopreserving ovarian tissue. Methodology used to cryopreserve human ovarian tissue was developed using three cryoprotectants: (1) dimethyl sulfoxide (DMSO), (2) ethylene glycol (EG), and (3) propanediol (PROH). Until recently there has been little assessment of vitrification of human ovarian tissue with only a single report of a live birth from tissue grafted at an orthotopic site [4]. This will not be discussed further.

The procedure reported by Gosden using DMSO continues unmodified today for the preservation of human ovarian tissue. The efficiency of this procedure in preservation of human primordial follicles has not been evaluated, but evidence of follicle preservation, development, and ovulation was established with


Zsolt Peter Nagy et al. (eds.), *Cryopreservation of Mammalian Gametes and Embryos: Methods and Protocols*, Methods in Molecular Biology, vol. 1568, DOI 10.1007/978-1-4939-6828-2_12, © Springer Science+Business Media LLC 2017

cryopreserved human ovarian tissue grafted in immunodeficient mice (xenografting) [5]. A number of live birth have been reported following transplantation of ovarian tissue cryopreserved with this method [6–11] to an orthotopic site (ovarian). Controversy as to whether these pregnancies have arisen from the cryopreserved tissue or reactivation of the follicles within the ovary continues [12]. However, it is clear from the collection of oocytes from cryopreserved tissue grafted at heterotopic sites that these are capable of fertilization and subsequent embryo development [13–15].

Again no efficiency has been reported for the EG method, but there is similar evidence of a number of live births from tissue grafted at orthotopic sites [16, 17] and development of two embryos from oocytes recovered from heterotopic sites [18].

A number of cryopreservation regimens using the cryoprotectant PROH [19, 20] were initially evaluated at both the light and electron microscopy level for evidence of normal ovarian morphology. The most efficient method is reported below; this resulted in a high proportion (85 %) of intact oocytes within primordial follicles with normal morphology (normal mitochondria and organelles), and 74 % of the pre-granulosa cells were intact and had normal morphology [19]. The method was subsequently assessed for preservation of developmental potential by xenografting the cryopreserved human tissue. Antral follicle development [21], metaphase II oocytes, and ovulation clearly established preservation of function of the primordial follicles cryopreserved by this method [22]. Subsequent reproducibility of the procedure was established with cryopreserved tissue from nine patients and assessment of over 400 antral follicles that developed [23]. The longevity of function using this method has been established in one patient with regular cycling for over 5 years following grafting of less than a third of an ovary [24]. Embryo development is relatively efficient with oocytes collected from either the orthotopic site (13/21, 61.9 % per MII oocyte) or a heterotopic site (15/17, 88 %) although the total numbers are still low. The unequivocal evidence of successful preservation of primordial follicles using this method was achieved with a twin birth following recovery of two oocytes from a heterotopic site and the transfer of the subsequent embryos which developed in a patient who had a bilateral oophorectomy at the time of collection of tissue for cryopreservation [25].

Although previously offered by only a few clinical centers, fertility preservation for young women about to undergo cytotoxic treatment for a wide variety of conditions has now moved to being a widely accepted treatment. This change is as a result of a number of babies born following the grafting of cryopreserved ovarian tissue [26]. Associated with this general acceptance is the issue that, for the vast majority of ART centers receiving ovarian tissue for cryopreservation, many have no experience of the appropriate procedures to cryopreserve this tissue.

## 2 Materials

Prepare all solutions in a laminar flow hood into sterile disposable tissue culture plastic ware (e.g., Falcon®, Becton Dickinson). Filter sterilize through 0.2 μm membrane (PALL Life Sciences).

### 2.1 Preparation of Solutions for Processing Tissue, Freezing, and Thawing

1. Basal buffer: Quinn's Advantage® Medium with HEPES—SAGE® In Vitro Fertilisation, CooperSurgical Company for collection and processing of ovarian tissue (*see* **Note 1**).
2. Quinn's Advantage Fertilisation Medium—SAGE® In Vitro Fertilisation, CooperSurgical Company (*see* **Note 2**).
3. Sucrose: BioXtra, ≥99.5 % (GC) (Sigma). FW 342.3.
4. 1,2-Propanediol (PROH): ACS reagent, ≥99.5 % (Sigma-Aldrich). FW 76.09; density 1.036 g/mL.
5. Human serum albumin (albumin) stock solution (100 mg/mL in isotonic saline)—SAGE® In Vitro Fertilisation, CooperSurgical Company (*see* **Note 3**).

### 2.2 Preparation of Solutions

1. Collection solution: Basal buffer with 4 mg/mL albumin. To 192 mL of basal buffer, add 8 mL of albumin stock, and store at 4 °C.
2. Holding medium: Quinn's Advantage Fertilisation Medium with 4 mg/mL albumin. To 96 mL of Quinn's Advantage Fertilisation Medium, add 4 mL of albumin stock, and store at 4 °C.
3. Pre-freeze solution: Basal buffer with 10 mg/mL albumin. To 180 mL of basal buffer, add 20 mL of albumin stock, and store at 4 °C.
4. Freeze solution (1.5 M PROH/0.1 M sucrose/10 mg/mL albumin): To 6.84 g of sucrose, add 152.0 mL basal buffer. When sucrose has dissolved, add 22.0 mL PROH (*see* **Note 4**) and 20.0 mL of albumin stock. Mix, filter, sterilize, and store at 4 °C (*see* **Note 5**).
5. Rehydration solution A (1.0 M PROH/0.2 M sucrose/10 mg/mL albumin): To 3.42 g of sucrose, add 39.5 mL of basal buffer; when dissolved add 3.7 mL PROH and 5 mL of albumin stock. Mix, filter, sterilize, and store at 4 °C.
6. Rehydration solution B (0.5 M PROH/0.2 M sucrose/10 mg/mL albumin): To 3.42 g of sucrose, add 41.5 mL of basal buffer, and then add 1.85 mL of PROH and 5 mL albumin stock. Mix, filter, sterilize, and store at 4 °C.

7. Rehydration solution C (0.2 M sucrose/10 mg/mL albumin): To 3.42 g of sucrose, add 43 mL of basal buffer and 5 mL of albumin stock. Mix, filter sterilize, and store at 4 °C.
8. Solution D (basal buffer/10 mg/mL albumin): To 180 mL of basal buffer, add 20 mL of albumin stock. Mix, filter, sterilize, and store at 4 °C.
9. Storage of solutions (*see* **Note 6**).

### 2.3 Equipment and Consumables

1. Programmable freezing machine—KRYO 360—1.7/MRV (Planer Products), attached to pressurized liquid nitrogen source. Configured with canes for vials.
2. Liquid nitrogen storage vessel (*see* **Note 7**).
3. Biohazard Class II cabinet.
4. Binocular dissecting microscope with working range approximately ×5–60.
5. Sterile gloves.
6. Sterile plastic pipettes (e.g., Falcon®) for dispensing solutions.
7. Sterile specimen collection container (50 and 200 mL, *see* **Note 8**).
8. Sterile petri dishes (large; 150350 Nunclon™ and small; 353002 Falcon®).
9. Sterile cell strainers (352350 70 μm Nylon Falcon®, *see* **Note 9**).
10. Sterile 6-well plate (140675 Delta Surface Nunclon™).
11. Sterile forceps and scissors (*see* **Note 10**).
12. Sterile double-edge stainless steel razor blades (L056 ProSci-Tech, *see* **Note 11**).
13. Vials (363401 Cryotube 1.8 mL round, Nunc, *see* **Note 12**) and Inserts (9375930 Nunc Cryo Colour Coders).
14. Labels (Brady Thermal Labelling System, *see* **Note 13**).
15. Large cotton swab (for manual seeding, 22-555 Multigate, *see* **Note 14**).
16. Storage boxes (*see* **Note 15**).
17. Insulated Dewar (*see* **Note 16**).
18. Water at 90–100 °C.
19. Large forceps.
20. Sterile flat end needle holders (15-2808 830 UU8, Pilling CE, Germany).
21. 6.0 Vicryl coated suture (J489 G, P-1 11 mm 3/8C, Ethicon).
22. Timer.
23. Portable oven (*see* **Note 17**).

## 3 Methods

The freezing machine should be programmed as follows:

Set pause temperature at +16 °C and hold at this temperature until run is initiated.

1st ramp: Cool from +16 to −7 °C at 2 °C/min.

2nd ramp: Hold at −7 °C for 10 min (to allow manual seeding).

3rd ramp: Cool from −7 °C to −30 °C at 0.3 °C/min.

4th ramp: Cool from −30 °C to −150 °C at 50 °C/min.

5th ramp: Hold at 150 °C for approximately 30 min (to allow for removal of vials and transfer to storage).

### *3.1 Preparation of Tissue*

1. In theater, the tissue is removed without the use of diathermy. The scrub nurse or surgeon places approximately half an ovary or a whole ovary into the sterile pot containing 30 mL of collection solution at 37 °C (*see* **Note 8**). This is transported to the laboratory at room temperature (*see* **Note 17**).
2. Using forceps the tissue is removed from the pot and placed in a large petri dish containing 20 mL of collection solution at 37 °C (*see* **Note 18**). All subsequent preparation is performed under a dissecting microscope with a heated stage. The tissue is examined for atypical appearance (*see* **Note 19**), and the weight, size and appearance are recorded (*see* **Note 20**). A cross section of ovary containing both medulla and cortex is removed and placed immediately into 10 % formalin for pathological assessment of follicular density and malignancy. A whole ovary is bisected longitudinally for examination of both the medulla and cortex, and at least two areas are removed with scissors for pathology (*see* **Note 21**, Fig. 1).
3. The following preparation of ovarian tissue is for adult tissue, and modifications to the preparation are necessary with prepubertal ovarian tissue (*see* **Note 22**). With the cortex on the bottom of the dish and, holding the medulla with forceps, starting at the cut edge, snip at the junction between the medulla and cortex with the fine point of the scissors. Continue to snip at the junction pulling the medulla back as it becomes free (*see* **Note 23**, Fig. 2). Place medulla aside (*see* **Note 24**).
4. The piece of thin cortex (1 mm in thickness) is moved to a clean petri dish of collection solution. The cortex is then cut into thin strips 5 mm in width (*see* **Note 25**).
5. Working with one strip at a time, with the trimmed surface up, hold the strip on the bottom of the dish with forceps. Bring the razor blade down on the tissue with one firm motion of the blade straight and horizontal (*see* **Note 26**, Fig. 3). Check the

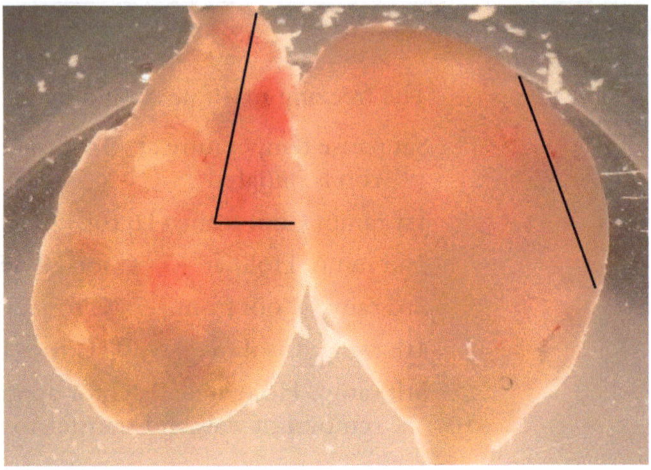

**Fig. 1** Bisected ovary with medulla surface (*left*) and cortex surface (*right*). Both surfaces are unremarkable in appearance; therefore, the bloody medulla region (*L*) was removed for pathology together with the small area to the *right of the line*

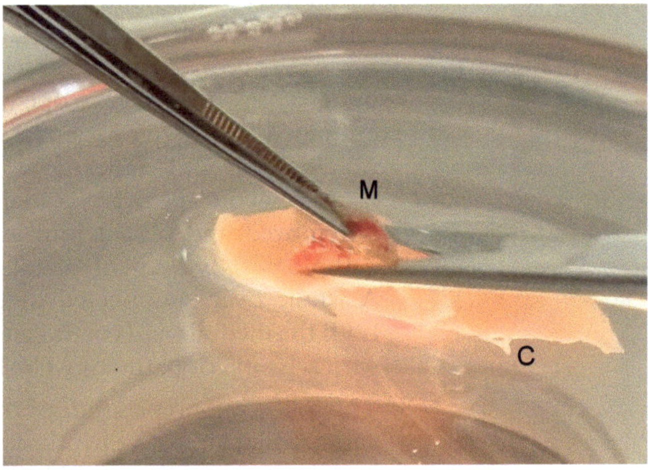


**Fig. 2** Cutting at the junction of the medulla (*M*) and cortex (*C*) while peeling back the medulla tissue

dimensions of the slices and move the slice away from the cutting area (*see* **Note 27**).

6. Continue slicing until the strip is completed. Move the slices out of the dish into a small petri dish containing the strainer and holding medium (*see* **Note 28**, Fig. 4). Record number of slices (*see* **Note 29**) and return the small dish to the gassed incubator.
7. Slice all strips and move slices into the holding medium (*see* **Note 30**).

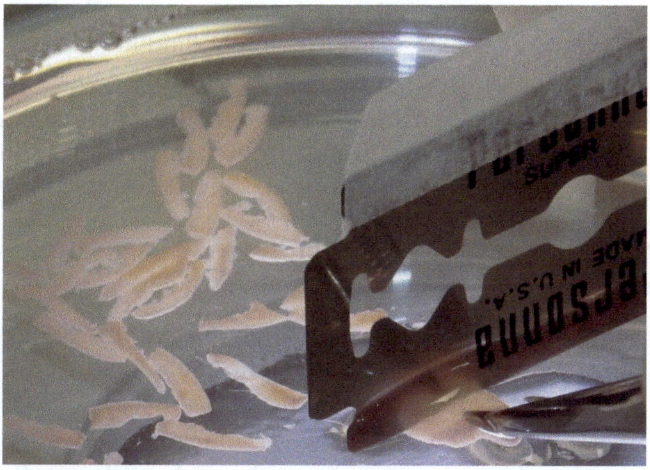


**Fig. 3** Slicing the strip of tissue with a double-sided blade

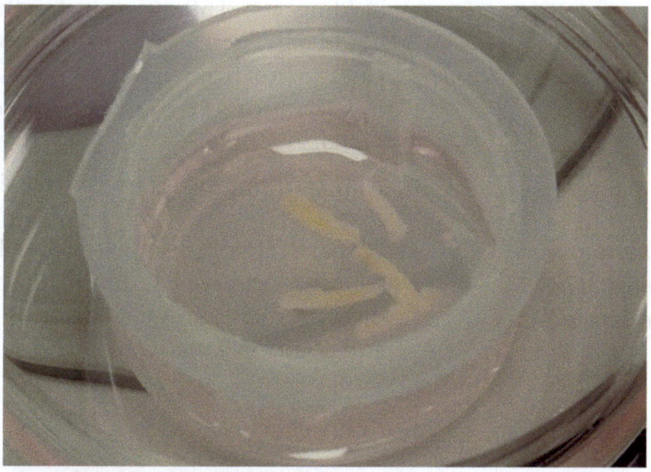

**Fig. 4** Strainer containing slices

### *3.2 Dehydration of Tissue Slices*

1. The dehydration plate (6 well) is prepared with 8 mL of pre-freeze solution in odd numbered wells and 8 mL of freeze solution in even numbered wells. Dehydration is performed at ambient temperature.
2. When all slices are in holding dishes, move each strainer using forceps to the pre-freeze well. Dip strainer in pre-freeze to remove any carryover medium, drain (*see* **Note 9**), and move to freeze solution well. Agitate to dislodge any air bubbles trapped under the strainer. Leave in freeze solution for 80 min, periodically agitate strainer to stop clumping of slices.
3. Label and prepare vials (*see* **Note 13**) containing 0.5 mL of freezing solution.

4. Using forceps to move slices to the vials (*see* **Note 28**), check slices are under fluid in the vial and all have been removed from the strainer (*see* **Note 31**).
5. Load vials on canes in freezing machine (*see* **Note 14**). At 90 min from transfer to freezing solution, turn off pause on freezing machine. Run the above program.

### 3.3 Running the Freezing Program

Personal protective equipment must be worn at all times when handling liquid nitrogen:

1. After 12 min, ensure that the freezing machine has entered the "Hold" ramp and is maintaining a temperature of −7 °C before commencing manual seeding.
2. Seed by holding the pre-cooled swab (soaked in liquid nitrogen) on the vial at the surface of the liquid; seeding has occurred when ice crystals have formed (*see* **Note 32**, Fig. 5).
3. When the freezing program is complete and the temperature is holding at −150 °C, transfer the vials to a Dewar containing liquid nitrogen. Transport this to the storage vessel. Remove storage box from vapor storage vessel, quickly place vials in the box (*see* **Note 15**), and record position. Return box to storage vessel.
4. Run warming program on freezing machine.

### 3.4 Thawing and Rehydrating Ovarian Tissue Slices

1. Prepare rehydration plate (6 well) with a strainer and 8 mL of solution A in well 1. Eight mL of each rehydration solution is dispensed as follows: solution B in well 2, solution C in well 3, and solution D in wells 4 and 5.

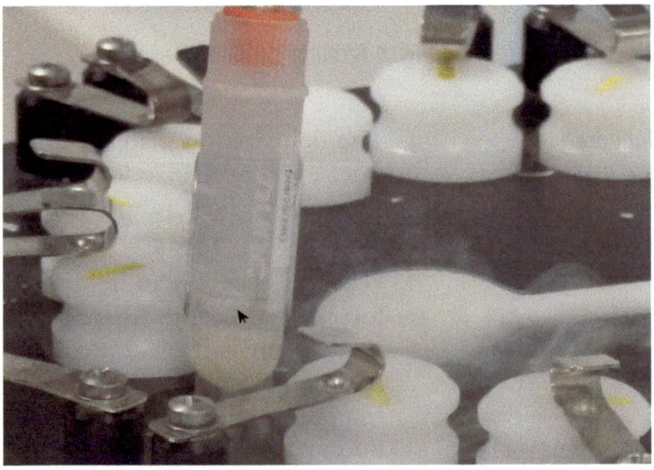

**Fig. 5** Seeding vials. Ice formation can be seen at the surface of the liquid (*arrow*) in the vial

**Fig. 6** A suture tied in a loop containing slices of ovarian tissue. Note needle has been removed

2. Remove vials from storage vessel. Place in Dewar containing liquid nitrogen for transport.
3. Hold vial with long forceps in a beaker containing water at 90–95 °C. Do not submerge the vial but hold so that screw cap seal is above water surface. Move vial through water to disperse thermal differences. Once thawed (*see* **Note 33**), remove the vial from the water. Using sterile forceps, remove slices from the vial and place in strainer in well 1. Check correct number of slices. Leave in rehydration solution A for 5 min (*see* **Note 34**).
4. Drain and transfer strainer to solution B for 5 min. Repeat for solution C and solution D twice, each step for 5 min.
5. Transfer slices to a large petri dish containing 20 mL of solution D at 37 °C. The above procedure is repeated for each vial to be thawed.
6. Under the dissecting microscope, thread slices onto a suture (*see* **Note 35**, Fig. 6).
7. Move the completed suture to a fresh large petri dish containing 20 mL of solution D at 37 °C. The dish is then placed in the portable oven (*see* **Note 16**) on a trolley ready for transport to theater.
8. Multiple sutures are prepared depending on the volume of tissue to be grafted (*see* **Note 36**).

#### 3.5 Collection of Cumulus Oocyte Complexes (Optional)

During the preparation of tissue, a number of small antral follicles may be observed, and, although this is experimental, some oocytes will subsequently mature. The developmental potential of these is unknown:

1. At Subheading 3.1, **step 3** follicles present can be punctured. A 26 gauge needle is used to pierce the follicle and pressure is applied with forceps to expel fluid and cells from the follicle. The medium is then scanned for the cumulus-enclosed oocyte (COC) (*see* **Note 37**). The COC is then moved to a culture plate containing 0.5 mL of holding medium with an overlay of 300 μL of oil and returned to a gassed incubator.
2. All dishes used for the preparation of the tissue are scanned for COC, and the COC collected are moved to the culture plate.
3. When all COC are collected, they are moved to individual 50 μL drops of IVM medium (*see* **Note 38**) with an 8 mL overlay of oil and incubated for 48 h.
4. At 48 h the cumulus is stripped away from the oocyte with a 0.146–0.155 mm pipette and mature oocytes vitrified.

## 4 Notes

1. A buffer that maintains pH in atmospheric conditions should be used for collection and preparation of tissue and in freezing and thawing solutions. Due to collection of cumulus oocyte complexes during the preparation of ovarian tissue and the preservation of follicles containing oocytes, a basal buffer used in ART has been used throughout our procedure. However, other groups use Leibovitz's L-15 medium [8]. A comparison of these two media used during preparation of human ovarian tissue showed follicle loss following xenografting of the tissue was lower in tissue prepared in Quinn's Advantage® Medium with HEPES at 37 °C than in Leibovitz's L-15 medium at 4 °C (unpublished data).
2. This step has been shown to decrease ischemic damage, but this requires a carbon dioxide (6 %), oxygen (5 %), and nitrogen (89 %) gas environment. If not available, continue in atmospheric buffer solution.
3. Alternative products are available, but it is preferable to use an albumin source that is approved for human ART use.
4. PROH is a viscous liquid. To ensure accurate delivery, it is important after expelling the PROH out of the pipette to wait for the remaining PROH to run down the side of the pipette and then expel this.
5. Due to the unpredictable nature of tissue freeze patients, we find having a supply of freeze solution always present is a benefit. Freeze solutions can be made in advance without albumin. Sterilize and store at 4 °C and the albumin can be added on the day of tissue collection.

6. When solutions are opened for use, sterility of the remaining solution should be maintained by ensuring that aliquots are dispensed in a laminar flow hood. Our general policy is for the shelf life of solutions containing sucrose and or albumin to be 1 month at 4 °C.
7. To reduce contamination risk during storage of vials containing tissue, vials should only be stored in the vapor phase of liquid nitrogen and not submerged in the liquid.
8. For whole adult ovaries, it may be necessary to collect the specimen in the larger sterile pot (250 mL).
9. To reduce handling damage and possible loss of the tissue slices, slices are loaded into cell strainers submerged in holding medium. All subsequent change of solution requires draining of excess medium by holding the strainer against side of dish or well and then moving the strainer to the next solution.
10. Our preference for preparing the tissue is to use fine dissecting scissors and forceps. These are washed in tissue culture detergent after use, rinsed, and gas sterilized.
11. The double-sided razor blade provides a thin, extremely sharp flat blade that facilitates in preparing thin slices of the dense, firm cortex of the ovary. These blades are carefully removed from packaging using forceps and submerged in 90 % alcohol to remove any solvents and oils used in manufacturing. They are rinsed with water a number of times to remove the alcohol and dried. One sharp edge of the blade is then covered with a piece of autoclave tape; these are then packaged individually and gas sterilized. Other groups use scalpel blades, but these have either a pointed tip or round-shaped cutting edge which is less suited to making straight vertical cuts (*see* Fig. 2).
12. A colored insert is placed in the cap of each vial and vial number written on the colored insert. The inserts are optional but aid in quickly finding the required patient and vial number to be removed from a storage box when thawing later. All vials from one patient are the same color, and a different color to the previous patient in the storage box is used to distinguish.
13. Printed adhesive labels containing three points of identification are placed on the vials. Be aware that many of the labels sold for straws do not stick on the smooth surface of vials. We place an ID label over the white label on the vial and then place another two labels end to end to make a complete ring of labels.
14. Place vials on the canes in freezing machine with ID label against the canes so liquid surface can be visualized when seeding.
15. Leave the corners vacant and an occasional other space in the storage boxes to reduce swelling. We find that the lids are

almost impossible to remove from completely full boxes and the boxes swell in storage racks. Due to this difficulty, the time vials that are out of the storage tank will be increased when taking vials out to be thawed.

16. In our situation, the laboratory and theater complex are a large distance apart, and, therefore, the dishes containing the prepared sutures are placed in a heating oven at 37 °C for transport and during time in theater.
17. The collection specimen pot initially contains warmed (37 °C) collection medium. This is transported in an insulated bag, but, due to the time involved in collecting and transporting the specimen, the temperature will reduce to ambient. Other groups, because of the cryoprotectant used in freezing procedure (i.e., DMSO or EG), will transport and maintain the tissue during preparation at 4 °C.
18. All preparation and slicing of tissue are performed in medium at 37 °C.
19. The typical cortical surface of the ovary should be smooth without adhesions, white in color and even in density. Any atypical areas should be removed for pathological examination; these include areas not in the contour of the cortex surface, firm solid-appearing areas. The importance of this examination was emphasized by the removal of an atypical area in a patient with Hodgkin's lymphoma which confirmed the presence of malignancy associated with the ovary [27].
20. Medulla is also examined for the presence of follicles and corpora lutea. A pathology specimen is taken of every ovarian tissue sample for cryopreservation regardless of the patient condition or quantity of tissue available for cryopreservation. If the cortex appears normal, then our preference for the pathology sample is to remove a piece of the ovary that has both the cortex and medulla from an area with high blood content in the medulla.
21. In the situation of a whole ovary, the larger tissue volume allows for multiple sampling for pathology across the ovary. For suspected ovarian cancers, the ovary is examined by the pathologist using sterile instruments, the area thought to be normal is removed, and frozen sections of this are examined. If no infiltration of malignancy is detected, this normal area is taken to the laboratory for cryopreservation. The pathologist subsequently examines the remaining ovary in more detail, and, if malignancy is detected, the cryopreservation is abandoned and all tissue returned to the pathologist.
22. A number of modifications to the preparation of the tissue are necessary when cryopreserving prepubertal ovarian tissue. The medulla is only removed if the thickness of the intact ovarian

tissue is greater than 2 mm; if less than this, no separation is required (prepubertal ovary is less dense than adult). Both the medulla and cortex are sliced and cryopreserved for subsequent grafting. In the prepubertal ovary, a large number of follicles are also present in the medulla. Record which type of tissue is present in each vial. Due to the very large number of follicles present in the prepubertal ovary, the number of slices is reduced to 2–3 per vial. All other components of the procedure are the same as for the adult.

23. The analogy to this process is a quarter of an orange with the peel (cortex) on the bottom of the dish and cutting with the fine point of the scissors at the junction of the peel and flesh (medulla). Using the forceps, hold the flesh back and continue to cut across the edge until the flesh is removed in one piece (*see* Fig. 1). Following this method will reduce damage to the cortex and the time involved in obtaining cortex with a thickness of 1 mm. The cortex may still have some irregularities in thickness. Trimming off the excess medulla remaining can be achieved by holding the medulla with forceps and again snipping with fine point scissors at the junction.
24. A piece of the medulla is also cryopreserved. This is used for subsequent pathological or molecular testing.
25. A transparency of graph paper is placed under the dish, and, using the scissors, the cortex is cut into 5 mm wide strips.
26. Hold the razor blade vertical on the autoclave tape edge, bring the blade down straight, and press down on the tissue with firm pressure against the base of the dish. This will make a straight cut of the cortex. If still attached use a sawing motion of the blade back and forward on the base of the dish to complete the cut.
27. We check the size of a few slices, using the graph paper transparency, at the start of slicing. The aim is to have slices $1 \times 1 \times 5$ mm, but, due to this being a manual procedure, there is some variation in the size of the slices. With experience this variation is reduced.
28. The small petri dishes are prepared prior to collecting the tissue. Using sterile scissors the small tab is removed from the strainer and the strainer placed into the small petri dish before 8 mL of the holding medium is added. Dishes are numbered and returned to the gassed incubator until required. To move slices from one dish to another, the slices are not held by the forceps but in the fluid column conducted up the forceps. With the slice between the arms of the forceps, bring the forceps arms closer to the slice while raising the forceps out of the fluid; this will create a column of fluid containing the slice. Dip this into next solution and the slice will be released.

29. Ten to twelve slices are transferred to each strainer. For large amounts of tissue, the number of slices is increased to 20–25.
30. In one strainer place a minimum of 2–4 slices from different regions. This smaller number of slices is dehydrated and cryopreserved as with the others but can be thawed for repeat pathological assessment prior to grafting. A large piece of the medulla, also for subsequent testing, is placed in another strainer, dehydrated, and cryopreserved.
31. Since slices are difficult to see when on the rim of the strainer, swirl forceps around strainer—this will move slices into the center of the strainer.
32. We prefer to use large cotton swabs for seeding vials as they tend to be more efficient at holding the low temperature required for effective seeding.
33. The vial has thawed when contents change from opaque to clear; moving the vial during the thawing will help identify when contents of vial are liquid. This generally takes 40 s to occur.
34. All rehydration steps are performed at ambient temperature.
35. Holding the base of the needle with needle holder, the suture is removed from packaging and cut to approximately 10 cm in length. Place the suture attached to the needle holder in the large petri dish containing slices. A slice of tissue is placed over the point of the needle and pushed onto the needle. Care is taken in manipulating the slice so as to reduce any possible damage. Once 10–12 slices are on the needle, the needle holder is released, and the slices shuffled down the suture to approximately half the length. A loop is made with the ends of the suture and secured with two reef knots. The needle is cut off the suture and discarded. Slices are then careful pushed apart to make sure they are not clumped together.
36. Our preference has been to have a small amount of tissue (10–12 slices) on each suture and preparing multiple sutures. Our aim has been to place sutures at a number of sites in close proximity to each other instead of a single large amount of tissue which has a high risk of the tissue being clumped together and therefore higher ischemic damage. Details of the laparoscopic grafting procedure at heterotopic and orthotopic sites are in our publications (references).
37. The cumulus cells surrounding oocytes obtained from these small follicles are dense, tightly packed cells with no matrix. They are difficult to identify.
38. Media used by others for clinical IVM may be used for these COC.

## References


1. Parkes AS (1958) Factors affecting the viability of frozen ovarian tissue. J Endocrinol 17:337–343
2. Gosden RG, Baird DT, Wade JC, Webb R (1994) Restoration of fertility to oophorectomized sheep by ovarian autografts stored at −196 °C. Hum Reprod 9(4):597–603
3. Meirow D (1999) Ovarian injury and modern options to preserve fertility in female cancer patients treated with high dose radio-chemotherapy for hemato- oncological neoplasias and other cancers. Leuk Lymphoma 33 (1–2):65–76
4. Kawamura K, Cheng Y, Suzuki N, Deguchi M, Sato Y, Takae S, Ho CH, Kawamura N, Tamura M, Hashimoto S, Sugishita Y, Morimoto Y, Hosoi Y, Yoshioka N, Ishizuka B, Hsueh AJ (2013) Hippo signaling disruption and Akt stimulation of ovarian follicles for infertility treatment. Proc Natl Acad Sci USA 110(43):17474–17479. doi:10.1073/pnas.1312830110
5. Kim SS, Soules MR, Battaglia DE (2002) Follicular development, ovulation, and corpus luteum formation in cryopreserved human ovarian tissue after xenotransplantation. Fertil Steril 78(1):77–82
6. Demeestere I, Simon P, Buxant F, Robin V, Fernandez SA, Centner J, Delbaere A, Englert Y (2006) Ovarian function and spontaneous pregnancy after combined heterotopic and orthotopic cryopreserved ovarian tissue transplantation in a patient previously treated with bone marrow transplantation: case report. Hum Reprod 21(8):2010–2014
7. Demeestere I, Simon P, Emiliani S, Delbaere A, Englert Y (2007) Ongoing pregnancy after a second cryopreserved ovarian transplantation procedure. Hum Reprod 22(suppl 1):i43–i44
8. Donnez J, Dolmans MM, Demylle D, Jadoul P, Pirard C, Squifflet J, Martinez-Madrid B, van Langendonckt A (2004) Livebirth after orthotopic transplantation of cryopreserved ovarian tissue. Lancet 364(9443):1405–1410
9. Donnez J, Squifflet J, Jadoul P, Demylle D, Cheron AC, Van Langendonckt A, Dolmans MM (2011) Pregnancy and live birth after autotransplantation of frozen-thawed ovarian tissue in a patient with metastatic disease undergoing chemotherapy and hematopoietic stem cell transplantation. Fertil Steril 95 (5):1787.e1–1787.e4
10. Meirow D, Levron J, Eldar-Geva T, Hardan I, Fridman E, Zalel Y, Schiff E, Dor J (2005) Pregnancy after transplantation of cryopreserved ovarian tissue in a patient with ovarian failure after chemotherapy. N Engl J Med 353(3):318–321
11. Roux C, Amiot C, Agnani G, Aubard Y, Rohrlich PS, Piver P (2010) Live birth after ovarian tissue autograft in a patient with sickle cell disease treated by allogeneic bone marrow transplantation. Fertil Steril 93(7):2413.e15–2413.e19
12. Oktay K (2006) Spontaneous conceptions and live birth after heterotopic ovarian transplantation: is there a germline stem cell connection? Hum Reprod 21(6):1345–1348
13. Kim SS, Lee WS, Chung MK, Lee HC, Lee HH, Hill D (2009) Long-term ovarian function and fertility after heterotopic autotransplantation of cryobanked human ovarian tissue: 8-year experience in cancer patients. Fertil Steril 91(6):2349–2354
14. Oktay K, Buyuk E, Veeck L, Zaninovic N, Xu K, Takeuchi T, Opsahl M, Rosenwaks Z (2004) Embryo development after heterotopic transplantation of cryopreserved ovarian tissue. Lancet 363(9412):837–840
15. Piver P, Amiot C, Agnani G, Pech JC, Rohrlich PS, Vidal E, Aubard Y, Roux C (2009) Two pregnancies obtained after a new technique of autotransplantation of cryopreserved ovarian tissue. Hum Reprod 24(Suppl 1):i15
16. Andersen CY, Rosendahl M, Byskov AG, Loft A, Ottosen C, Dueholm M, Schmidt KL, Andersen AN, Ernst E (2008) Two successful pregnancies following autotransplantation of frozen/thawed ovarian tissue. Hum Reprod 23(10):2266–2272
17. Ernst E, Bergholdt S, Jorgensen JS, Andersen CY (2010) The first woman to give birth to two children following transplantation of frozen/thawed ovarian tissue. Hum Reprod 25 (5):1280–1281
18. Rosendahl M, Loft A, Byskov AG, Ziebe S, Schmidt KT, Andersen AN, Ottosen C, Andersen CY (2006) Biochemical pregnancy after fertilization of an oocyte aspirated from a heterotopic autotransplant of cryopreserved ovarian tissue: case report. Hum Reprod 21 (8):2006–2009
19. Gook DA, Edgar DH, Stern C (1999) Effect of cooling rate and dehydration regimen on the histological appearance of human ovarian cortex following cryopreservation in 1, 2- propanediol. Hum Reprod 14(8):2061–2068
20. Gook DA, Edgar DH, Stern C (2000) The effects of cryopreservation regimens on the

morphology of human ovarian tissue. Mol Cell Endocrinol 169(1–2):99–103

21. Gook DA, McCully BA, Edgar DH, McBain JC (2001) Development of antral follicles in human cryopreserved ovarian tissue following xenografting. Hum Reprod 16(3):417–422
22. Gook DA, Edgar DH, Borg J, Archer J, Lutjen PJ, McBain JC (2003) Oocyte maturation, follicle rupture and luteinization in human cryopreserved ovarian tissue following xenografting. Hum Reprod 18(9):1772–1781
23. Gook DA, Edgar DH, Borg J, Archer J, McBain JC (2005) Diagnostic assessment of the developmental potential of human cryopreserved ovarian tissue from multiple patients using xenografting. Hum Reprod 20(1):72–78
24. Stern CJ, Toledo MG, Hale LG, Gook DA, Edgar DH (2011) The first Australian experience of heterotopic grafting of cryopreserved ovarian tissue: evidence of establishment of normal ovarian function. Aust N Z J Obstet Gynaecol 51(3):268–275
25. Stern CJ, Gook D, Hale LG, Agresta F, Oldham J, Rozen G, Jobling T (2013) First reported clinical pregnancy following heterotopic grafting of cryopreserved ovarian tissue in a woman after a bilateral oophorectomy. Hum Reprod 28(11):2996–2999
26. Stoop D, Cobo A, Silber S (2014) Fertility preservation for age-related fertility decline. Lancet 384(9950):1311–1319. doi:10.1016/S0140-6736(14)61261-7
27. Bittinger SE, Nazaretian SP, Gook DA, Parmar C, Harrup RA, Stern CJ (2011) Detection of Hodgkin lymphoma within ovarian tissue. Fertil Steril 95(2):803.e3–803.e6

# Chapter 13

# Human Ovarian Tissue Vitrification


## Sherman Silber



## Abstract

Ovarian freezing and transplantation has garnered increasing interest as a potential way of preserving fertility in cancer patients as well as for women who just wish to delay childbearing. This chapter spells out our techniques of ovarian cortex vitrification and results for frozen compared to fresh ovarian cortex transplantation (in one single series from one center for the sake of consistency), as well as potentially provides insight into the mechanism behind ovarian follicle recruitment. This represents an effort to simplify and popularize an approach that has yielded favorable results (all cases recovered ovulation and 75% had successful spontaneous pregnancy) in one single, disciplined study. It should be clear that this is a review for the more general reader of our original scientific papers published in *Reproductive BioMedicine Online, New England Journal of Medicine, Fertility and Sterility, Human Reproduction, Molecular Human Reproduction, and Journal of Assisted Reproduction and Genetics (JARG).*



**Key words** Fertility, Preservation, Ovarian freezing, Transplant, Follicle recruitment


Patient characteristics and results are all summarized in Tables 1 and 2. All recipients of fresh or frozen ovarian transplants had the same robust return of FSH to premenopausal levels by 150 days, menstruation by 130 days, and massive AMH elevation by 170 days. The AMH then fell to below normal by 240 days and remained at that level for many years with normal ovarian cycling and hormonal function. We have reported seventeen babies resulting from 11 fresh and six frozen transplants. However there are even more pregnancies and babies that we will soon submit in another original paper. Grafts of one-third of an ovary cortex lasted as long as 8 years or more despite eventually low AMH. Our results support a hypothesis that ovary freezing and transplantation is clinically a robust method of preserving ovarian function, and it supports a hypothesis that ovarian stromal density gradient can account for the early meiotic arrest of fetal oogonia and also for resting follicle recruitment in the adult.

---


**Electronic supplementary material:** The online version of this chapter (doi:10.1007/978-1-4939-6828-2_13) contains supplementary material, which is available to authorized users.

Zsolt Peter Nagy et al. (eds.), *Cryopreservation of Mammalian Gametes and Embryos: Methods and Protocols*, Methods in Molecular Biology, vol. 1568, DOI 10.1007/978-1-4939-6828-2_13, © Springer Science+Business Media LLC 2017

**Table 1**
**Pregnancy and duration of function of fresh donor ovarian grafts from identical twin sister [9] or nonidentical sister allograft [2] (*see* Ref. [1] in References)**

| Days of ovarian function | Date of transplant | Ovarian function end date | Miscarriage | Baby born | Girl | Boy |
|---|---|---|---|---|---|---|
| 975 | 4/21/04 | 12/22/06 | 1 | 1 | 1 | |
| 3025 | 4/20/05 | 8/1/13 | | 3 | 1 | 2 |
| 2285 | 6/7/05 | 9/9/11 | | 2 | 1 | 1 |
| 1534 | 8/23/05 | 11/4/09 | 1 | 1 | | 1 |
| 1107 | 9/20/05 | 10/1/08 | | | | |
| 752 | 7/11/06 | 8/1/08 | | 1 | 1 | |
| 2410+ | 8/16/06 | n/a | 2 | 2 | | 2 |
| 1666 | 1/9/07 | 2/1/11 | | 1 | 1 | |
| 618 | 4/21/08 | 12/30/09 | 1 | | | |
| 263[a] | 7/21/10[b] | 4/10/11 | | | | |
| 1172+ | 2/11/11 | n/a | 1 | | | |
| Total | | | 6 | 11 | 5 | 6 |

**Table 2**
**Pregnancy after frozen ovary autograft (*see* Ref. [1] in References)**

| Days of ovarian function | Date of transplant | Date of first menstruation post OT | Diagnosis | Miscarriage | Baby born | Girl | Boy |
|---|---|---|---|---|---|---|---|
| 804 | 3/6/07 | 9/19/2008 | POF | | 1 | 1 | |
| 884 | 1/13/09 | 6/7/09 | Hodgkin's | | 1 | | 1 |
| 997 | 6/9/09 | 11/28/09 | POF | 1 | 0 | | |
| 1155+ | 6/17/11 | 11/5/11 | Hodgkin's | | 0 | | |
| 564+ | 10/12/12 | 3/2/13 | MS | | 1 | 1 | |
| 487+ | 3/29/13 | 7/5/13 | POF | | 1 | 1 | |
| 496+ | 4/5/13 | 12/27/13 | Brain cancer | | 0 | | |
| 502+ | 4/12/13 | 1/1/14 | Blood disorder | | 1 | | 1 |
| 328+ | 10/1/13 | 12/19/13 | Synovial sarcoma | | 0 | | |
| 294+ | 10/7/13 | 3/6/14 | All | | 0 | | |
| 306+ | 10/23/13 | 12/29/13 | Breast cancer | | 0 | | |
| Total | | | | 1 | 5 | 3 | 2 |

## 1 Methodology for Ovarian Cortex Cryopreservation

Successful fresh as well as frozen ovarian cortex transplants in humans were first published in 2004 and 2005 as a case reports, and many other case reports have subsequently followed [1–15]. There has now developed a rapidly growing interest in frozen ovarian cortical transplantation despite the absence of a clearly documented success rate for this procedure. The primary impetus for this procedure has been to cryopreserve ovarian tissue prior to sterilizing cancer treatment with the objective of transplanting the tissue back after cancer cure, after the proof of principle was reported in sheep in 1994 [16]. The possibility also loomed of preserving fertility and even hormonal function against the natural decline caused by aging [17, 18].

Oocyte freezing is often mentioned as another alternative for preserving fertility [19–21]. However, many different centers do it in different ways, and most have not verified long-term viability of the oocytes. There could therefore be many disappointed women in years to come. Among the pitfalls to oocyte freezing are too rapid and changing osmolality with dangerously quick osmotic shifts. Also, there is a common failure to ensure a rapid enough freeze and worse yet not a rapid enough thaw. The commercial kits that are designed to make this easy often fail in this regard, and closed freezing is more problematic than open freezing for rapid freeze and thaw. It was not until 2005 that a highly efficient method was published, which stimulated a huge wave of enthusiasm, for this approach, also known as the bridge technique (Fig. 1a–f), but it also is very user dependent, and none of the oocyte vitrification protocols are as foolproof and easy as freezing of ovarian tissue.

One extra benefit of ovary cryopreservation and transplantation compared to oocyte freezing is the putting off of hormonal menopause as well as preserving fertility. Although oocyte freezing is now mostly performed with vitrification, there is still controversy regarding the ovary tissue between slow freeze and vitrification. We prefer vitrification, and have healthy babies, and have discussed this in previous papers [22]. It is well known that unilateral oophorectomy does not diminish fertility and does not hasten menopause. Thus, it is speculatively possible that grafts taken from young women could be used to delay menopause in the future [23–25].

## 2 Ovarian Cortical Transplantation

The main benefit of ovary cryopreservation over oocyte cryopreservation is for cancer patients who need to start their chemotherapy and radiation promptly and cannot afford to wait around for two or three cycles of IVF, knowing the pregnancy rate per single egg is only about 4%. With ovary freezing, they can potentially get

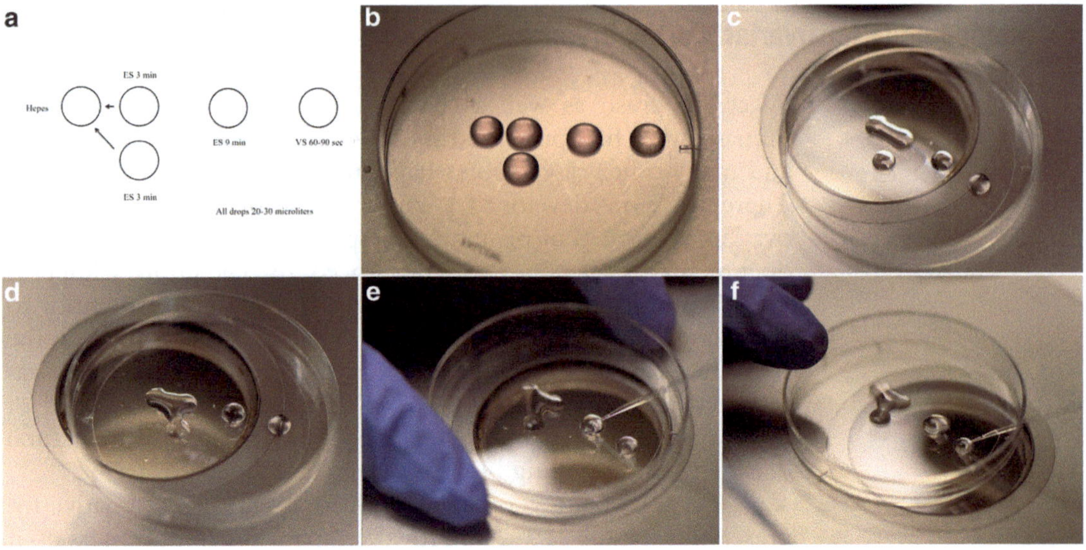


**Fig. 1** (**a–f**) (**a**) The bridge technique for oocyte freezing. (**b**) Setup for bridge equilibration. (**c**) First bridge between ES and isotonic HEPES media; 3 min. (**d**) Second bridge equilibration; 3 min (**e**) Transfer to full concentration ES; 9 min. (**f**) Transfer from ES to VS; 60–90 s. Caption: High cryoprotectant concentration is not toxic. It is just the rapid osmotic shifts that kill the egg or embryo and give the incorrect impression of toxicity. To avoid over rapid osmotic shifts (that are more poorly tolerated by the egg than the embryo), the original bridge technique is best. ES solution droplets are first bridged over to the isoosmotic solution the egg is in, and 3 min later, another droplet of ES solution is bridged over to the original solution very gradually and continuously raising the osmolality of the solution in which the oocyte is resting

pregnant naturally and preserve several hundred thousand eggs (Figs. 2, 3, and 4).

Successful fresh and cryopreserved ovarian cortex transplants in humans were first published in 2004 and 2005, as case reports, and many other case reports have subsequently followed [3, 5, 6, 10–15, 17, 18, 26–30]. The first human applications were preceded by a long history of animal experimentation. As far back as 1954, Deanesly showed in rats and, in 1960, Parrott showed in mice that ovarian tissue could be successfully frozen and autografted resulting in live births [31, 32]. Interest in human applications began after Gosden's report of successful pregnancies in sheep in 1994 [16]. One of the most interesting case reports in humans involved rejuvenating menopausal ovarian cortex with cryopreserved autotransplantation [33]. Interest in cryopreserved ovarian cortical transplantation is rapidly growing, although only two reports to date have clearly stated its success rate [8, 22]. No systematic report has been published from one center comparing fresh donor transplants and cryopreserved ovary autografts, and little has been gleaned from studies of these procedures on analysis of ovarian function and resting follicle recruitment prior to our reports [22, 34].

The primary impetus for this procedure has been to cryopreserve ovarian tissue before sterilizing cancer treatment, with the

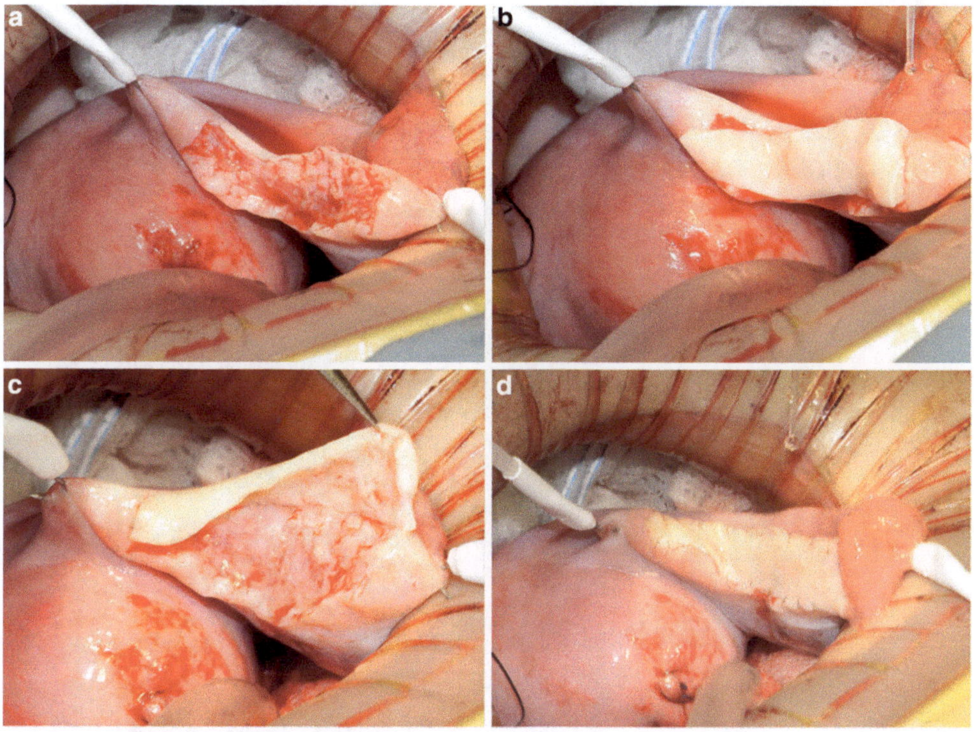


**Fig. 2** Steps in the procedure of ovarian transplantation between MZ twin sisters: (**a**) preparation of donor ovarian cortex by dissection in a Petri dish on ice, (**b**) preparation of recipient ovarian medulla, (**c**) attaching donor cortical tissue to recipient ovarian medulla, and (**d**) attaching thawed donor cortical tissue for re-transplant to the recipient medulla

objective of transplanting the tissue back after cancer has been cured, thus allowing patients to preserve their fertility. It is also possible that grafts taken from young women with cancer could be used to delay their menopause in the future [17, 18, 23–25, 35–42]. The possibility of preserving fertility and even hormonal function against the natural decline caused by aging has also been speculated but is considered of less importance [17–21].

Most published research in this field consists of case reports of cryopreserved transplants only. We conducted a worldwide survey of 37 babies born from cryopreserved transplants, but still could not establish a clear success rate [18]. Here, we summarize our report from a single series of both fresh and cryopreserved transplants from one center, carried out with the same technique and assessed uniformly over follow-up, with the aim of improving our understanding of resting follicle recruitment and demonstrating the clinical robustness of the procedure. All of the recipients had normal return of hormonal and menstrual ovarian function at about the same time after surgery (4.5 months) (Figs. 5 and 6).

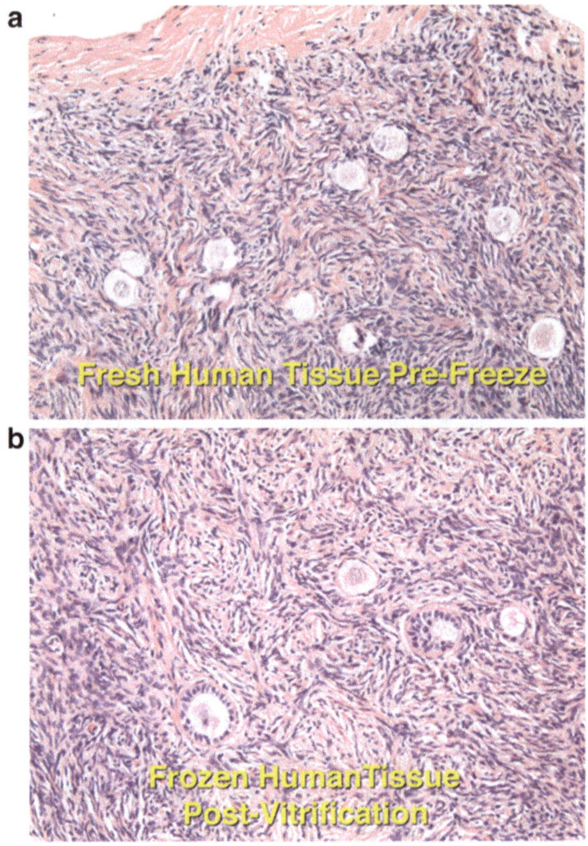


**Fig. 3** Histology (**a**) pre- and (**b**) post-vitrification of ovarian tissue

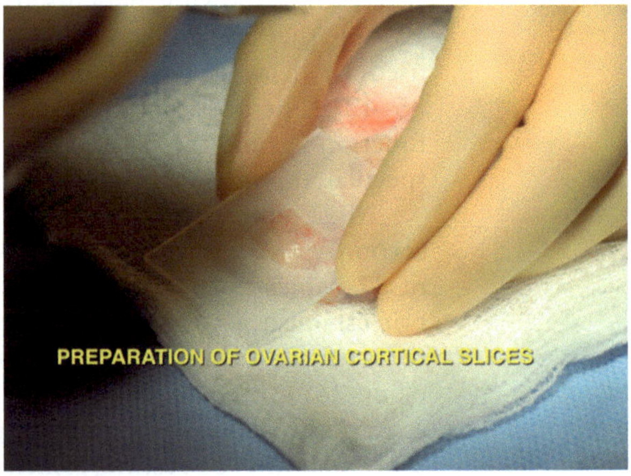


**Fig. 4** Ovarian tissue slice

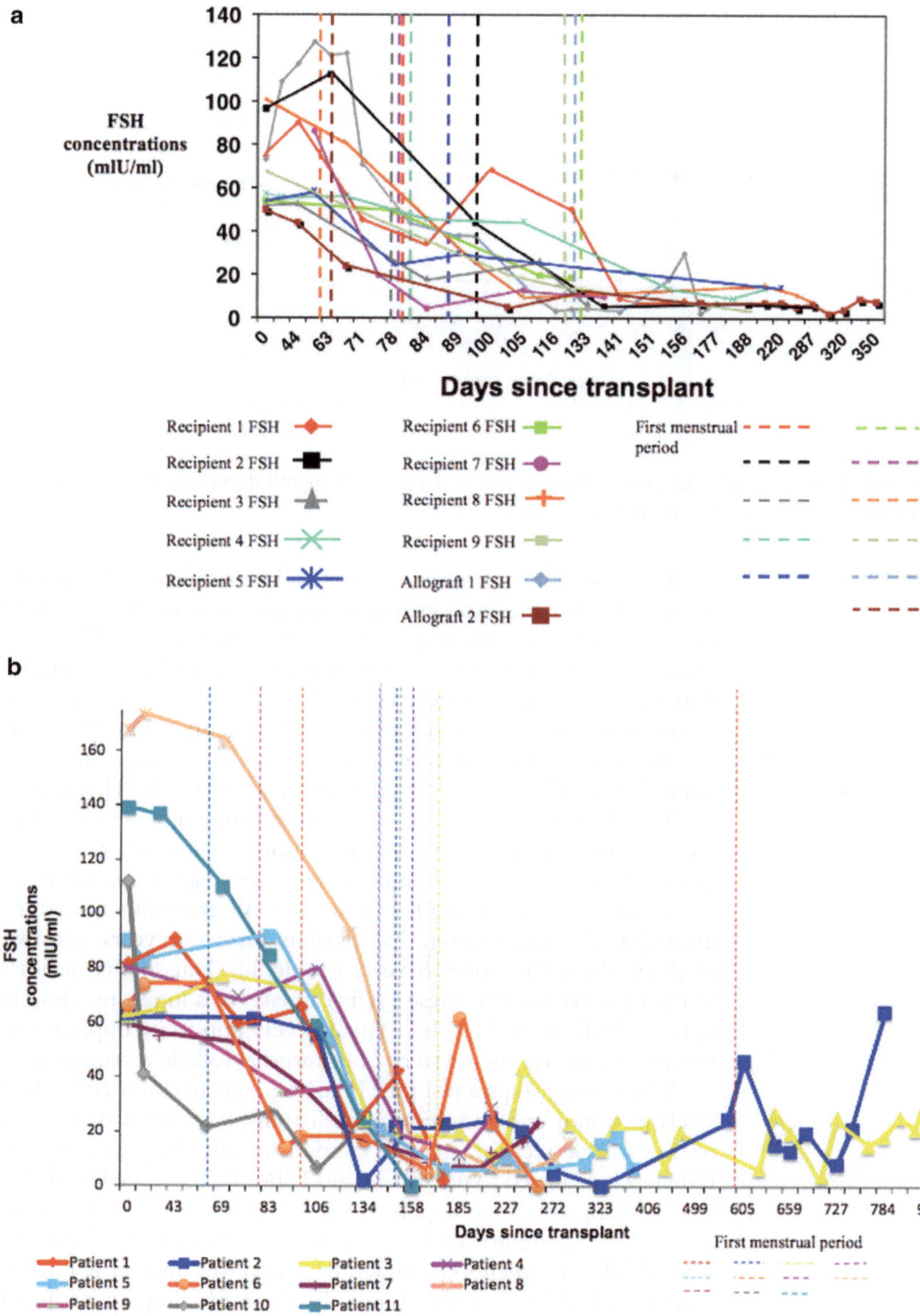


**Fig. 5** (**a**, **b**) Levels of FSH after fresh or cryopreserved ovary graft. The time of the first menstruation and the duration of time required for FSH levels to revert to normal levels, allowing for ovulation to occur, are represented for the 11 patients who received fresh ovary tissue transplants (**a**) and the 11 patients who received cryopreserved ovary tissue transplants (**b**) (*see* Ref. [1] in References)

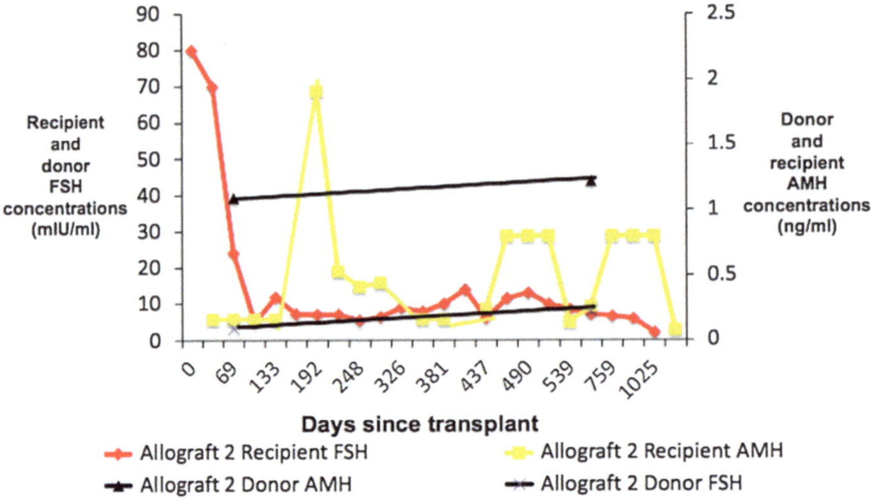


**Fig. 6** Levels of FSH and anti-Mullerian hormone levels in the second allograft donor and recipient. AMH = anti-Mullerian hormone (*see* Ref. [1] in Bibliography)

In all fresh and cryopreserved transplant recipients, day 3 FSH decreased from the menopausal range to normal levels in about 150 days, and menstrual cycling resumed by roughly 130 days. Since all developing follicles were destroyed or excluded from the thin (1-mm) donor ovarian cortical graft, this period seems to represent the number of days required for resting primordial follicles to be recruited and develop to the ovulatory stage, at which point they finally become sensitive to cyclic FSH and LH (Fig. 5).

The duration of function of fresh grafts was directly related to the original ovarian reserve of the donor. In all cases, only one-quarter to one-half of the donor ovary was transplanted, and most of the tissue cryopreserved for future use. All grafts functioned for more than 2 years, over one-half of them for over 6 years, and two of them already for over 8 years (Table 1). Thus if the donor's ovarian reserve is high, these grafts can last for a long time despite reduced AMH levels [17, 43]. That has very important implications for understanding the mystery of primordial follicle recruitment.

The relationships among FSH levels, menstruation and AMH levels in donor and recipient in fresh transplants are indicative of resting follicle recruitment and ovarian reserve [44]. As recipient FSH levels returned to normal within 130–170 days, the low AMH level of recipients then began to rise in response to an increasing number of mature gonadotropin-sensitive follicles (Figs. 5 and 6). The AMH of recipients continued to rise to well above the normal baseline AMH level of the donor. In the fresh allograft recipient shown in Fig. 6, although FSH decreased to normal levels by day 133 and normal menstrual cycling resumed, AMH levels rose far above normal (higher than the donor) shortly thereafter. Despite the transplanted graft continuing function, AMH then returned to low levels. This analysis of both donor and recipient data after transfer is rare and useful in its demonstration of no significant long-term change in donor AMH levels despite loss of an ovary. The rise of recipient AMH levels above donor AMH levels reflects over-recruitment of resting follicles in the recipient compared with the donor (Fig. 6). Contrary to what might have been expected based on earlier studies [45], no evidence was found of significant loss of follicles from transplantation ischemia, as each patient had FSH levels return to normal as AMH levels rose far above normal. Despite continued function of the transplanted graft, AMH then returns to low levels (Fig. 7a–c). Preservation of follicles was supported by our observation of substantial follicle recruitment with many eggs produced in the two patients who underwent IVF during the window of high AMH levels. Quantitative studies have also supported this finding in bovines [46].

The autotransplantation of cryopreserved ovary tissue yielded results almost identical to fresh transplantation (Fig. 5). This finding is consistent with the observed absence of histological damage from cryopreservation [43, 46, 47]. As with fresh ovary transplants, FSH levels returned to normal by about 150 days in all cases, and menstrual cycles resumed shortly before that. The return to normal did not differ between slow freeze cases and vitrification cases. In all of the cryopreserved cases, just as with fresh transplants, AMH rose to high levels shortly after FSH returned to normal, at around 130–170 days (Fig. 7a–c). Then, exactly as with fresh transplant cases, AMH dropped to a lower baseline level by about 240 days and remained at that lower level [17].

Eight of the 11 cryopreserved autotransplant recipients had a follow-up of over 1 year, which allowed the assessment of pregnancy potential. Seven of these eight patients spontaneously conceived, although one spontaneously aborted (Table 2). The other six were healthy singleton pregnancies. In fact there are now even 3 more healthy babies from 11 frozen transplants followed for over a year. Therefore, cryopreserved and fresh transplantation were similar in hormonal function and high pregnancy outcome. Functional hormonal results thus far have demonstrated a remarkable degree of repeatable concordance in all 22 cases of fresh ovary donor transplantation and cryopreserved ovary autotransplantation. Thus we had a spontaneous pregnancy rate overall of 75%.

In conclusion, no difference was found in clinical and functional results between fresh and cryopreserved ovary cortical grafts, and both demonstrated a high success rate in preserving fertility as well as endocrine function for long periods of time. Therefore, aside from the benefit of fertility preservation for cancer patients, this procedure offers the benefit of relief from menopause without having to resort to exogenous hormone replacement [17].

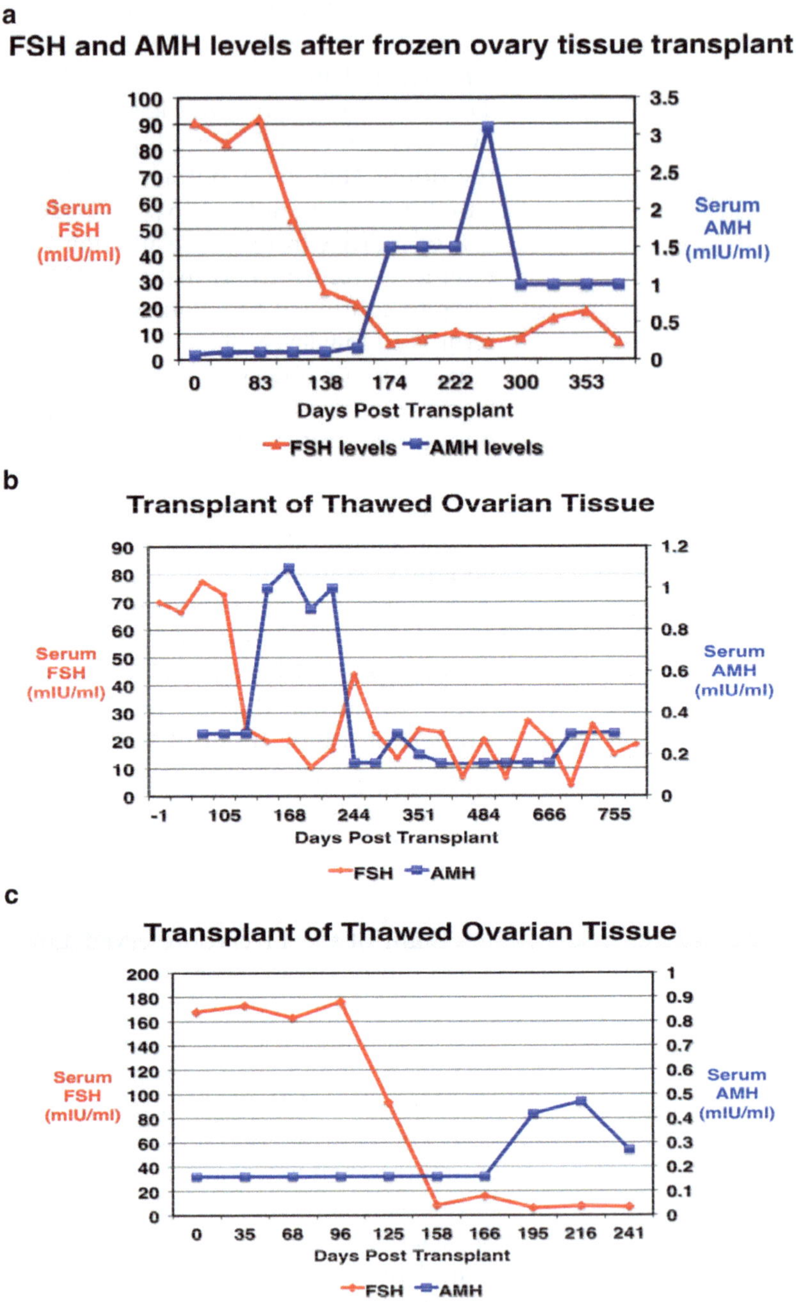


**Fig. 7** (**a–c**) Levels of FSH and anti-Mullerian hormone levels after ovary transplant. AMH = anti-Mullerian hormone (*see* Ref. [1] in References)

## 3 Ovarian Cryopreservation by Original Slow Freeze Technique

For slow freezing, after enucleating medullary tissue with a sharp scalpel dissection, the cortex was pared down manually to an ultra-thin translucent shell with a thickness of $\leq 1$ mm. Tissue for cryopreservation was divided into multiple strips and transferred to 1.5 ml cryovials after equilibration in 1.5 mol/l 1,2-propanediol and 0.1 mol/l sucrose at 37 °C for 30 min, followed by 1.5 mol/l 1,2 propanediol and 0.2 mol/l sucrose for 5 min, then cooled at a controlled rate, as described previously [16, 47]. Thawing was achieved rapidly by agitating the vials in a warmed water bath. If tissue had thickened by contraction after thawing, it was pared down again to <1 mm under an operating microscope with microsurgical scissors before transplantation [47].

However, we prefer vitrification for cryopreservation because of our in vitro viability analysis studies as well as in vivo transplant studies in the bovine [46]. A total of 16 cancer patients requesting fertility preservation by ovarian banking consented to an oocyte viability test and histologic review of a small sample (<10 %) of their fresh or preserved tissue. In eight cases, the tissue had been preserved by vitrification, six by a slow freezing protocol and in two only fresh tissue was analyzed. The goal of this in vitro study was to determine which method produced a higher cell survival rate [43].

The high viability (92%) of oocytes in both control (fresh) specimens and vitrified specimens indicated that disaggregation per se had only caused minimal damage to this cell type [43]. Overall, 2301 oocytes were examined from 16 specimens. The results within each of the three groups revealed no significant difference overall between fresh and vitrified tissue, but the viability of slow freeze-cryopreserved tissue was less than one-half (42%; $P<0.01$). Transmission electron microscopy also has been used to analyze ovarian tissue that had been either cryopreserved by slow freezing or vitrified by ultrarapid freezing, showing vitrification to be superior (Table 3) [48].

We know there have certainly been successes with slow freeze, but vitrification is easier and does less cryodamage.

**Table 3**
**Survival of small oocytes after enzymatic isolation from ovarian tissues following cryopreservation by vitrification or slow freezing**

| Group | No. of ovaries | No. of oocytes harvested | No. of surviving oocytes (%) |
|---|---|---|---|
| Fresh | 2 | 358 | 329 (91.9%)[a] |
| Vitrified | 8 | 1122 | 1000 (89.1%)[a] |
| Cryopreserved | 6 | 821 | 342 (41.7%)[b] |

[a,b] Groups with the same subscript are not significantly different ($P > 0.05$), whereas those with different superscripts are significantly different ($P < 0.01$)
*Silber. Fertility after ovary transplantation. Fertil Steril 2010*

## 4 Cryoinjury

Most of the stroma cells in the slow freeze-cryopreserved specimen were lysed and their nuclei compressed between dense bundles of extracellular fibers. The same cells were generally intact after vitrification. Small follicles were found in each specimen, all of which were intact within their basement membranes. The high viability (92 %) of oocytes in control (fresh) specimens indicated that disaggregation per se had only caused minimal damage to this cell type. Overall, 2301 oocytes were examined from 16 specimens. Results within each of the three groups were consistent and revealed no significant difference overall between fresh and vitrified tissue, although the viability of slow freeze-cryopreserved tissue was less than one-half (42 %) and highly significant ($P < 0.01$; Table 3).

## 5 Cryopreservation

For slow freezing, after enucleating medullary tissue with sharp scalpel dissection, the cortex was pared down manually to an ultra-thin translucent shell with a thickness of ≤1 mm. Tissue for cryopreservation was divided into multiple strips and transferred to 1.5 mL cryovials after equilibration in 1.5 mol/L 1,2-propanediol and 0.1 mol/L sucrose at 37 °C for 30 min, followed by 1.5 mol/L 1,2-propanediol and 0.2 mol/L sucrose for 5 min, and then cooled at a controlled rate, as described previously [16, 47]. Thawing was achieved rapidly by agitating the vials in a warmed water bath. If tissue had thickened by contraction after thawing, it was pared down again to <1 mm under an operating microscope with microsurgical scissors before transplantation. For vitrification, details have been described elsewhere [43, 46].

## 6 Tissue Analyses

Small samples of cortical tissue from donor organs were assessed: (1) on a semiquantitative scale of relative follicle density in histology slides (0, +, ++, or +++, in order of increasing abundance) and (2) by testing viability after enzymatic disaggregation. Also, all original cortical tissue from transplant recipients who were in ovarian failure was removed and prepared histologically after excision to verify that total follicular depletion had occurred. For viability testing, tissues were incubated and pipetted in type I collagenase (1 mg/mL) for 10 min to isolate the small follicles and visualize their oocytes. The cells were briefly incubated in Hoechst 33342 and propidium iodide and then washed before viewing by fluorescence microscopy, for a total of 2301 oocytes from 16 patients. Transmission electron microscopy was also used to analyze ovarian tissue that had been either cryopreserved by slow freezing or vitrified by ultrarapid freezing [48].

## 7 Ovarian Tissue Vitrification

Cortex tissue of each ovary is cut into slices of 1 × 0 × 10 mm. The precise 1-mm tissue thickness was guaranteed with a tissue slicer designed explicitly for this purpose (Kitazato BioPharma, Japan). The tissue slicer is put on the surface of the ovary. Then another plate is placed over the tissue slicer, and the ovary is cut between the slicer and the surface of the ovary using a sharp blade. The cortical ovarian tissue is thus cut into 1 × 10 × 10 mm pieces. The ultra-thinness of the tissue is crucial, not only for the cryopreservation but also for rapid revascularization after grafting.

Ovarian tissues are initially equilibrated in 7.5 % ethylene glycol (EG) and 7.5 % dimethyl sulfoxide (DMSO) in handling medium (HM:HEPES-buffered TCM-199 solution) supplemented with 20 % (v/v) synthetic serum substitute (SSS; Irvine Scientific, Santa Ana, CA, USA) for 25 min followed by a second equilibration in 20 % DMSO with 0.5 mol/l sucrose for 15 min. Ovarian tissues are then placed in a minimum volume solution (virtually "dry") onto a thin metal strip (Cryotissue, Kitazato BioPharma, Fujinomiya, Japan) and submerged directly into sterile liquid nitrogen [49], following which the strip was inserted into a protective container and placed into a liquid nitrogen storage tank.

For thawing, the protective cover is removed, and the Cryotissue metal strip is immersed directly into 40 ml of 37 °C HM solution supplemented with 1.0 mol/l sucrose for 1 min. Then, the ovary tissues are transferred into 15 ml of 0.5 mol/l sucrose HM solution for 5 min at room temperature and washed twice in HM solution for 10 min before viability analysis or transplantation. No ice crystal formation occurs during any of these vitrification procedures [46]. This approach is remarkably easier than slow freeze and also more effective.

## 8 Cortical Ovarian Tissue Transplantation Technique

The recipients were prepared by minilaparotomy via a 3.5-cm incision above the pubis. For cortical tissue transplantation, recipient ovarian cortex was resected to completely expose medullary tissue; hemostasis was controlled with micro-bipolar forceps, and irrigation with heparinized saline was performed to avoid adhesion formation or micro-hematomas between donor and recipient tissues. This technique may be the most important reason for a minimal ischemic loss of oocyte viability. The tissue graft was trimmed to the dimensions of the exposed surface of the recipient and attached using 9–0 [43, 46, 50–56]. Since 1996, we have frozen ovary tissue for 100 young women with solid organ cancer, or with a great risk for POF, of whom 16 had spare frozen tissue subjected to detailed viability testing before cryopreservation and after thawing. All had histologic review by a variety of pathologists. Only one had ovarian metastasis, a young woman with widespread breast cancer metastasis throughout her entire body. Otherwise, none of our other cases had any tumor cells in their ovary. Andersen has also noted a complete lack of ovarian metastasis, even in the majority of leukemia cases. The reason for the remarkable absence of ovarian metastasis might possibly be due to the fibrous avascular nature of the ovarian cortex [57]. In fact, the reason why fetal ovarian tubules (which in the fetal male become seminiferous tubules) invade the fibrous cortex and become follicles is that the dense fibrous tissue of the cortex (which in the fetal and adult testis is just tunica albuginea) is needed to suppress the resting follicles from developing all at once prematurely.

With the classical slow freeze technique, our in vitro viability testing showed only 41 % of oocytes survived [43, 46, 47, 58]. However, with vitrification of the ovarian tissue, there was no difference between fresh unfrozen controls and frozen tissue. It seems likely, therefore, that vitrified ovarian tissue would give better results after transplantation than tissue cryopreserved by slow freeze, but it is too early to state that with any certainty.

Although the clinical results seem promising from this large series of fresh and frozen ovarian transplants, the functional results may be even more interesting (Figs. 5a, 6, and 7a–c). Day 3 FSH came down from menopausal range to normal or near normal levels in all eleven cases by 150 days, and menstrual cycling resumed in all eleven recipients by 130 days. Since all developing follicles were destroyed or excluded from the thin (1 mm) donor ovarian cortical graft, this would represent the number of days required for resting follicles to be recruited and develop to the ovulatory stage when they would finally become sensitive to cyclic FSH and LH.

The relationship between FSH levels, menstruation, and AMH levels in these fresh transplants is indicative of resting follicle recruitment. As the FSH returned to normal by 130 days, the very low AMH level of the recipient began to rise in response to an increasing number of mature gonadotropin-sensitive follicles [43, 44]. The AMH of the recipient then continued to rise to well above the normal baseline AMH of the donor (Figs. 5a, 6, and 7a–c). Contrary to what might have been expected from earlier studies [45], this indicated no significant loss of follicles from the transplantation, which was also indicated by ultrasonographic examination of stimulated tissues at this period, which showed a very robust follicular response to gonadotropin stimulation when IVF was attempted (third identical twin case). Then by 240 days, the AMH of the recipient descended to well below baseline levels and then remained steady at this level usually for many years (Fig. 6) [43, 44, 46–48].

The autotransplantation of frozen ovary tissue yielded results almost identical to fresh (as might have been guessed by failure to notice any histological damage from freezing) [43, 46–48]. As with fresh ovary transplants, the FSH always returned to normal in these cases by 150 days without exception, and menstrual cycles resumed shortly before that. The return to normal was no different for slow freeze cases than for vitrification cases. In all of these cases, just like with fresh transplants, the AMH rose to very high levels shortly after the FSH returned to normal, at about 170 days. Then the AMH, exactly as with fresh transplant cases, came down to a lower baseline level by 240 days and remained at that lower indefinitely.

Eight of the 11 frozen cases were over 1 year ago and thus suitable to look at pregnancy potential. However, one of these eight has chosen not to get pregnant yet and is using barrier contraception. Of the remaining seven, six have gotten pregnant although one miscarried. Thus frozen results were similar to fresh in terms of hormonal function and pregnancy outcome. Quite remarkable was the repeatable concordance of functional results thus far seen in all 22 cases of fresh ovary donor transplantation and frozen ovary autotransplantation.

## 9 Future Perspective

Firstly, ovarian freezing and transplantation should no longer be considered experimental moving forward. Secondly, this could provide a unifying theory for adult recruitment and fetal arrest of primordial follicles. Without a steady controlled release of resting follicles in the adult, women would run out of eggs prematurely. Similarly, if fetal oocyte arrest did not occur after meiotic activation, there would be no oocytes remaining at birth. Similarly, if there were not a slow, steady, monthly release of primordial (resting or better, arrested) follicles from the resting phase, without any hormonal regulation, related to tissue pressure gradients in the ovarian cortex (the same tissue as the dense stroma in the tunica albuginea of the male), all the adult follicles also would soon be gone (Fig. 8a–c). Thus we propose a unifying theory for both fetal oocyte arrest and adult resting follicle recruitment.

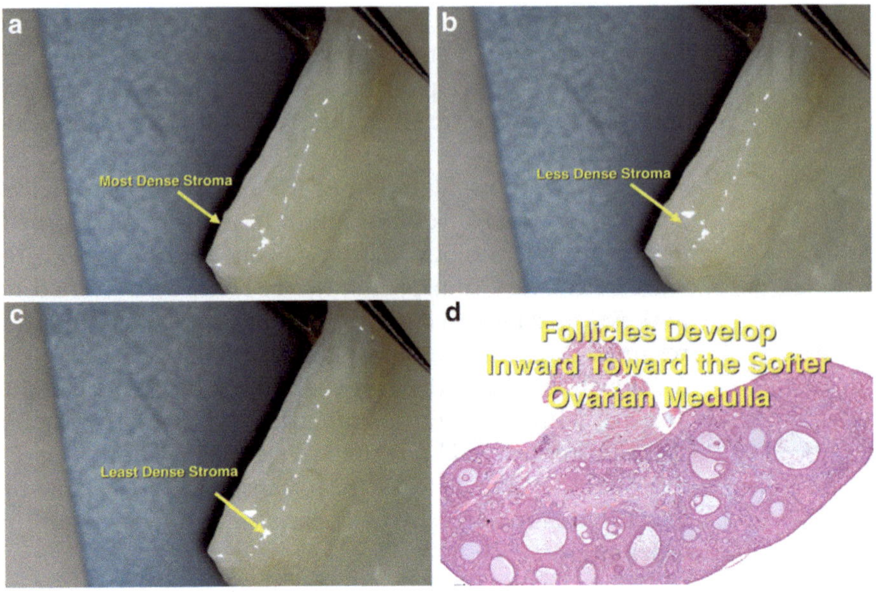


**Fig. 8** (**a–d**) Stromal density gradient in the ovarian cortex. (**d**) Follicles develop inward toward the softer ovarian medulla

## References


1. Silber S, Pineda J, Lenahan K, DeRosa M, Melnick J (2015) Fresh and cryopreserved ovary transplantation and resting follicle recruitment. Reprod Biomed Online 30 (6):643–650
2. Silber S (2015) Unifying theory of adult resting follicle recruitment and fetal oocyte arrest. Reprod Biomed Online 31(4):472–475
3. Silber SJ, Lenahan KM, Levine DJ, Pineda JA, Gorman KS, Friez MJ, Crawford EC, Gosden RG (2005) Ovarian transplantation between monozygotic twins discordant for premature ovarian failure. N Engl J Med 353(1):58–63
4. Silber SJ (2012) Ovary cryopreservation and transplantation for fertility preservation. Mol Hum Reprod 18(2):59–67
5. Demeestere I, Simon P, Englert Y, Delbaere A (2003) Preliminary experience of ovarian tissue cryopreservation procedure: alternatives, perspectives and feasibility. Reprod Biomed Online 7(5):572–579
6. Demeestere I, Simon P, Moffa F, Delbaere A, Englert Y (2010) Birth of a second healthy girl more than 3 years after cryopreserved ovarian graft. Hum Reprod 25 (6):1590–1591
7. Donnez J, Dolmans MM, Demylle D, Jadoul P, Pirard C, Squifflet J, Martinez-Madrid B, van Langendonckt A (2004) Livebirth after orthotopic transplantation of cryopreserved ovarian tissue. Lancet 364(9443):1405–1410
8. Donnez J, Dolmans MM, Pellicer A, Diaz-Garcia C, Sanchez Serrano M, Schmidt KT, Ernst E, Luyckx V, Andersen CY (2013) Restoration of ovarian activity and pregnancy after transplantation of cryopreserved ovarian tissue: a review of 60 cases of reimplantation. Fertil Steril 99(6):1503–1513
9. Donnez J, Silber S, Andersen CY, Demeestere I, Piver P, Meirow D, Pellicer A, Dolmans MM (2011) Children born after autotransplantation of cryopreserved ovarian tissue. a review of 13 live births. Ann Med 43(6):437–450
10. Meirow D, Levron J, Eldar-Geva T, Hardan I, Fridman E, Zalel Y, Schiff E, Dor J (2005) Pregnancy after transplantation of cryopreserved ovarian tissue in a patient with ovarian failure after chemotherapy. N Engl J Med 353 (3):318–321
11. Poirot C, Abirached F, Prades M, Coussieu C, Bernaudin F, Piver P (2012) Induction of puberty by autograft of cryopreserved ovarian tissue. Lancet 379(9815):588
12. Sanchez M, Alama P, Gadea B, Soares SR, Simon C, Pellicer A (2007) Fresh human orthotopic ovarian cortex transplantation: long-term results. Hum Reprod 22 (3):786–791

13. Silber SJ, DeRosa M, Pineda J, Lenahan K, Grenia D, Gorman K, Gosden RG (2008) A series of monozygotic twins discordant for ovarian failure: ovary transplantation (cortical versus microvascular) and cryopreservation. Hum Reprod 23(7):1531–1537
14. Silber SJ, Gosden RG (2007) Ovarian transplantation in a series of monozygotic twins discordant for ovarian failure. N Engl J Med 356(13):1382–1384
15. Silber SJ, Grudzinskas G, Gosden RG (2008) Successful pregnancy after microsurgical transplantation of an intact ovary. N Engl J Med 359 (24):2617–2618
16. Gosden RG, Baird DT, Wade JC, Webb R (1994) Restoration of fertility to oophorectomized sheep by ovarian autografts stored at −196 degrees C. Hum Reprod 9(4):597–603
17. Andersen CY, Silber SJ, Bergholdt SH, Jorgensen JS, Ernst E (2012) Long-term duration of function of ovarian tissue transplants: case reports. Reprod Biomed Online 25(2):128–132
18. Stoop D, Cobo A, Silber S (2014) Fertility preservation for age-related fertility decline. Lancet 384(9950):1311–1319
19. Homburg R, van der Veen F, Silber SJ (2009) Oocyte vitrification–women's emancipation set in stone. Fertil Steril 91(4 Suppl):1319–1320
20. Nagy ZP, Chang CC, Shapiro DB, Bernal DP, Kort HI, Vajta G (2009) The efficacy and safety of human oocyte vitrification. Semin Reprod Med 27(6):450–455
21. Cobo A, Domingo J, Perez S, Crespo J, Remohi J, Pellicer A (2008) Vitrification: an effective new approach to oocyte banking and preserving fertility in cancer patients. Clin Transl Oncol 10(5):268–273
22. Silber S (2016) Ovarian tissue cryopreservation and transplantation: scientific implications. J Assist Reprod Genet
23. Lass A, Paul M, Margara R, Winston RM (1997) Women with one ovary have decreased response to GnRHa/HMG ovulation protocol in IVF but the same pregnancy rate as women with two ovaries. Hum Reprod 12(2):298–300
24. Zhai A, Axt J, Hamilton EC, Koehler E, Lovvorn HN III (2012) Assessing gonadal function after childhood ovarian surgery. J Pediatr Surg 47(6):1272–1279
25. Gosden RG, Telfer E, Faddy MJ, Brook DJ (1989) Ovarian cyclicity and follicular recruitment in unilaterally ovariectomized mice. J Reprod Fertil 87(1):257–264
26. Anderson EL, Baltus AE, Roepers-Gajadien HL, Hassold TJ, de Rooij DG, van Pelt AM, Page DC (2008) Stra8 and its inducer, retinoic acid, regulate meiotic initiation in both spermatogenesis and oogenesis in mice. Proc Natl Acad Sci U S A 105(39):14976–14980
27. Dittrich R, Mueller A, Maltaris T, Hoffmann I, Magener A, Oppelt PG, Beckmann MW (2009) Hormonal and histologic findings in human cryopreserved ovarian autografts. Fertil Steril 91(4 Suppl):1503–1506
28. Revel A, Laufer N, Ben Meir A, Lebovich M, Mitrani E (2011) Micro-organ ovarian transplantation enables pregnancy: a case report. Hum Reprod 26(5):1097–1103
29. Revelli A, Marchino G, Dolfin E, Molinari E, Delle Piane L, Salvagno F, Benedetto C (2013) Live birth after orthotopic grafting of autologous cryopreserved ovarian tissue and spontaneous conception in Italy. Fertil Steril 99 (1):227–230
30. Stern CJ, Gook D, Hale LG, Agresta F, Oldham J, Rozen G, Jobling T (2013) First reported clinical pregnancy following heterotopic grafting of cryopreserved ovarian tissue in a woman after a bilateral oophorectomy. Hum Reprod 28(11):2996–2999
31. Deanesly R (1954) Immature rat ovaries grafted after freezing and thawing. J Endocrinol 11(2):197–200
32. Bruce HM, Parrott DM (1960) Role of olfactory sense in pregnancy block by strange males. Science 131(3412):1526
33. Kawamura K, Cheng Y, Suzuki N, Deguchi M, Sato Y, Takae S, Ho CH, Kawamura N, Tamura M, Hashimoto S, Sugishita Y, Morimoto Y, Hosoi Y, Yoshioka N, Ishizuka B, Hsueh AJ (2013) Hippo signaling disruption and Akt stimulation of ovarian follicles for infertility treatment. Proc Natl Acad Sci U S A 110(43):17474–17479
34. Oktay K, Economos K, Kan M, Rucinski J, Veeck L, Rosenwaks Z (2001) Endocrine function and oocyte retrieval after autologous transplantation of ovarian cortical strips to the forearm. JAMA 286(12):1490–1493
35. Aydin Y, Celiloglu M, Koyuncuoglu M, Ulukus C (2010) Follicular dynamics and apoptosis following unilateral oophorectomy. Syst Biol Reprod Med 56(4):311–317
36. Faddy MJ, Gosden RG, Gougeon A, Richardson SJ, Nelson JF (1992) Accelerated disappearance of ovarian follicles in mid-life: implications for forecasting menopause. Hum Reprod 7(10):1342–1346
37. Kaaijk EM, Hamerlynck JV, Beek JF, van der Veen F (1999) Clinical outcome after unilateral oophorectomy in patients with polycystic ovary syndrome. Hum Reprod 14(4):889–892

38. Kim SS (2003) Ovarian tissue banking for cancer patients. To do or not to do? Hum Reprod 18(9):1759–1761
39. Koskas M, Uzan C, Gouy S, Pautier P, Lhomme C, Haie-Meder C, Duvillard P, Morice P (2011) Fertility determinants after conservative surgery for mucinous borderline tumours of the ovary (excluding peritoneal pseudomyxoma). Hum Reprod 26(4):808–814
40. Meredith S, Dudenhoeffer G, Butcher RL, Lerner SP, Walls T (1992) Unilateral ovariectomy increases loss of primordial follicles and is associated with increased metestrous concentration of follicle-stimulating hormone in old rats. Biol Reprod 47(2):162–168
41. Saiduddin S, Rowe RF, Casida LE (1970) Ovarian follicular changes following unilateral ovariectomy in the cow. Biol Reprod 2 (3):408–412
42. Yasui T, Hayashi K, Mizunuma H, Kubota T, Aso T, Matsumura Y, Lee JS, Suzuki S (2012) Factors associated with premature ovarian failure, early menopause and earlier onset of menopause in Japanese women. Maturitas 72 (3):249–255
43. Silber S, Kagawa N, Kuwayama M, Gosden R (2010) Duration of fertility after fresh and frozen ovary transplantation. Fertil Steril 94 (6):2191–2196
44. La Marca A, Giulini S, Tirelli A, Bertucci E, Marsella T, Xella S, Volpe A (2007) Anti-Mullerian hormone measurement on any day of the menstrual cycle strongly predicts ovarian response in assisted reproductive technology. Hum Reprod 22(3):766–771
45. Baird DT, Webb R, Campbell BK, Harkness LM, Gosden RG (1999) Long-term ovarian function in sheep after ovariectomy and transplantation of autografts stored at −196 °C. Endocrinology 140(1):462–471
46. Kagawa N, Silber S, Kuwayama M (2009) Successful vitrification of bovine and human ovarian tissue. Reprod Biomed Online 18 (4):568–577
47. Newton H, Aubard Y, Rutherford A, Sharma V, Gosden R (1996) Low temperature storage and grafting of human ovarian tissue. Hum Reprod 11(7):1487–1491
48. Keros V, Xella S, Hultenby K, Pettersson K, Sheikhi M, Volpe A, Hreinsson J, Hovatta O (2009) Vitrification versus controlled-rate freezing in cryopreservation of human ovarian tissue. Hum Reprod 24(7):1670–1683
49. Fuentes, F., Dubettier, R. (2004). Air Separation Method and Plant, United States Patent. Patent No. US 6,776,005 B2.
50. Bleyer WA (1990) The impact of childhood cancer on the United States and the world. CA Cancer J Clin 40(6):355–367
51. Ries, L. A. G., and SEER Program (National Cancer Institute (U.S.)). (1999). Cancer incidence and survival among children and adolescents: United States SEER program 1975–1995//edited by Lynn A. Gloecker Ries ... et al.], Bethesda, MD: National Cancer Institute, SEER Program
52. Larsen EC, Muller J, Schmiegelow K, Rechnitzer C, Andersen AN (2003) Reduced ovarian function in long-term survivors of radiation- and chemotherapy-treated childhood cancer. J Clin Endocrinol Metab 88(11):5307–5314
53. Anderson RA, Themmen AP, Al-Qahtani A, Groome NP, Cameron DA (2006) The effects of chemotherapy and long-term gonadotrophin suppression on the ovarian reserve in premenopausal women with breast cancer. Hum Reprod 21(10):2583–2592
54. Lee SJ, Schover LR, Partridge AH, Patrizio P, Wallace WH, Hagerty K, Beck LN, Brennan LV, Oktay K, American Society of Clinical, O (2006) American society of clinical oncology recommendations on fertility preservation in cancer patients. J Clin Oncol 24(18):2917–2931
55. Anderson RA, Cameron DA (2007) Assessment of the effect of chemotherapy on ovarian function in women with breast cancer. J Clin Oncol 25(12):1630–1631. author reply 1632
56. Jeruss JS, Woodruff TK (2009) Preservation of fertility in patients with cancer. N Engl J Med 360(9):902–911
57. Greve T, Clasen-Linde E, Andersen MT, Andersen MK, Sorensen SD, Rosendahl M, Ralfkiaer E, Andersen CY (2012) Cryopreserved ovarian cortex from patients with leukemia in complete remission contains no apparent viable malignant cells. Blood 120(22):4311–4316
58. Gook DA, Edgar DH, Stern C (1999) Effect of cooling rate and dehydration regimen on the histological appearance of human ovarian cortex following cryopreservation in 1, 2-propanediol. Hum Reprod 14(8):2061–2068

# Chapter 14

# Establishing an Oocyte Cryobank Network

## James Graham, Josh Lim, and Michael Tucker

## Abstract

Improvements in oocyte cryopreservation has prompted wider acceptance of this technology leading to its use for several reasons. Notably, in addition to elective and medically driven reasons for oocyte cryostorage for fertility preservation, donor oocyte cryobanking is beginning to gain traction, potentially replacing fresh oocyte donation in assisted reproduction. Donor "egg banking," while not totally analogous to donor sperm banking, does provide strong benefits in terms of scheduling flexibility and improved clinical efficiencies, while providing a wider immediate inventory choice. The development of a successful cryobank "network" and subsequent growth into a full-access donor egg bank are only possible through adoption of a series of key steps involving establishment of a repeatable vitrification protocol with a strong clinical record, incorporation of a comprehensive database and quality management system, and strict control over the logistics of inventory and shipping and receiving, to establish a flawless chain from donor to recipient. Confidence will grow in this potentially difficult process over time.

**Key words** Oocyte, Vitrification, Cryopreservation

Oocyte cryopreservation is a proven method of storing oocytes for future use for attempting pregnancy [(1, 2);] there are various reasons for cryopreservation of oocytes or "banking."

## 1 Why Start an "Egg Bank"?

Banking of oocytes could involve many aspects of the term banking. The three types of banking are:

- Fertility preservation for social reasons
- Fertility preservation for oncology patients/medical reasons
- The banking of oocytes from anonymous donors to be distributed to donor egg recipients

The focus of this chapter will be on banking for donor egg recipients.

Zsolt Peter Nagy et al. (eds.), *Cryopreservation of Mammalian Gametes and Embryos: Methods and Protocols*, Methods in Molecular Biology, vol. 1568, DOI 10.1007/978-1-4939-6828-2_14, 

There are many benefits to banking donor eggs [3, 4]. The primary advantage is it gives your patients and your practice more flexibility in their cycle and reduces the time your clinical team needs for donor-recipient synchronization. Failed linings can just postpone the oocyte thaw, delaying the recipients' transfer days not weeks. The donor's eggs can be shared with several recipients throughout your network, maximizing your pool of donors. This also minimizes the problem of a large number of extra embryos cryopreserved that a recipient may never use. The selection of donors will be broader not just limited to the availability of donors that are available to cycle concurrent to your recipient.

### 1.1 Why Start an Egg Bank Network?

The establishment of a donor egg bank network has many advantages. A network can provide a larger selection of donors available to your patients. An egg bank network supplies more ethnically diverse donors. A recipient looking for a specific ethnicity or, even more difficult, a specific mix of ethnicities may be more challenging in a given geographic location. Donors can be easily shared among several recipients and with recipients in different geographic locations. Simply put, a patient will have more choice.

Giving a patient a choice to make a selection from different geographic regions can be an added benefit. The perception exists that if you choose a donor from your local area, you are more likely to have offspring or other recipient's offspring nearby. While this may not be true, the perception exists. A clinic that stores their inventory locally gives your patient a feeling of exclusivity.

There are also financial benefits to belonging to a network. One practice may have a surplus of highly desirable donors of specific ethnicity. This practice can cycle their donors and store the donor oocytes and then supply areas of the network that are in need of donors of that specific ethnicity. Another financial benefit is in the coordination of donors. The coordination of a fresh donor with recipients requires clinical manpower that many clinics do not posses. Also trying to create an individual bank with diversity can incur high recruiting expenses. Belonging to a network allows those with limited resources the opportunity to deal with surplus or deficits of donors or recipients.

## 2 Establishing a Vitrification Protocol

Both slow freeze [5–7] and vitrification methods [8, 9] have been developed. While slow freeze methods have been employed successfully, the use of vitrification for cryopreserving oocytes is a more common and successful approach [10, 11]. Spindle depolarization occurs in both but spindle recovery is similar in both approaches [ (12).] Reviewing protocols for oocyte vitrification needs to include carrier options. While there are many carriers commercially available, the efficacy of the carrier is important. HSV (CBS) straws, micropipettes, and sealed straws are just some of the closed systems available. A closed system may be preferable, but to date these closed systems have not been shown to have as consistent results as open systems. Many open systems that are available (i.e., Cryotop, Cryolock, and Cryotech) are very similar in design and are used in a similar manner. The mass of the actual tip of the unit is important for rapid cooling and warming. A carrier with a lower tip mass appears to ensure more rapid and consistent cooling/warming rates. There are several commercial vitrification kits available all with oocyte warming/cooling protocols that included many with positive results. A practice needs to evaluate these devices and to decide what protocol works best for them.

## 3 Protocol Evaluation

Choosing a commercially available oocyte vitrification kit or commercially available media that is easily available to all participating network clinics is important. This will minimize workloads and variables in performing oocyte vitrification. This is important when troubleshooting suboptimal outcomes in the embryology labs within a network. A user-friendly protocol that can be easily standardized and implemented in multiple locations should be used. The protocol should be thoroughly tested prior to being implemented.

Using discard IVM oocytes from consented patients is a good choice to test and refine your protocol. Using IVM oocytes will give the embryologists practice familiarizing themselves with the vitrification protocol. This also allows for labs to evaluate potential issues that can arise in various lab settings. Often, when oocytes are vitrified and then warmed, survival or the appearance of survival will occur. While this is encouraging, it can be misleading. Subsequent fertilization and early cleavage may have the appearance of progressing normally as well. Embryo development from Day 3 may still be poor if there is a poor vitrification/warming technique. A true clinical test of a protocol is necessary for validation of the protocol. Vitrifying oocytes and subsequent warming for use in an IVF cycle is necessary to truly validate an oocyte vitrification protocol.

## 4 Preclinical Study

A challenging precursor of implementation is the preclinical study. Refining protocol and doing test vitrification/warmings are effective tools to hone a vitrification protocol, but an actual clinical study is needed to ensure true efficacy of the protocol prior to implementation. There are numerous ways to approach this. Autologous patients may be enrolled in a study to have a portion of their oocytes vitrified and warmed during their IVF cycle. Recipients that are scheduled to receive fresh donor eggs might also be enrolled to do a similar test. Donor eggs could be banked and then thawed for a vitrified donor egg cycle, perhaps vitrification of some or all of a total cohort of donor oocytes and then warming for a subsequent donor oocyte IVF cycle.

At our clinic we chose to vitrify donor oocytes, then matched them to recipients in our IVF refund program, and gave patients the opportunity to participate in the study with no risk to the number of cycles they would be allowed. We then warmed the vitrified oocytes in advance of their normally scheduled ICSI. This would allow us to get immediate results from our egg vitrification protocol. Ten patients were enrolled and a 46 % pregnancy rate was obtained, with a 42 % delivery rate. This result gave us confidence that the protocol was effective.

Our practice has had a successful shared donor oocyte program for several years. Typically 75 % of our donors shared their cohort among three synchronized recipients. The minimum number of M2 oocytes for this fresh program is 4×M2 per recipient. Given that during the preclinical egg vitrification study, survival rates were around 85 % and in order to try to model the "egg bank" after this successful fresh donor oocyte share program, the minimum lot size was therefore set at 5×M2 per patient.

## 5 Preclinical and Clinical "Proof of Concept" (Single Site)

### *5.1 Analyze Study Data*

By looking at the data, parameters that should be evaluated are warming survival rates, fertilization rates, cleavage rates, blastocyst formation rates, and implantation/clinical pregnancy rates, in addition to overall embryo utilization rates. If the preestablished benchmarks are met, then offering clinical application can commence.

## 6 Offer Limited Clinical Application

Once the preclinical trial is successful, a limited clinical offering may begin. This will allow you to develop confidence in the protocol, while increasing your experience as you work toward offering "egg banking" at additional programs. This may extend the time until you build your donor egg bank network, but it will allow your embryology and clinical teams time to gain experience and for your "home" team to be able to be the expert resource for the "egg bank" network programs. It is imperative that highly selected/desirable donors be used in the initial phase of the clinical application preferably with proven track records of fresh donation, so that they will be chosen quickly, and the vitrified oocytes can be used and clinical results be assessed. The following table shows our early experience:

| | |
|---|---|
| Oocytes thawed | 934 |
| Oocytes survived | 817 |
| 2PN embryos | 667 (82 %) |
| Cycles | 143 |
| Embryo transfers | 143 |
| Day 3 ET/# of embryos | 48/1.75 |
| Day blastocyst ET/# of embryos | 95/1.96 |
| D-3 pregnancy %/ET | 52 % |
| D-5 pregnancy %/ET | 57 % |
| Cycles with surplus blastocyst cryo | 65 (44 %) |
| Blastocyst utilization | 66 % |
| Clinical pregnancy | 55 % |
| SAB | 6 |
| **Delivered** | **51 %** |

## 7 Data Collection

Data collection is a vital part of any clinical medicine program. This is especially true with an "egg bank" network. Being able to evaluate embryologist and clinicians in their roles throughout your "egg bank" network will ensure their success. Evaluation of the performance with both vitrification and warming techniques needs to be done for each individual performing the task. As the number of egg lots are banked and then warmed, the more apparent variations in technique per embryologist become. Thus, regular data tabulation is necessary to ensure consistent outcomes and to identify potential outliers and thereby any protocol deviations.

A stimulation protocol review and assessment of outcomes per physician stimulation should be evaluated according to stimulation approach. A clinic that continually produces poor performing egg lots may not have an embryology performance problem, but possible ovarian stimulation issues that need to be remediated.

## 8 Establishing a Network

Careful selection of partner clinics is of paramount importance. Unlike with a donor "sperm bank," having the technical expertise on site to even warm eggs for clinical use is challenging. Minimum criteria should be established as to which clinics are interviewed and then ultimately selected to participate in a donor "egg bank" network. Clinic geography, clinic success, resources, and embryology talent should all be part of the decision process. A smaller clinic with exceptional pregnancy rates may not have the necessary personnel for recruiting and screening donors. The clinic's ability or inability to manage the inventory and shipping may also be a limiting factor. A large clinic may have the resources, but not the embryology staffing and talent, or may have poor clinical outcomes. A well-run clinic with adequate resources, talent, and outcomes should all factor in deciding in taking on a partner in your "egg bank" network.

## 9 Identifying Partner Sites

A well-run clinic with adequate resources (both in staffing and facilities) is necessary to participate in an egg bank network. The first criterion should be that the network clinic is a SART reporting clinic. This allows easy access to historical data for the clinic as the first step in evaluation. While non-SART clinics may be considered, this does pose the problem of having comparable data to evaluate that a potential particular clinic's overall performance. Though CDC ART clinic data may suffice for comparison.

Geography: Is this clinic in an underserved area? Are there donors that can be accessed? Would their donors be desirable? Do they compete with a clinic that you already have an agreement with? A clinic that has access to desirable donors that does not compete with another partner clinic may be a good choice. A last area of concern is if the practice has adequate resources: embryologists, donor coordinators, and donor screening staff. There are many areas to evaluate before extending/establishing a partnership.

## 10 Logistics

After partner clinics have joined your network and have been trained, there is the often overlooked issue of logistics. Donor-recipient matching, verifying, shipping and receiving, and thaw coordination all involve systems that may not exist in your clinic or potential egg bank network partner clinics. There can be a central location for storage of oocytes, similar to commercial donor "sperm bank," or alternatively the oocytes may be stored at their place of origin. If stored onsite at the clinic of origin, an online web-based or "virtual egg bank" inventory can be used. This allows potential recipients access to the available inventory, while minimizing repeated shipping.

## 11 Training

An extensive embryologist training program is essential. The embryologist training program is necessary to ensure protocol adherence and excellent outcomes. Even with extensive training, success can still be limited by a technician's skill set or later "protocol drift." Many times an experienced embryologist, even one that has had previous embryo vitrification experience, will have difficulty following the oocyte vitrification protocol appropriately. Typically this is not deliberate, but a tendency to return to a protocol they have used more routinely or to treat the oocyte similar to an embryo when vitrifying.

An embryologist should be trained at your facility and then allowed to return to their own facility to practice the technique. This allows for the workflow to be established prior to any "live" cases. It also allows for the embryologist to familiarize themselves with the protocol working through any technical difficulties. The trained embryologist can then vitrify some discard oocytes and send to your central facility for validation. This will not only test their newly learned vitrification technique but also allow for identification of any technical shipping issues. It is also important for an onsite visit with the newly trained embryologist by a vitrification trainer. This can be done in conjunction with batched donor retrieval cycles to allow for onsite observation of the newly trained embryologist. The trainer can provide support for the first egg vitrification cases or even help if the newly trained embryologist becomes overwhelmed. Follow-up site visits are also important to identify any potential "protocol drift." It is recommended that a semiannual or even quarterly site visit coordinated with donor egg retrievals that are to be banked, as an ongoing quality assurance plan. This is vitally important in the first year of a newly trained embryologist, as this is when "protocol drift" is most likely to occur. "Banking" multiple lots of oocytes with poor outcomes because of "protocol drift" will be a particularly discouraging experience.

## 12 Data Collection

Data collection should occur in real time. Many eggs may sit in a storage tank for weeks or months after collection unused, and therefore feedback may be limited. Getting outcome data in a timely manner is difficult. Given this delay, it is important to collect and analyze the data as soon as possible. Reviews of ovarian stimulations are included in the data that is collected. Cryosurvival, fertilization, embryo cleavage, blastocyst formation rate, implantation, as well as cumulative pregnancy rates should all be collected per "freeze clinic", as well as for the clinic where the oocyte warming ultimately occurs. A clinic that has poor outcomes during the oocyte warming may have had several lots that performed poorly at other clinics. Conversely, a "warming site" clinic that has received oocytes that performed poorly but that had sibling lots do well at other "warming sites" may warrant a site visit, to troubleshoot any problems they may have occurred technically.

## 13 QA/QC: QMS (Quality Management System)

While initial training and quality programs may be instituted on a quarterly or semiannual basis, distance quality programs have limitations. Thresholds for performance for clinical as well as laboratory performance need to be established. Clinics that deviate from the mean need quick intervention and continuous monitoring. Establishing a minimum threshold of performance in key areas is critical for consistent outcomes. Measuring survival rates, blastocyst formation, and ultimately clinical pregnancy rates are key benchmarks in a QMS. This also ensures the longevity of the program. Lack of these quality programs will discourage the use of a donor "egg bank" due to inconsistent clinical outcomes, with clinical staff as well as patients losing confidence in the program.

### References

1. Tucker MJ, Morton PC, Wright G, Sweitzer CL, Massey JB (1998) Clinical application of human egg cryopreservation. Hum Reprod 13:3156–3159
2. Tucker M, Morton P, Liebermann J (2004) Human oocyte cryopreservation: a valid alternative to embryo cryopreservation? Eur J Obstet Gynecol Reprod Biol 113S:S24–S27
3. Cobo A, Meseguer M, Remohi J et al (2010) Use of cryobanked oocytes in an ovum donation programme: a prospective, randomized, controlled, clinical trial. Hum Reprod 25:2239–2246
4. Nagy ZP, Chang CC, Shapiro DB et al (2009) Clinical evaluation of the efficiency of an oocyte donation program using egg cryo-banking. Fertil Steril 92:520–526
5. Larman M et al RBM Online 2007 Citation, Borini A, Lagalla C, Bonu MA et al (2006) Clinical outcome of oocyte cryopreservation after slow cooling with a protocol utilizing a high sucrose concentration. Hum Reprod 21:512–517
6. De Santis L, Cino I, Rabellotti E et al (2007) Oocyte cryopreservation: clinical outcome of slow-cooling protocols differing in sucrose concentration. Reprod Biomed Online 14:57–63
7. Borini A, Lagalla C, Bonu MA et al (2006) Clinical outcome of oocyte cryopreservation after slow cooling with a protocol utilizing a high sucrose concentration. Hum Reprod 21:512–517
8. Cobo A, Kuwayama M, Perez S et al (2008) Comparison of concomitant outcome achieved with fresh and cryopreserved donor oocytes vitrified by the cryotop method. Fertil Steril 89:1657–1664
9. Cobo A, Diaz C (2011) Clinical application of oocyte vitrification: a systematic review and meta-analysis of randomized controlled trials. Fertil Steril 96:277–285

10. Bianchi V, Lappi M, MA B et al (2012) Oocyte slow freezing using a 0.2-0.3M sucrose concentration protocol: is it really the time to trash the cryopreservation machine? Fertil Steril 97:1101–1107
11. Cil AP, Bang H, Oktay K (2013) Age-specific probability of live birth with oocyte cryopreservation: an individual patient data meta-analysis. Fertil Steril 100:492
12. Cobo A, Perez S, MJ D l S, Zulatequi J, Domingo J, Remohi J Effect of different cryopreservation techniques on the metaphase II spindle in human oocytes. Reprod Biomed Online 17:350–359

# Chapter 15

# Development of a Nationwide Network for Ovarian Tissue Cryopreservation

**Jana Liebenthron and Markus Montag**

## Abstract

Ovarian tissue cryopreservation is gaining much interest since the publication of the first live birth after retransplantation of frozen-thawed tissue in 2004 (Donnez et al., Lancet 364:1405–1410, 2004). In contrast to cryopreservation of gametes and embryos, ovarian tissue freezing is a complex requiring a proper approach in order to make this a viable option for fertility preservation of cancer patients. Due to the need in terms of laboratory space, equipment, personnel, and adequate logistics, an ovarian tissue cryobank is most economic if managed as a centralized service unit that interacts with numerous clinics covering the surgical part. Transportation of ovarian tissue under appropriate conditions from the surgical unit to the cryobank for subsequent preparation and freezing has been shown to have no impact on cryo-survival (Schmidt et al., Hum Reprod 18:2654–2659, 2003; Isachenko et al., Fertil Steril 91:1556–1559, 2009). Several children have been born after retransplantation of such tissue that was derived from the cryobank in Bonn, Germany (Homepage FertiPROTEKT. http://www.fertiprotekt.de). This cryobank is one of the largest in the world with more than 1300 tissue samples that were frozen from 2003 until today. It is integrated in the network FertiPROTEKT (Homepage FertiPROTEKT. http://www.fertiprotekt.de) and is served by 108 surgical centers that are located all over Germany. The concept of this cryobank is a blueprint for success and has recently been used for another regionally centralized cryobank in Beijing, China. In this chapter the most important topics that need to be considered while creating a centralized cryobank within a national or regional network are highlighted.

**Key words** Fertility protection, Ovarian tissue, Cryopreservation, Transplantation, Cryobanking, Cancer

## 1 Introduction

Fertility protection in male cancer patients is well established; however, this does not hold true for the female side. Up to the year 2004, the public awareness of the possibilities to preserve fertility in women was scarce. Only few oncologists and gynecologists were aware of adequate therapeutic options. Until then this was mostly freezing of embryos generated by in vitro fertilization. However, especially young women do not necessarily live already in a stable partnership or are married; thus, the source of sperm to generate an

Zsolt Peter Nagy et al. (eds.), *Cryopreservation of Mammalian Gametes and Embryos: Methods and Protocols*, Methods in Molecular Biology, vol. 1568, DOI 10.1007/978-1-4939-6828-2_15, 

embryo was a major issue. Cryopreservation of oocytes in the pre-vitrification era was accomplished by slow freezing, and retrospectively it is obvious that it was not an established routine with acceptable success rates [1]. Few specialists did ovarian tissue freezing, but with no case that has been successfully published, the method was considered highly experimental and thought to hold more hope than a real promise [2]. Up to then they were usually highly educated women who actively asked for information on how they could preserve their fertility and tried by all means to get relevant information [3].

In 2004 the first publication of a case report that involved ovarian tissue freezing, retransplantation, and live birth by the group of Jacques Donnez in Belgium stirred global interest [4]. It was the foundation to shape the further development of this technology for fertility protection in women. Reproductive as well as oncological medical societies took up the topic and started to initiate task forces and interest groups in order to evaluate how ovarian tissue freezing could be brought into clinical practice—still under the assumption of being experimental and requiring more proof prior to further adoption on a broader scale [5].

It became evident though that fertility protection in women has several angles that all need to be considered, including the multidisciplinary approach especially in oncological cases but also the fact that ovarian tissue freezing is one treatment option among others [6]. Ovarian tissue freezing is often considered as being part of routine IVF technology; however, it is definitely not. Taking the example of Germany, the development of a nationwide network for ovarian cryopreservation will be shown.

## 2 Establishing a Network

The foundation of a network for ovarian cryopreservation in Germany was mainly influenced by two facts. One was the knowledge, which was available on ovarian tissue freezing based on research conducted at the Universities of Erlangen, Cologne, and Bonn [3, 7–9] and which resulted in the start of a cryobank with a clear focus on ovarian tissue freezing at the University of Bonn in 2003. The other was the foundation of a fertility protection network called FertiPROTEKT by the Universities of Bonn and Heidelberg [3, 6, 10]. The possibility to integrate the service of a national operating cryobank like the one in Bonn or a regional cryobank, like the one at the University of Erlangen, into such a network allowed every member of the network to offer ovarian freezing as a treatment option [10]. Furthermore, the professional service offered by such cryobanks made ovarian tissue freezing available to almost every woman who wanted to opt for this therapy.

### 2.1 Centralized Cryobanking Facility

The cryobank for ovarian tissue freezing at the University of Bonn was initiated on the base of personnel expertise on ovarian cryopreservation and on logistics in running such a cryobank as a centralized service [3, 9, 11–14]. The requirements are manifold and regulatory, and financial issues are in clear favor of having a centralized cryobanking facility—dependent on the expected volume. The calculations for Germany were based on the number of women that may be threatened by cancer in the reproductive age period. Based on a population of 80 million people, the corresponding figure is approximately 2000 women per year [10]. This figure should be considered as the number of consultations and not necessarily the number of initiated therapies and under the assumption that numerous therapies are offered—not only ovarian tissue freezing.

Some women will not go for any therapy, either because of the severity of their disease (i.e., leukemia) or because they already have a child or because they do not want to take the burden of a treatment, which has implications they cannot overlook at the time of their disease. Another factor is that not all women will be informed about the possibility to preserve fertility. In some cases the oncologists may not grant the time needed for the therapy; in other cases a gonadotoxic therapy was initiated in the past leaving an already impaired or interrupted reproductive function of the ovaries. And finally one needs to mention that there is a choice between different treatment options [6]. Ovarian tissue freezing is one, but others include hormonal stimulation followed by follicle puncture and freezing of oocytes or embryos, in vitro maturation [15], GnRH analogue treatment, and repositioning of ovaries to avoid gonadotoxic effects of radiotherapy [6].

All this results in the fact that up to 2014 the number of cases with ovarian tissue freezing amounted to 2080 in Germany and more than 60 % of the cases were covered by the central cryobanking facility at the University of Bonn and all remaining were mostly served by local cryobanks in various regions of the country [10].

### 2.2 Organizational Requirements

Organizing a centralized cryobanking facility requires a variety of topics that need to be considered. Ovarian tissue banking is not comparable to cryopreservation and storage of embryos in the context of an IVF treatment. For most women this tissue is the last chance to restore fertility by retransplantation. The storage period may easily exceed 10–15 years and hence the working live span of the team that initially started such a cryobank.

The most important topics are described below in more detail.

### 2.3 Laboratory Facilities, Material, and Equipment

Ovarian tissue will be used for retransplantation, and thus transplantation law will in most countries cover the procedure. Transplantation will be performed by stitching the tissue to a remaining part of an ovary or by placing tissue pieces into a

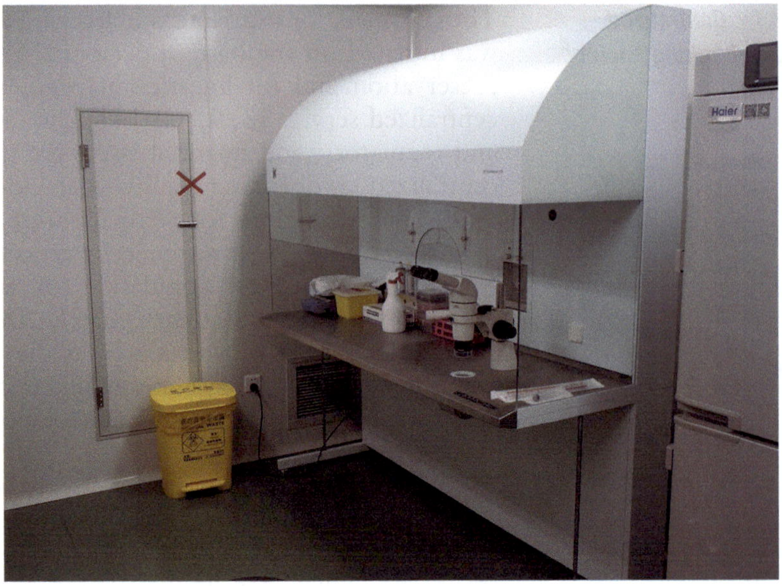


**Fig. 1** Workbench for preparation of material located within the tissue preparation clean room

peritoneal pocket [4, 13]. Both procedures require the highest standard of care in regard to sterility and tissue integrity.

Thus clean room environment should be considered by using a lamina flow bench specified as class A device and a background environment of class B/C for the tissue preparation room (Figs. 1–3).

Once the tissue pieces have been allocated into the cryotubes and these are closed, further requirements to special air quality during transport to the freezing facility and during the actual freezing procedure and the subsequent storage period are not required. Storage of frozen tissue in suitable storage containers is often done in the same room where the freezing device is located, as both, the storage tank and the freezer, require liquid nitrogen. The entire setup is shown in Fig. 4. Due to this the location has to be equipped with an oxygen sensor that constantly measures the oxygen concentration and is coupled to an acoustic and optical warning system outside the storage room, which signals low atmospheric oxygen and fatal conditions in the storage room. The floor in the freezing and storage room should be covered with a material that is inert to nitrogen in liquid or gaseous phase, i.e., stainless steel covers do serve that purpose very well and avoid cracks in the floor lining.

The freezing device should allow for controlled-rate freezing and possess the possibility for easy manual seeding or even automatic seeding [16, 17]. The freezer has to be connected to a suitable nitrogen dewar that provides liquid nitrogen for at least

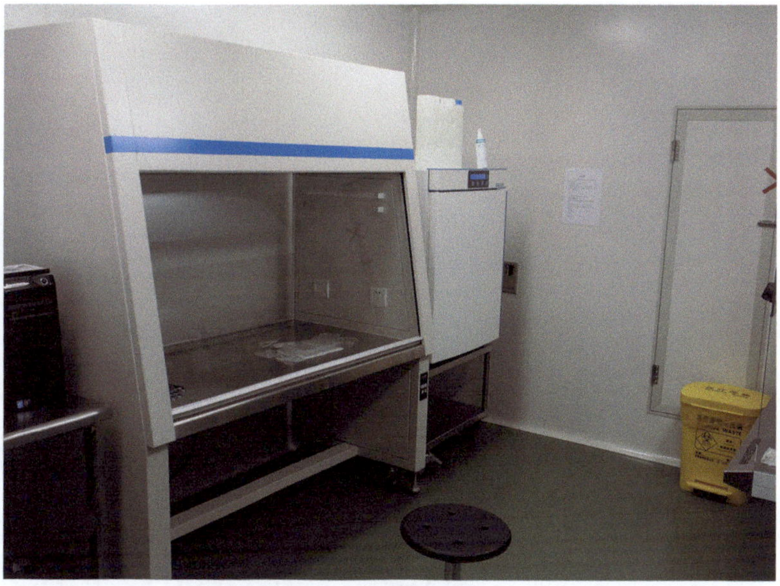

**Fig. 2** Class A lamina workbench for preparation of ovarian tissue pieces located inside the tissue preparation clean room

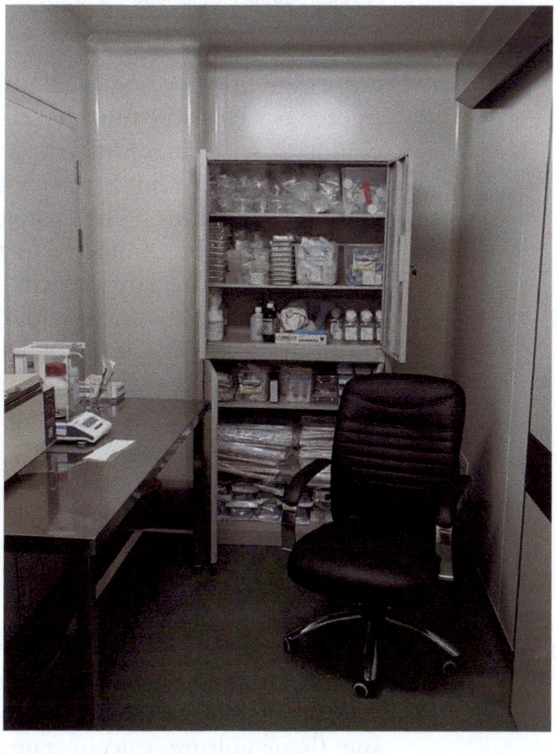

**Fig. 3** Storage facility located in a separate area within the clean room area but outside the tissue preparation room

**Fig. 4** Separate ovarian tissue freezing room within the clean room area. The freezing device is located on the *left side* and connected to a nitrogen supply tank. The cryotank for long-term storage (*right side*) is connected to a separate nitrogen supply tank located in the middle

more than one freezing cycle and that holds a sufficient reserve. The freezer as well as the unit that controls the freezing process must be equipped with an uninterrupted power supply that can support enough power to finish a cryopreservation project even with complete power failure. The storage container should be sufficiently large and must be connected to an automatic liquid nitrogen supply tank with an adequate capacity. As storage is expected to be long term, the equipment should be chosen accordingly. Economic consumption of liquid nitrogen is an important factor, and storage tanks that operate with storage in a gaseous phase offer a good deal. The storage tank should contain a build-in battery that enables to overcome power failure for up to 24 h at least. Despite this, the storage container should be connected to an external alarm system that notifies alarms and informs via modem.

All safety requirements have to be fulfilled, service agreements need to be in place, and regular service has to be documented.

### 2.4 Personnel

The personnel performing ovarian tissue freezing and thawing must be trained and experienced in sterile working, preparation of freezing media, ovarian tissue preparation, thawing of ovarian tissue, tissue culture, calcein staining technology, and histology. This can be best achieved by either hospitation at an ovarian tissue cryobank or by in-house training through an adequate specialist at the own equipment and setting. Furthermore, technical background knowledge like operation of the freezing device, the nitrogen supply, and storage tanks, microscopes, and lamina flow bench is required. All personnel should have an adequate background in medical or biomedical education according to national requirements and specifications. Continuous education and training are required in this fast-forward moving field.

### 2.5 Optimal Management of Tissue Transportation

One important aspect of centralized cryobanking is the transportation of the ovarian tissue to the cryobank for further processing, freezing, and storage [11, 13, 18]. This requires a suitable transportation system, adequate instructions, and proper logistics even for short in-house transportation from surgery to the tissue preparation unit of the cryobank. In-house transportation can be done using warm media (37 °C) and room temperature or in the cold (4 °C) provided a suitable (pH-stable) transportation medium, and a proper container that maintains a constant transportation temperature is used.

Long-distance transportation can be accomplished by overnight transport at 4–10 °C. This requires a suitable transportation container that maintains the required temperature and that can be locked and is isolated for temperature stability and to warrant shock resistance.

It is important to note that the feasibility of overnight transportation in regard to preserving tissue potential for achieving successful therapy outcome is proven [13, 19].

### 2.6 Getting Started Clinically

Once all requirements have been fulfilled, the equipment is installed and in place, safety issues are covered, and the personnel is trained accordingly, and the cryobank needs to run several experimental cycles prior to getting started with the first real clinical cases.

This should initially involve the freezing device, as it is important that the tissue pieces are frozen according to a defined temperature profile [16, 17]. How a defined temperature profile including proper seeding is obtained depends on the type of freezer. The setup is best accomplished by using cryo-solution in a reference cryotube without tissue and by comparing the temperature profile of the chamber with the profile that is recorded through the temperature-measuring device in the reference tube. The freezing curve needs to be reproducible in several runs, and only then freezing of tissue samples can be considered.

For experimental test runs, ovarian tissue can be obtained from patients that undergo ovarectomy and agree to donate part of the removed tissue for research via informed consent. The tissue should be exactly treated as if it would be a clinical case for real fertility preservation. Following simulation of proper in-house as well as external/overnight transportation, the tissue has to be prepared for freezing, pre-equilibrated in freezing solution, and loaded into cryotubes. The preparation process is shown in Figs. 5 and 6.

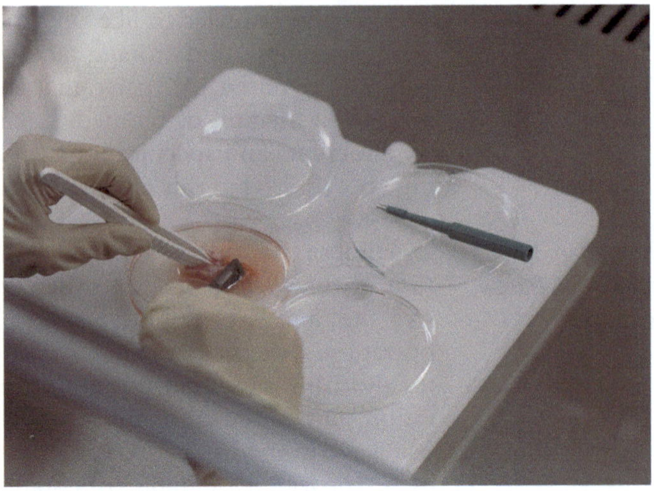

**Fig. 5** Preparation of the ovarian cortex within the class A lamina workbench. The preparation is performed on a cooling plate. Two dishes are for preparation and rinsing of the tissue (*bottom*), one dish is for the prepared tissue pieces for freezing (*top left*), and one dish is for the tissue pieces that are prepared for viability assay, which is performed before and after cryopreservation (*top right*)

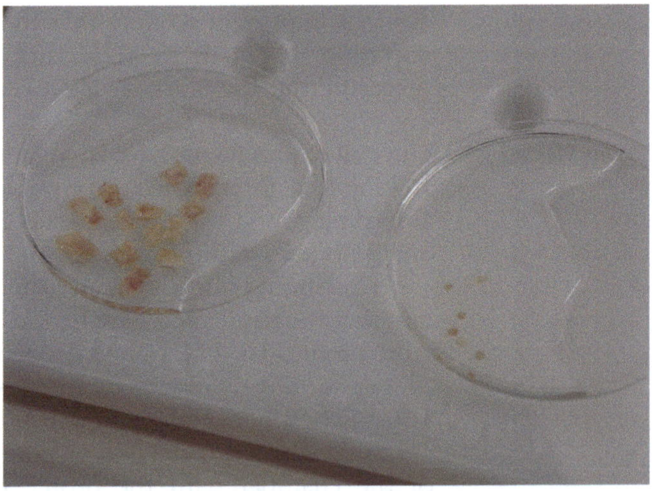

**Fig. 6** Cortical tissue stripes after final preparation and prior to freezing (*left dish*) and standard-sized cortical tissue pieces for viability staining (*right dish*)

Usually it is best to place the tubes with the tissue into the freezer and proceed to the start temperature (usually 2–4 °C) to finish the equilibration time. Following cooling to the seeding temperature, either manual or automatic seeding has to be initiated (Fig. 7), the increase of the temperature after seeding has to be verified, followed by a subsequent slow cooling slope at −0.3 °C per minute to −42 °C and final rapid cooling to −100 °C [16, 17]. Tubes can

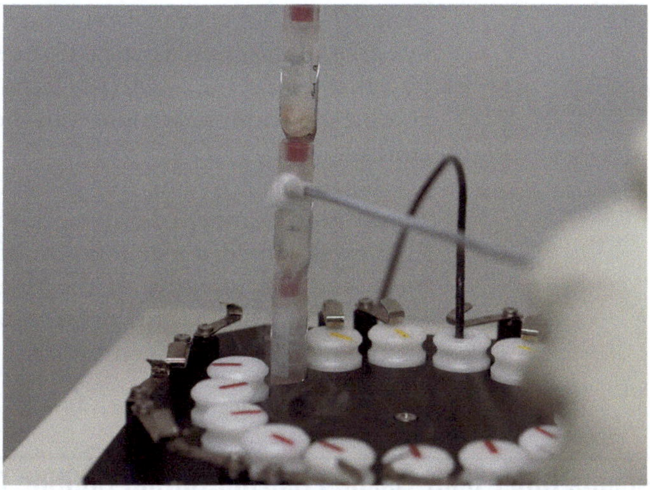

**Fig. 7** Manual seeding of cryotubes at the proper seeding temperature

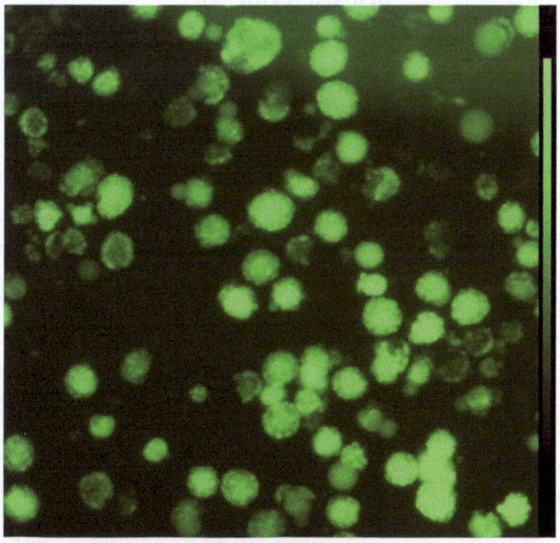

**Fig. 8** Viability stain of primordial follicles after combined collagenase digestion and calcein staining shown under fluorescence light

then be transferred into liquid nitrogen and into the appropriate storage container that are finally placed in the storage device. This has to be performed under optimal temperature control, so that the samples do not get warm. It is important to perform several test runs and to check the quality of the process. This is best accomplished with a viability assay like calcein staining comparing the follicle count and cellular integrity of a fresh part of the incoming tissue with a frozen-warmed tissue piece (Figs. 6 and 8) [19]. However, tissue samples should be stored for at least one day prior to thawing. The result of the viability assay, expressed by the rate of viable follicles after calcein staining, must not differ by more than 10–20 % between fresh and thawed tissue. In addition the integrity of the entire cortical tissue should be occasionally assessed by other assays (i.e., glucose/lactate assay; [19]). Only then the cryopreservation procedure can be considered as being under control.

Finally, at least one to two samples from ovaries with a sufficiently high number of viable primordial follicles must be successfully transplanted under the skin of ovariectomized female SCID mice, and the integrity of the tissue as well as the initiation of follicular growth in the tissue transplant has to be verified by histology [20]. Only then the tissue bank has all necessary proof to go live.

## 3 Aspects Other than Tissue Freezing During Routine Operation

Cryopreservation of ovarian tissue is only one aspect of a somehow holistic concept.

Prior to cooperation with a surgical unit, a contract should be signed that clarifies the responsibilities and duties of both parties.

Although it may sound strange, the cryobank should interact with the surgeons regarding tissue requirements for optimal results. The main issue is that bipolar scissors cannot be used as they damage up to 50 % of the tissue, especially the primordial follicles, which are located in the periphery (Fig. 9).

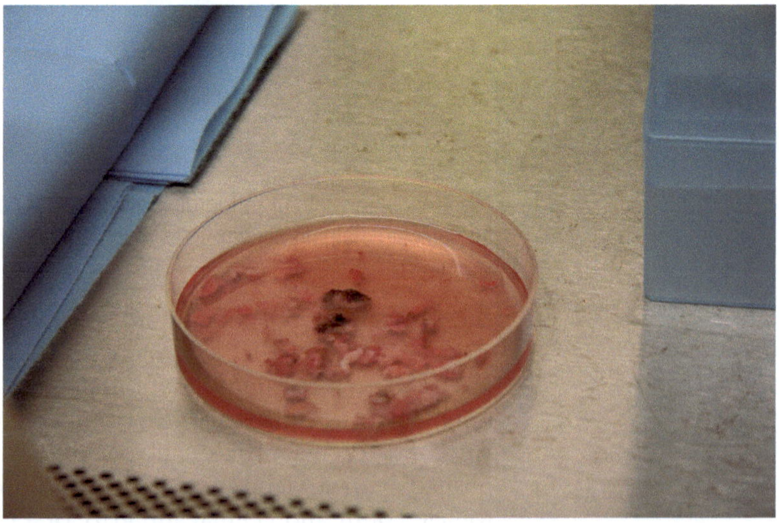

**Fig. 9** Ovarian cortex with partial burned/coagulated tissue (*dark part in the middle of the dish*) due to the use of bipolar scissors during excision of the tissue. All follicles located within this area as well as in the surrounding areas will be irreversibly damaged

Another advice is to retrieve tissue from the inactive ovary and not from the region with the corpus luteum, as follicle density in that part of the cortex is very low and the thin tissue is prone to rupture and not well suited for retransplantation.

Further, it is absolutely important to know the potential of the tissue—fresh as well as after freeze-thaw. The density of viable primordial follicles determines if the tissue can be stored and how many tissue pieces may be used later for retransplantation. So for every case, a patho-histological examination of a representative part of the biopsied tissue and a viability assay (life-dead assay) for fresh and frozen-thawed tissue must be performed. Assessing the AMH level prior to surgery and prior to retransplantation is suggested in order to generate data that may be used in the future for patient counseling, although the value of AMH after transplantation is under discussion [21]. All this should be initiated and coordinated by the cryobank, and all data should be stored in a database at the cryobank.

It should be also noted that an ovarian cryobank can coordinate adequate research that is related to all aspects of ovarian tissue. Patients do welcome "patient-related research" as they do see the need for further basic research and the shortcomings in our current knowledge if it comes to find the best modalities for protecting their fertility. Numerous such examples are available in the literature and document the important role of ovarian cryobanks in that regard [14, 19, 20, 22].

### 3.1 Key Performance Indicators

A professional cryobank service has to include KPIs for the temperature control during transport with a maximum drop between the start temperature and the incoming temperature of the transportation medium. The viability count of follicles in the fresh tissue must be used in relation to the viability count after a freeze-thaw procedure and should not differ by more than 10–20 %. Prior to routine large-scale clinical kickoff, tissue integrity and the potential to allow for growth of primordial follicles of retransplanted tissue into SCID mice must be proven [20].

### 3.2 Quality Management of a Cryobank

Quality management of an ovarian cryobank includes that the relevant authorities approve the cryobank. All procedures that are used have to be defined as SOPs in a handbook and cooperation contracts with surgical units, and informed consent forms and contracts with patients have to be prepared. Medium used for cryopreservation has to be tested and evaluated prior to clinical use. All procedures and lot numbers of material and consumables used must be documented. If local transplantation law specifies detailed conditions, these need to be followed. In general ovarian tissue banking should in all aspects follow GMP guidelines. Constant records of assays for tissue integrity and viability have to be kept. It is crucial to perform statistical analysis of patient-related data and success rates of the freeze-thaw procedures as well as of retransplantation, including the outcome and the health of the children born. Last but not least, an analysis of current receipts and expenses has to be done to document and determine long-term calculations.

### 3.3 Documentation and Patient Management

A centralized cryobank service will require proper patient documentation and management. This includes documentation of the medical record of the patient, contract details, as well as informed consent. For the cryopreservation process, the temperature profile during freezing protocol and the storage information are mandatory. Adequate software solutions are commercially available from several sources, and these basics are essential for running such a cryobank. The most important factor is the contact to the patients. To our experience, women having cryopreserved ovarian tissue are highly sensitized toward the value of their tissue and all aspects that relate to safe storage and professionalism in interaction [3, 23]. Patients do welcome a yearly reminder for prolonging the storage period and have the impression that the cryobank personnel do take care for them. Although such a concept requires administrative work, it helps to reduce dropout rates in regard to patient contact. The cryobank in Bonn receives from almost more than 95 % of all patients an annual feedback. And for some patients, the information that they are deceased is also obtained.

## 4 Examples from the Cryobank Bonn, Germany

### 4.1 Cryopreservation

From 2000 to 2015, 1300 ovarian biopsies were stored after cryopreservation in the cryobank in Bonn. The mean age of patients who cryopreserved their tissue for fertility protection is 28 years (2–45 years). The number of prepubertal tissue (girls aged 2–12 years and without menstrual cycle at the time of harvesting) is 13 biopsies. The majority of patients of the Cryobank Bonn are women diagnosed with breast cancer, followed by female patients diagnosed with Hodgkin lymphoma (Table 1).

On average, nine cortical tissue pieces (sized $8 \times 4 \times 1$ mm$^3$) were prepared and cryopreserved. Most of the ovarian cortical biopsies that are stored at the Cryobank Bonn were obtained from collaboration partners, mainly clinics in Germany and Austria. The transportation of tissue to the cryobank is organized via special transplantation media and transportation boxes, which are able to maintain temperature conditions between 4 and 8 °C [19]. The transportation equipment including the medium is provided by the Cryobank Bonn. After surgical retrieval of the cortical tissue, the return transportation to the cryobank has to be organized by the surgical clinic, mostly by overnight transportation. The logistic aim is to get the cortical tissue on the following day at 9 a.m. at the

**Table 1**
**Overview of cryopreserved ovarian biopsies at the Cryobank Bonn that are stored for fertility protection**

| | | |
|---|---|---|
| Ratio of diagnoses | Total number: 1300 | Patients with breast cancer: 624<br>Patients with Hodgkin lymphoma: 397<br>Patients with other malignant diseases: 156<br>Patients with other benign diseases: 20<br>Patients with unknown diagnoses: 103 |
| Ratio of overnight transportation procedures | Total number: 1125 | Optimal (4–8 °C): 1029<br>Not optimal (>8 °C): 91<br>Not optimal (<2 °C): 5 |
| Ratio of direct transportation procedures | Number of ovarian tissue biopsies obtained in Bonn: 124<br>Number of ovarian tissue biopsies obtained in a collaborating clinic: 51 | Optimal (4–8 °C): 51 |
| Ratio of tested ovarian tissue biopsies before cryopreservation (since 2008) | Total number: 1200 | Viable: 1147<br>Nonviable: 7<br>Reduced viability: 8<br>No results after testing: 8<br>Not analyzed: 30 |

Cryobank Bonn. In the majority of cases, the transportation procedure was realized within the specified temperature and time frame (Table 1).

Since 2008, a viability test based on calcein staining is routinely performed to ensure that the incoming tissue is not damaged during laparoscopy or during transportation. The results until today are also shown in Table 1. Since 2013 the viability test was refined, and the potential of the ovarian cortex is since then analyzed with standardized three 2 mm biopsies prepared from different parts of the cortical tissue. These are stained with calcein and subjected to collagenase digest under defined conditions to determine the primordial follicle count, which is representative of the potential of the ovarian cortex.

### 4.2 Thawing and Retransplantation

According to the homepage of the network FertiPROTEKT, 43 female patients have been documented that received a laparoscopic retransplantation of ovarian cortical tissue in Germany, Austria, and Switzerland during 2008 and 2013 [10]. Ovarian cortex pieces were predominantly retransplanted into a peritoneal pocket. From 24 out of these 43 patients, the tissue was initially cryopreserved at the Cryobank Bonn, 22 after overnight transportation, and two after direct transportation on the same day. Fourteen of these 43 documented retransplantations were accompanied through the Cryobank Bonn: two retransplantations were performed at the University of Bonn and 12 at the same seven collaborating clinics, which initially retrieved the ovarian tissue prior to gonadotoxic therapy. The reproductive biologist of the Cryobank Bonn performed the thawing procedure. From these 24 cases, nine pregnancies were achieved in eight patients after thawing, and the first retransplantation resulted in six healthy babies, and another pregnancy ended in a miscarriage. Four pregnancies were still ongoing in June 2015. The mean age of the patients at the time of cryopreservation of ovarian tissue was 31 years, and the mean age at the time of retransplantation is 35 years.

By the time of April 2015, the number of cases with active participation of the Cryobank Bonn during retransplantation have doubled within two years ($n = 27$, four cases in Bonn, 23 in eight cooperation clinics); in Germany, Austria, and Switzerland, 84 cases are documented [24]. The success rates in Germany, Austria, and Switzerland amount to 15 pregnancies from 13 patients after the first retransplantation; eight healthy babies were born and five pregnancies are ongoing. One pregnancy ended in a miscarriage and one pregnancy was documented as biochemical pregnancy. The success rates of the Cryobank Bonn are shown in Table 2.

**Table 2**
**Success rates of the Cryobank Bonn**

| Success rates after retransplantation from the Cryobank Bonn (updated: April 2015) | | | | |
|---|---|---|---|---|
| **No.** | **Diagnosis** | **Retransplantation** | **Patient age at surgery/at retransplantation** | **Outcome** |
| *1* | *Hodgkin lymphoma* | *06/2010* | *27/32* | *Delivery 10/2011* |
| *2* | *Hodgkin lymphoma* | *06/2012* | *35/37* | *Delivery 12/2013* |
| ***3*** | ***Breast cancer*** | ***08/2012*** | ***36/37*** | ***Delivery 11/2013*** |
| *4* | *Hodgkin lymphoma* | *10/2013* | *33/37* | *Delivery 12/2014* |
| ***5*** | ***Breast cancer*** | ***12/2013*** | ***33/36*** | ***Delivery 12/2014*** |
| *6* | *Fibrosis of ovarian cortex* | *07/2013* | *20/27* | *Delivery 04/2015* |
| **7** | **Breast cancer** | **09/2014** | **?/38** | **Ongoing pregnancy** |
| **8** | **Breast cancer** | **07/2014** | **30/36** | **Ongoing pregnancy** |
| *9* | *Hodgkin lymphoma* | *12/2012* | *22/29* | *Ongoing pregnancy* |
| *10* | *Breast cancer* | *05/2013* | *28/34* | *Miscarriage* |
| *11* | *Breast cancer* | *05/2013* | *28/34* | *Ongoing pregnancy* |

All cases in black were cryopreserved at the Cryobank Bonn after overnight transportation (1–6, 9, and 10/11). Retransplantations that were accompanied by personnel from the Cryobank Bonn are marked in bold (Nos. 3, 5, 7, and 8). Cases written in black cursive were cryopreserved at the Cryobank Bonn and retransplanted in external collaborating clinic (Nos. 1, 2, 4, 6, 9, and 10/11). Nos. 10 and 11 are the same patient, who achieved two pregnancies after the first retransplantation, where the first ended as a miscarriage

## 5 Knowledge and Experience Transfer: Creating a New Cryobank

The concept of the ovarian tissue bank at the University of Bonn has been transferred to a recently created new cryobank at the Gynecology and Obstetrics Hospital of the Capital Medical University of Beijing, China. The process was completely supervised by external experts (JL and MM) and involved blueprint drafts for the clean room laboratories (tissue preparation room, storage room) and an adjacent documentation room, a complete list for the required equipment, training of personnel at the University of Bonn and Heidelberg, on-site training in Beijing with the equipment, and technical properties of the local environment. The first clinical cases were done under guidance of the external experts in January 2015. The cryobank in Beijing will initially serve the local area, which comprises more than 20 million people, and then expand to nearby regions.

## 6 Conclusions

Centralized cryobanking for ovarian tissue freezing in conjunction with a national or regional network is no longer fiction but reality. Transportation of ovarian tissue overnight can be carried out under appropriate conditions and full control. The increasing number of reports on successful retransplantation of cryopreserved-thawed ovarian tissue—both from in-house sources and from transported tissue—is encouraging. A centralized approach that is based on a reasonable number of ovarian cryobank facilities is a major backbone of national as well as international fertility protection registries that have evolved over the last few years. The responsible implementation of this concept may soon allow changing the classification of this treatment option from "experimental" to "implementation into clinical routine." Transferring cryobanking concepts from one location and region to another is feasible and will further enhance the spreading of this technology for the health benefit of cancer patients.

## Acknowledgment

The authors acknowledge Prof. Xiangyang Ruan (Capital Medical University of Beijing) for sharing the common experience on knowledge transfer and Prof. Hans van der Ven (University of Bonn) and Prof. Thomas Strowitzki (University of Heidelberg) for their constant support of educating and training others.

## References


1. Oktay K, Cil AP, Bang H (2006) Efficiency of oocyte cryopreservation: a meta-analysis. Fertil Steril 86:70–80
2. Gosden RG (2000) Low temperature storage and grafting of human ovarian tissue. Mol Cell Endocrinol 163:125–129
3. Wollenschein M, van der Ven K, Montag M, Rohde A (2008) Protection of fertility in female cancer patients: Cryobank Bonn- consultation, implementation and the impact on affected women. Geburtshilfe Frauenheilkd 68:262–267
4. Donnez J, Dolmans MM, Demylle D et al (2004) Livebirth after orthotopic transplantation of cryopreserved ovarian tissue. Lancet 364:1405–1410
5. ESHRE Task Force on Ethics and Law (2004) Taskforce 7: ethical considerations for the cryopreservation of gametes and reproductive tissues for self use. Hum Reprod 19:460–462
6. von Wolff M, Montag M, Dittrich R et al (2011) Fertility preservation in women—a practical guide of preservation techniques and therapeutic strategies in breast cancer, hodgkin's lymphoma and borderline ovarian neoplasms. Arch Gynecol Obstet 284:427–435
7. Maltaris T, Dragonas C, Hoffmann I et al (2006) Simple prediction of the survival of follicles in cryopreserved human ovarian tissue. J Reprod Dev 52:577–582
8. Isachenko E, Isachenko V, Rahimi G, Nawroth F (2003) Cryopreservation of human ovarian tissue by direct plunging into liquid nitrogen. Eur J Obstet Gynecol Reprod Biol 108:186–193
9. Isachenko V, Montag M, Isachenko E et al (2006) Effective method for in-vitro culture of cryopreserved human ovarian tissue. Reprod Biomed Online 13:228–234
10. Homepage FertiPROTEKT. http://www.fertiprotekt.de
11. Isachenko E, Isachenko V, Nawroth F et al (2009) Effect of long-term exposure at suprazero temperatures on activity and viability of human ovarian cortex. Fertil Steril 91:1556–1559
12. Von Wolff M, Donnez J, Hovatta O et al (2009) Cryopreservation and autotransplantation of human ovarian tissue prior to cytotoxic therapy—a technique in its infancy but already successful in fertility preservation. Eur J Cancer 45:1547–1553
13. Dittrich R, Lotz L, Keck G et al (2012) Live birth after ovarian tissue autotransplantation following overnight transportation before cryopreservation. Fertil Steril 97:387–390
14. Liebenthron J, Köster M, Drengner C et al (2013) The impact of culture conditions on early follicle recruitment and growth from human ovarian cortex biopsies in vitro. Fertil Steril 100:483–491
15. Von Wolff M, Strowitzki T, van der Ven H, Montag M (2006) In vitro maturation is an efficient technique to generate oocytes and should be considered in combination with cryopreservation of ovarian tissue for preservation of fertility in women. J Clin Oncol 24:5336–5337
16. Isachenko V, Isachenko E, Reinsberg J et al (2007) Cryopreservation of human ovarian tissue: comparison of rapid and conventional freezing. Cryobiology 55:261–268
17. Isachenko V, Isachenko E, Reinsberg J et al (2008) Cryopreservation of human ovarian tissue: effect of spontaneous ice formation. Reprod Biomed Online 55:336–345
18. Schmidt KL, Ernst E, Byskov AG et al (2003) Survival of primordial follicles following prolonged transportation of ovarian tissue prior to cryopreservation. Hum Reprod 18:2654–2659
19. Bastings L, Liebenthron L, Westphal JR et al (2014) Efficacy of ovarian tissue cryopreservation in a major European center. J Assist Reprod Genet 31:1003–1012
20. Lotz L, Liebenthron L, Nichols-Burns SM et al (2014) Spontaneous antral follicle formation and metaphase II oocyte from a non-stimulated prepubertal ovarian tissue xenotransplant. Reprod Biol Endocrinol 12:41
21. Janse F, Donnez J, Anckaert E et al (2011) Limited value of ovarian function markers following orthotopic transplantation of ovarian tissue after gonadotoxic treatment. J Clin Endocrinol Metab 96:1136–1144
22. Lotz L, Montag M, van der Ven H et al (2011) Xenotransplantation of cryopreserved ovarian tissue from patients with ovarian tumors into SCID mice—no evidence of malignant cell contamination. Fertil Steril 95:2612–2614
23. Wollenschein M, Montag M, van der Ven K et al (2008) Cancer female patients with desire to have children: significance of cryopreservation of the ovary tissues: an example of Cryobank Bonn. Geburtshilfe Frauenheilkd 68:188–189
24. Liebenthron J, Dittrich R, Toth B et al. (2015) Orthotopic ovarian tissue transplantation—results in relation to experience of the transplanting centers, overnight tissue transportation and transplantation into the peritoneum. Hum Reprod. Abstract book ESHRE in Press—to be updated at proof stage

# Part IV

## Advances in Cryotechnology, Research, and Animal Models

# Chapter 16

# Directional Freezing of Ovarian Tissue and Freeze-Drying of Stem Cells for Fertility Preservation

**Amir Arav and Yehudit Natan**

## Abstract

Fertility preservation is practiced today for women in their fertile years who wish or need to postpone or prolong their reproductive era. Currently women have two main options to preserve their fertility either by oocyte vitrification or by freezing of ovarian cortical slices. In this chapter we will describe the use of directional freezing for cortical slices and freeze-drying of stem cells. These procedures open the way for the preservation of reproductive cells and tissue for future use in fertility preservation treatments.

**Key words** Fertility preservation, Cryopreservation, Freezing, Directional freezing, Freeze-drying, Ovary, Stem cells, Cortical slices

## 1 Introduction

Fertility preservation is practiced today for women in their fertile years who for medical reasons (e.g., undergoing chemotherapy or radiation) or lifestyle reasons (career choices, finding a partner) wish to postpone their pregnancy [1–5].

The main way to preserve women's fertility is either by cryopreserving ovarian cortical slices or by vitrification of oocytes [6–10].

There are three methods for cryopreservation: slow freezing, directional freezing, and vitrification. Vitrification is a process in which ice crystal formation is avoided. This occurs when the viscosity of the sample is high (using high concentrations of cryoprotectants (CPs)), the volume is very low, and the cooling rate is very high (at the range of 20,000°C/min) [11]. Vitrification is done by exposing the tissue to high concentration of CPs and rapidly cooling into liquid nitrogen (LN).

In conventional slow freezing, the tissue is cooled to temperature below the freezing point with or without seeding and then slowly frozen in a controlled cooling rate to a temperature between −35°C and −130°C before plunging into LN [12].

Zsolt Peter Nagy et al. (eds.), *Cryopreservation of Mammalian Gametes and Embryos: Methods and Protocols*, Methods in Molecular Biology, vol. 1568, DOI 10.1007/978-1-4939-6828-2_16, © Springer Science+Business Media LLC 2017

The alternative technique to slow cooling is known as directional solidification or directional freezing; after the initial seeding stage, the sample is advanced at a constant velocity through a linear temperature gradient. This allows precise control over the freezing process, thus facilitating very accurate and homogenous cooling of the sample. The constant sample propagation also results in continuous seeding, thus controlling ice crystal growth in a direction opposite to the direction in which the sample is moved down the temperature gradient. This directional growth of ice crystals produces lamellae between which the cells in the sample are trapped. Dissipation of heat from the sample is achieved primarily in two ways. The highly heat-conductive metal block snugly surrounding the sample efficiently removes heat away as it freezes. Because of the temperature gradient and the directional movement of the sample, heat also continuously moves away from the ice front into the unfrozen fraction of the sample that is still located at the higher temperature end of the gradient. This controlled and directional heat dissipation protects the ice front from melting, thus avoiding damages to the cells. All these processes result in highly controlled and "cell-friendly" ice crystal morphology that contributes to reducing mechanical damages [13–15].

The efficient heat removal and controlled ice crystal propagation make it possible to use this technology to freeze small and large volumes alike [16].

Cryopreservation of ovarian tissue is now a well-established method for preserving female fertility [17, 18]. This can be achieved through the excision and banking of the entire organ or of fragments of the cortical region. However, whereas no live births or pregnancies have been achieved following the cryopreservation of whole human ovaries, 24 live babies were obtained after transplantation of frozen-thawed ovarian cortical fragments [19]. This clearly suggests that the use of cortical fragments is the method of choice even if, at present, the average functional life span of these samples ranges from 2 to 5 years [19, 20]. We have recently performed a direct comparison of sheep ovarian tissue viability between organs cryopreserved with a convectional programmable slow-freezer or a directional freezing apparatus [21]. Our results indicate that ovarian structures and functions were significantly better preserved by the directional freezer.

Freeze-drying of cells is the holy grail many scientists wish for the field of cryopreservation of cells. It will allow the long-term storage of cell and tissue samples at elevated temperatures with no need for freezers or cryogenic tanks, thus simplifying storage and transportation of samples as well as lowering costs.

Scientists have achieved to successfully freeze-dry sperm cells [22, 23] and mononuclear cells from umbilical cord blood which were able to grow and differentiate after rehydration with water [24] (*see* Fig. 1). A recent development showed that stem cells can differentiate into gametes (spermatozoa and oocytes) [25].

**Fig. 1** Freeze-dried human MNC derived from umbilical cord blood before and after rehydration with DDW

Therefore, the ability to preserve stem cells in a dry state will be of great importance to the field of fertility preservation as well as other regenerative medicine applications.

## 2 Materials

All solution should be prepared at room temperature unless otherwise stated and with analytical grade reagents.

### 2.1 Saline Solution

Nine grams (9 g) of sodium chloride is dissolved in 100 ml of ultrapure water (prepared by purifying deionized water to attain a sensitivity of 18 MΩ cm at 25°C) (*see* **Note 1**).

### 2.2 Freezing Solution for Cortical Slices

Ten percent (v/v) of fetal bovine serum (FBS) and 10 % (v/v) of dimethyl sulfoxide (DMSO) in medium 199-Hepes solution (*see* **Note 2**).

### 2.3 IMT-2 Freezing Solution for Freeze-Drying

0.945 mg/ml epigallocatechin gallate (EGCG) having >98 % purity and 0.1M trehalose dissolved in $Ca^{2+}$- and $Mg^{2+}$-free PBS (*see* **Note 3**).

### 2.4 Thawing Solutions

Three thawing solutions are used; all three are based on the medium used for the freezing solution, medium 199-Hepes solution, supplemented with 0.25 M, 0.125 M, and 0 M sucrose.

### 2.5 Rehydration Solution

Rehydration was done either by double-distilled water (DDW) or by medium 199.

## 3 Methods

All work is done at room temperature unless written otherwise. In addition, when working with liquid nitrogen, make sure to follow liquid nitrogen safety guides.

### 3.1 Cortical Slice Preparation

Sheep female reproductive tracts collected at the local slaughterhouse are transported to the laboratory in cold (0–4°C) saline solution.

In the laboratory cortical fragments (10 × 5 × 1 mm) are sliced, using forceps and scalpel, from the cortical region and immersed into the freezing solution described in Subheading 2.2 above.

### 3.2 Directional Freezing Procedure for Cortical Slices

1. Cortical samples prepared as described above are inserted into a 16 mm diameter, 80 mm long, glass test tube containing about 10 ml of the same freezing solution.
2. Close the samples using a special silicon cup (Fertilesafe Ltd., Israel) (*see* **Note 4**).
3. Prepare the directional freezing apparatus MTG-1314 (Fertilesafe, Ltd., Israel). The two thermal blocks were set as follows: the first block was set to 4 °C and the second block was set to −10 °C and −70 °C, respectively, thereby imposing a temperature gradient. Freezing tubes were pushed along the thermal gradient, and the velocity was set to 0.01 mm/s resulting in a cooling rate of 0.3 °C/min (*see* **Note 5**).
4. At the end of the freezing process, remove the samples (using long forceps) from the device and plunge them into liquid nitrogen for storage (*see* **Note 6**).

### 3.3 Thawing Procedure

1. Remove samples from liquid nitrogen storage tank using long forceps (*see* **Note** 7).
2. Insert the test tubes into a water bath heated to 68°C, and make sure the cup is left outside the water bath, for 20 s while moving the test tubes from side to side in the water bath.
3. After 20 s put the sample into a water bath warmed to 37 °C for 2 min until samples are completely thawed.

### 3.4 Preparation of Stem Cell Samples for Freeze-Drying

1. Stem cells are harvested according to each lab procedure. When we worked with mononuclear cells (MNC) from human umbilical cord blood, cells were isolated using histopaque-1077 solution and then washed twice using PBS. To the received pellet, we then added 2 ml of IMT-2 (*see* Subheading 2.3 for preparing this solution).
2. The cell suspension is then transferred into a 16 mm diameter, 80 mm long, glass test tubes (*see* **Note 4** for further description of the test tubes). Test tubes are closed using a special silicon cup. Make sure samples are protected from light, e.g., using an aluminum foil or a dark box.

### 3.5 Directional Freezing Procedure for Freeze-Drying Stem Cells

1. Stem cell samples prepared as described above are inserted into a 16 mm diameter, 80 mm long, glass test tube containing about 10 ml of the same freezing solution.
2. Prepare the directional freezing apparatus MTG-1314 (Fertilesafe, Ltd., Israel); two thermal blocks are set as follows: the first block was set to 5 °C, and the second block was set to −10 °C and −70 °C, respectively, thereby imposing a temperature gradient. Freezing tubes were pushed along the thermal gradient, and the velocity was set to 0.2 mm/s resulting in a cooling rate of 5.1 °C/min, and samples are rotated during freezing at 60 RPM. When the samples enter the cold block (set to −10 °C), seeding is performed by holding the tip of the test tubes inside the second block for 30 s before allowing to continue at 0.2 mm/s (*see* **Note 8**).
3. At the end of the freezing process, remove the samples (using long forceps) from the device and plunge them into liquid nitrogen for storage or open the test tubes and put inside the lyophilizer.

### 3.6 Drying and Rehydration of Stem Cells

1. Prepare the lyophilizer device according to user guide manual. We used the VirTis Genesis (Millrock Technology Inc., Kingston, NY, USA) lyophilizer. The device was set to a shelf temperature of −50 °C and vacuum was set to 5 mTorr.
2. Remove test tubes from LN tanks or from the freezing device and remove the silicon cups to open the test tubes (*see* **Note 9**).
3. Wrap each test tube opening with gauze strip and insert samples into the lyophilizer (already cold and ready for samples).
4. Samples are left to dry for 24–72 h.
5. After the above time period, remove samples and rehydrate with 1.9 ml of DDW or medium 199 pre-warmed to 37 °C.

## 4 Notes

1. Any saline for injection that is purchased can be used as well.
2. Instead of the 199-Hepes medium, other media can be used such as Leibovitz L-15 medium, 1314-RPMI, or University of Wisconsin (UW) solution.

   After the solution is ready, it should be kept protected from light, e.g., using an aluminum foil.
3. The glass test tubes are made of quartz glass having a thickness of 1 mm and a specially designed silicon cap (Fertilesafe, Ltd., Israel).
4. When inserting the cortical slices to the test tube, some freezing solution should be there prior to the insertion (about

1–2 ml), and after inserting the cortical slices, the test tube should be filled with the freezing solution as much as possible.

5. The freezing apparatus should be prepared ahead in order to minimize the time samples that are waiting in the freezing solution to be frozen. In addition, the freezing tubes containing the cortical slices should wait in a refrigerator (or in the first block of the MTG-1314 device if they have reached 4 °C) until the temperatures of the freezing device are set and the device is ready for the freezing process.
6. The MTG-1314 can freeze simultaneously up to five test tubes at the same freezing cycle.
7. When holding the test tubes upon removal from LN tank, be sure that the cup is facing downward to allow access LN that enters the test tube to evaporate.
8. The MTG-1314 device has LCD screen controlled in which all freezing parameters are entered (block temperature, sample velocity, RPM, and seeding waiting time) according to the device user manual guide.
9. Opening of the samples should be performed in nitrogen vapor, i.e., above LN, in order to avoid warming of the samples.

## References

1. Cobo A, Domingo J, Pérez S, Crespo J, Remohí J, Pellicer A (2008) Vitrification: an effective new approach to oocyte banking and preserving fertility in cancer patients. Clin Transl Oncol 10(5):268–273
2. Gosden R (2011) Cryopreservation: a cold look at technology for fertility preservation. Fertil Steril 96(2):264–268
3. Silber SJ (2011) Ovary cryopreservation and transplantation for fertility preservation. Mol Hum Reprod 18(2):59–67
4. Silber SJ (2009) Fresh ovarian tissue and whole ovary transplantation. Semin Reprod Med 27 (6):479–485
5. Stoop D, Cobo A, Silber S (2014) Fertility preservation for age-related fertility decline. Lancet 384(9950):1311–1319
6. Andersen CY, Silber SJ, Bergholdt SH, Jorgensen JS, Ernst E (2012) Long-term duration of function of ovarian tissue transplants: case reports. Reprod Biomed Online 25(2):128–132
7. Cobo A, Garcia-Velasco JA, Domingo J, Remohí J, Pellicer A (2013) Is vitrification of oocytes useful for fertility preservation for age-related fertility decline and in cancer patients? Fertil Steril 99(6):1485–1495
8. Nagy ZP, Chang CC, Shapiro DB, Bernal DP, Elsner CW, Mitchell-Leef D, Toledo AA, Kort HI (2009) Clinical evaluation of the efficiency of an oocyte donation program using egg cryobanking. Fertil Steril 92(2):520–526
9. Noyes N, Labella PA, Grifo J, Knopman JM (2010) Oocyte cryopreservation: a feasible fertility preservation option for reproductive age cancer survivors. J Assist Reprod Genet 27 (8):495–499
10. Rienzi L, Cobo A, Paffoni A, Scarduelli C, Capalbo A, Vajta G, Remohí J, Ragni G, Ubaldi FM (2012) Consistent and predictable delivery rates after oocyte vitrification: an observational longitudinal cohort multicentric study. Hum Reprod 27(6):1606–1612
11. Arav A (1992) Vitrification of oocyte and embryos. In: Lauria A, Gandolfi F (eds) New trends in embryo transfe. Portland press, Cambridge, pp 255–264
12. Morris GJ, Acton E, Faszer K, Franklin A, Yin H, Bodine R, Pareja J, Zaninovic N, Gosden R (2006) Cryopreservation of murine embryos, human spermatozoa and embryonic stem cells using a liquid nitrogen-free, controlled rate freezer. Reprod Biomed Online 13 (3):421–426

13. Arav A, Gavish Z, Elami A, Natan Y, Revel A, Silber S, Gosden RG, Patrizio P (2010) Ovarian function 6 years after cryopreservation and transplantation of whole sheep ovaries. Reprod Biomed Online 20:48–52
14. Arav A, Natan D (2012) Directional freezing of reproductive cells and organs. Reprod Domest Anim 47(Suppl 4):193–196
15. Arav A, Natan Y (2009) Directional freezing: a solution to the methodological challenges to preserve large organs. Semin Reprod Med 27:438–442
16. Arav A, Revel A, Nathan Y, Bor A, Gacitua H, Yavin S, Gavish Z, Uri M, Elami A (2005) Oocyte recovery, embryo development and ovarian function after cryopreservation and transplantation of whole sheep ovary. Hum Reprod 20:3554–3559
17. Donnez J, Dolmans MM (2013) Fertility preservation in women. Nat Rev Endocrinol 9 (12):735–749
18. Grynberg M, Poulain M, Sebag-Peyrelevade S, le Parco S, Fanchin R, Frydman N (2012) Ovarian tissue and follicle transplantation as an option for fertility preservation. Fertil Steril 97(6):1260–1268
19. Donnez J, Dolmans MM, Pellicer A, Diaz-Garcia C, Sanchez Serrano M, Schmidt KT, Ernst E, Luyckx V, Andersen CY (2013) Restoration of ovarian activity and pregnancy after transplantation of cryopreserved ovarian tissue: a review of 60 cases of reimplantation. Fertil Steril 99(6):1503–1513
20. Bromer JG, Patrizio P (2009) Fertility preservation: the rationale for cryopreservation of the whole ovary. Semin Reprod Med 27 (6):465–471
21. Maffei S, Pennarossa G, Brevini TA, Arav A, Gandolfi F (2014) Beneficial effect of directional freezing on in vitro viability of cryopreserved sheep whole ovaries and ovarian cortical slices. Hum Reprod 29(1):114–124
22. Liu JL, Kusakabe H, Chang CC, Suzuki H, Schmidt DW, Julian M, Pfeffer R, Bormann CL, Tian XC, Yanagimachi R, Yang X (2004) Freeze-dried sperm fertilization leads to full-term development in rabbits. Biol Reprod 70 (6):1776–1781
23. Kusakabe H, Yanagimachi R, Kamiguchi Y (2008) Mouse and human spermatozoa can be freeze-dried without damaging their chromosomes. Hum Reprod 23(2):233–239
24. Natan D, Nagler A, Arav A (2009) Freeze-drying of mononuclear cells derived from umbilical cord blood followed by colony formation. PLoS One 4(4):e5240
25. Easley CA IV, Phillips BT, McGuire MM, Barringer JM, Valli H, Hermann BP, Simerly CR, Rajkovic A, Miki T, Orwig KE, Schatten GP (2012) Direct differentiation of human pluripotent stem cells into haploid spermatogenic cells. Cell Rep 2(3):440–446

# Chapter 17

# Sterile Plate-Based Vitrification of Adherent Human Pluripotent Stem Cells and Their Derivatives Using the TWIST Method

Julia C. Neubauer, Frank Stracke, and Heiko Zimmermann

## Abstract

Due to their high biological complexity, e.g., their close cell-to-cell contacts, cryopreservation of human pluripotent stem cells with standard slow-rate protocols often is inefficient and can hardly be standardized. Vitrification that means ultrafast freezing already showed very good viability and recovery rates for this sensitive cell system, but is only applicable for low cell numbers, bears a high risk of contamination, and can hardly be implemented under GxP regulations. In this chapter, a sterile plate-based vitrification method for adherent pluripotent stem cells and their derivatives is presented based on a procedure and device for human embryonic stem cells developed by Beier et al. (Cryobiology 66:8–16, 2013). This protocol overcomes the limitations of conventional vitrification procedures resulting in the highly efficient preservation of ready-to-use adherent pluripotent stem cells with the possibility of vitrifying cells in multi-well formats for direct application in high-throughput screenings.

**Key words** Sterile plate-based vitrification, Adherent cells, Stem cells, GMP compliant

## 1 Introduction

Pluripotent stem cells (PSCs), regardless of whether naturally occurring, like embryonic stem cells (hESCs), or artificially induced, like induced pluripotent stem cells (hiPSC), are currently the key research topic in developmental biology, regenerative medicine, and pharmaceutical industry. Especially their ability to differentiate into all cell types of the human body [1] and their high telomerase activity [2] theoretically preventing cell aging and programmed cell death make them a promising cell model for organ and cell replacement therapies. Besides, with the possibility to artificially produce PSCs from patients carrying specific mutations by epigenetic reprogramming techniques, these cells have the potential to accelerate drug discovery and drug development [3, 4]. But particularly for applications with therapeutic focus and for high-throughput screenings of new target drugs, a readily

Zsolt Peter Nagy et al. (eds.), *Cryopreservation of Mammalian Gametes and Embryos: Methods and Protocols*, Methods in Molecular Biology, vol. 1568, DOI 10.1007/978-1-4939-6828-2_17, © Springer Science+Business Media LLC 2017

available and reliable supply of high numbers of stem cells or stem cell-derived cells with good quality is crucial [5].

Until now, slow-rate cryopreservation of PSCs in suspension is not efficient probably due to their sensitivity to cell-to-cell contact disruption [6] and prevents direct usage of PSCs in screening assays because of the prolonged recovery time after thawing. Vitrification approaches, typically by immersion in liquid nitrogen combined with a very high concentration of cryoprotective agents, already showed high recovery rates for hESC cultures [7]. This procedure instantly transforms the cytoplasm of cells into a glasslike structure, preventing the formation of intracellular ice crystallization [8] that is the main reason for cell death during slow-rate freezing of cells sensitive to cellular damage [9].

But current vitrification techniques are based on using detached cells or cell clumps in small volume straws for reaching the required high cooling rate, just enabling vitrification of up to ten cell clumps [10]. Additionally, the vitrification and thawing procedure is quite labor-intensive and cumbersome, increasing the risk of contamination and significantly impeding the implementation of GxP workflows.

Hence, we have developed a surface-based vitrification method that can overcome the limitations of small cell numbers and tricky handling [11]. Adherent cells immobilized on a plastic coverslip show a very high cell viability and recovery and maintain their stem cell characteristics after vitrification and thawing. But for this method, direct contact of the adherent cells with liquid nitrogen is necessary, still struggling with the risk of contamination and GxP regulations.

Based on these results, we have developed a sterile plate-based vitrification method using the so-called TWIST-substrate [12] which is reviewed by this chapter, enabling vitrification of adherent cells in a closed and sterile environment without direct contact with liquid nitrogen. This technology facilitates the highly efficient cryopreservation and storage of large cell numbers for therapeutic applications. For the evaluation of vitrification protocols as well as for the analysis of possible cryoinjury mechanisms, the vitrification process can be mimicked in a particular microscopy chamber and observed in situ. To this end we performed multiphoton laser scanning microscopy (cryo-MPLSM) enabling specific labeling of cellular structures and three-dimensional resolution even in the solid body of a vitrified sample (or of a frozen sample in the unfavorable case).

Further development of the TWIST-substrate to a multi-well-plate design, e.g., 96-well plate, will help to provide adherent stem cells or their derivatives with high quality in a format ready to use for high-throughput screenings in the field of drug discovery and drug development. Currently, also specific computer-controlled freezing devices are developed for the efficient vitrification of cells in multi-well plates based on standard cooling systems using liquid nitrogen or also electronic solutions.

## 2 Materials

***2.1 Components of the TWIST-Substrate for Sterile Vitrification***

1. IBIDI μ-dish, high (35 mm diameter, IBIDI, Martinsried, Germany).
2. Two-component glue.

***2.2 Cultivation Components for H1 Human Embryonic Stem Cells on Inactivated MEF***

1. H1 human embryonic stem cells (WiCell, Madison, WI, USA).
2. Treated cell culture dishes (60 mm diameter, Thermo Scientific, Nunc, Schwerte, Germany).
3. PBS (without Mg and Ca; Gibco/Life Technologies, Darmstadt, Germany).
4. H1 culture medium: add 0.1 mM β-mercaptoethanol (Sigma-Aldrich, Taufkirchen, Germany), 20% syntactical serum replacer, 2 mM L-glutamine, 1% nonessential amino acids, 4 ng/mL human recombinant bFGF, 100 U/mL penicillin, and 100 μg/mL streptomycin to DMEM/F12 (all Gibco/Life Technologies, Darmstadt, Germany).
5. Mitotically inactivated MEF cells, CF-1 strain (Merck Millipore, Darmstadt, Germany).
6. MEF culture medium: add 10% heat-inactivated FCS and 1% nonessential amino acids to DMEM (all Gibco/Life Technologies, Darmstadt, Germany).
7. Coating solution for hESCs and MEFs: dissolve 0.1% gelatin from porcine skin (Sigma-Aldrich, Munich, Germany) in water and autoclave it.

***2.3 Components for sterile, plate-based vitrification***

1. Vitrification solution 1 (VS1): add 10% $Me_2SO$ (WAK-Chemie Medical, Steinbach, Germany) and 10% ethylene glycol (Sigma-Aldrich, Munich, Germany) to H1 culture medium.
2. Vitrification solution 2 (VS2): add 300 mM sucrose (Sigma-Aldrich, Munich, Germany), 20% $Me_2SO$, and 20% ethylene glycol to H1 culture medium.
3. Warming solution 1 (WS1): add 200 mM sucrose to H1 culture medium.
4. Warming solution 2 (WS2): add 100 mM sucrose to H1 culture medium.
5. 37 °C warm water.
6. A Dewar vessel with liquid nitrogen.

***2.4 Components for Analysis of Cell Viability and Recovery of Adherent hESC Colonies***

1. Live/dead staining solution: add 50 μL ethidium bromide (stock solution, 1 mg/mL in $ddH_2O$; Molecular Probes/Life Technologies, Darmstadt, Germany) and 8 μL fluorescein diacetate (stock solution, 5 mg/mL in acetone; Molecular

Probes/Life Technologies, Darmstadt, Germany) to 5 mL PBS (without Mg and Ca).

2. Stereo fluorescence microscope (SMZ 1500, Nikon, Japan).

*2.5 Components for Analysis of Cell Morphology and Membrane Integrity of Adherent hESC Colonies During Vitrification*

1. Calcein AM (Molecular Probes/Life Technologies, Darmstadt, Germany) as cytosol dye.
2. Hoechst 33324 (Molecular Probes/Life Technologies, Darmstadt, Germany) for staining of the cell nuclei.
3. Circular microscopy slides, diameter 18 mm (VWR, Darmstadt, Germany).

*2.6 Setup of the Cryo-MPLSM for Analysis of Cryoinjury Mechanisms During the Vitrification Process*

1. A laser scanning microscope with at least two detection channels equipped for multiphoton excitation. Herein we used a Zeiss LSM 510 META NLO laser scanning microscope (Carl Zeiss, Jena, Germany).
2. A mode-locked laser in the NIR spectral range that delivers pulses on the 100 femtosecond order and nJ pulse energy. Herein we used a Coherent Chameleon Ultra mode-locked laser (Coherent, Santa Clara USA) with effective pulse durations below 200 fs.
3. An excitation wavelength of 780 nm was chosen to uniformly pump both stains and the evoked blue and green fluorescence was split into two detection channels by means of appropriate filter arrangements.
4. A Linkam MDS 600 cryo-microscopy chamber (Linkam, Waterfield UK) without the outer observation window. The MDS 600 is designed for upright operation. As the scanning microscope is operated in inverted mode, an objective inverter (LSM Tech, Etters, USA) was applied. The objective inverter is mounted into a thread of the microscope's objective turret, and the objective is mounted to the endpiece of the inverter in upright direction. The objective inverter lets both the excitation and the fluorescent light make a U-turn. It translates all lateral scanning movement of the laser beam 1:1 by creating an intermediate image at the half of its optical path (Fig. 1).

## 3 Methods

*3.1 Assembly of the TWIST-Substrate and Preparation for Vitrification*

1. Remove the cultivation foil of one IBIDI μ-dish, high, from the plastic rim.
2. Glue the remaining plastic rim upside-down onto the bottom of an intact IBIDI μ-dish, so that a sealable cultivation chamber and a chamber for liquid nitrogen are formed (Fig. 2).
3. Let the glue evaporate for at least 4 days prior to cell plating.

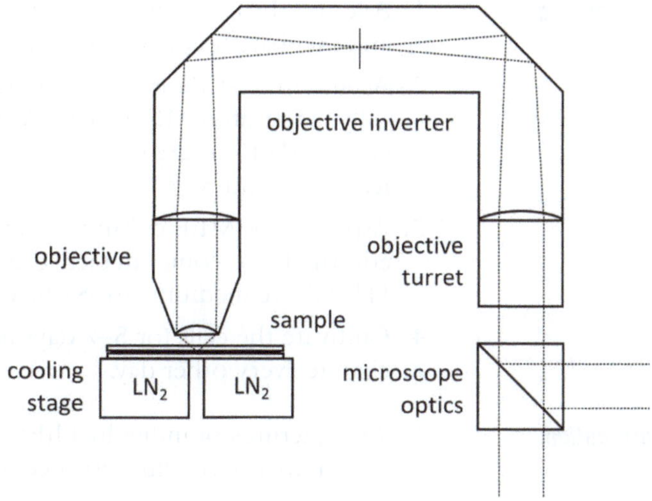


**Fig. 1** Setup of the cryo-MPLSM

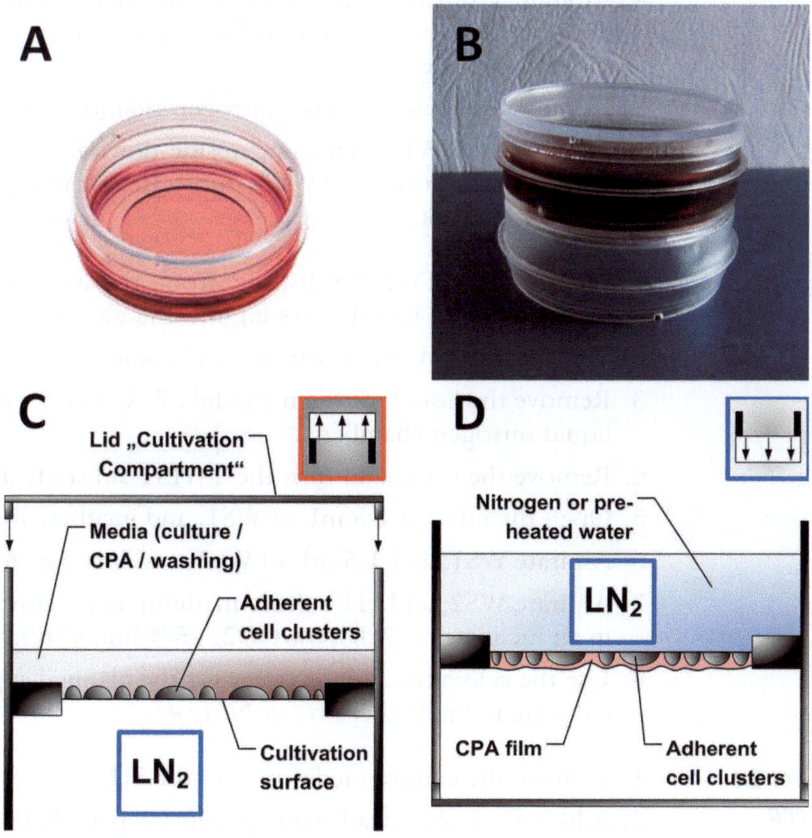


**Fig. 2** The TWIST-substrate for sterile vitrification (**A**) IBIDI μ-dish, high (35 mm diameter) [13]; (**B**) the assembled TWIST-substrate consisting of a sealable cultivation and an open liquid nitrogen chamber, (**C**) schematic drawing of the TWIST-substrate during cultivation, and (**D**) schematic drawing of the TWIST-substrate during vitrification

### 3.2 Cell Seeding

1. Add 1 mL 0.1% gelatine solution to the cultivation chamber and incubate it at 37 °C overnight.
2. Aspirate the 0.1% gelatine solution, plate $1.25 \times 10^5$ mitotically inactivated MEFs per TWIST-substrate and cultivate the cells in MEF culture medium in an incubator with 95 % humidity, 37 °C and 5% $CO_2$ for 24 h.
3. Aspirate the MEF culture medium, wash once with PBS, and add the H1 colonies detached using your standard protocol in H1 culture medium (*see* **Note 1**).
4. Cultivate the cells for 5–7 days in an incubator with a medium change every other day.

### 3.3 Vitrification

1. Take pictures of individual hESC colonies with 7.5× magnification on a stereo fluorescence microscope before vitrification.
2. Remove the culture medium from the H1 colonies, add 1.5 mL VS1, and incubate the cells for 1 min.
3. Aspirate VS1, add 1.5 mL VS2, and incubate the cells for 5 s.
4. Aspirate VS2 (*see* **Note 2**), close immediately the cultivation chamber with the corresponding lid, and turn the TWIST-substrate upside-down.
5. Fill liquid nitrogen into the nitrogen chamber (*see* **Note 3**).
6. Transfer the TWIST-substrate upside-down to a nitrogen storage tank and store the TWIST-substrate in the vapor phase of the storage tank.

### 3.4 Thawing

1. Remove the TWIST-substrate from the storage tank and immediately fill liquid nitrogen into the nitrogen chamber.
2. Transfer the TWIST-substrate to a laminar flow.
3. Remove the liquid nitrogen and fill 37 °C warm water into the liquid nitrogen chamber.
4. Remove the water and turn the TWIST-substrate around.
5. Open the lid, add 1.5 mL of WS1, and incubate for 1 min.
6. Aspirate WS1, add 1.5 mL of WS2, and incubate for 5 min.
7. Aspirate WS2, add H1 culture medium, and cultivate the cells in an incubator (37 °C, 5% CO2, 95 % humidity).
8. Use the cells after 24 h recovery for the planned application or cultivate them further on (*see* **Note 4**).

### 3.5 Analysis of the Cell Viability and Recovery of Adherent hESC Colonies After Vitrification

1. Remove the culture medium from the H1 colonies.
2. Add 1.5 mL live/dead staining solution per TWIST-substrate.
3. Incubate cells in the dark at room temperature for 5 min.
4. Aspirate the live/dead staining solution and add 1.5 mL PBS (without Mg and Ca) per TWIST-substrate.

5. Take pictures of the same hESC colonies as before vitrification using the stereo fluorescence microscope with 7.5× magnification with FITC and Texas Red filter (*see* Fig. 3).

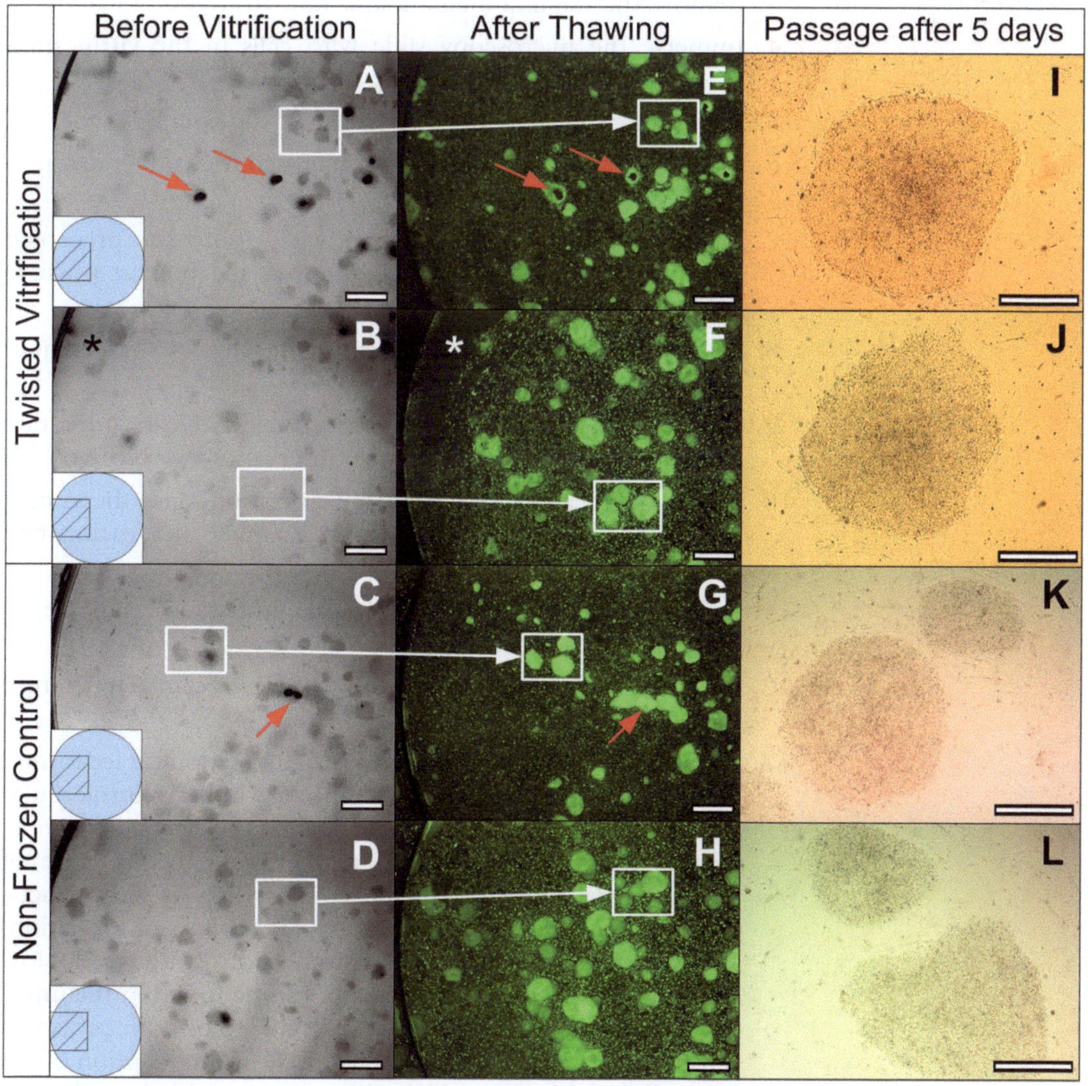


**Fig. 3** Representative images of hESC colonies after sterile vitrification in the TWIST-substrate (**A–D**) hESC colonies before cryopreservation; (**E, F**) hESC colonies after thawing; (**G, H**) non-frozen control. Inlays show the position of the images on the complete cultivation surface of the TWIST-substrate; (**E, G**) cells were stained with live/dead staining solution immediately after thawing; (**F, H**) cells were stained with live/dead staining solution 24 h after thawing. Corresponding cell colonies are marked with white squares and connecting arrows. (**I-L**) Thawed hESC colonies after passage and 6 days of cultivation; (**I, K**) cells were passaged immediately after thawing; (**J, L**) cells were passaged after a 24 h post-thawing cultivation in the TWIST-substrate. Note the thicker, possibly differentiated areas of the colonies (*red arrows*) lost after the vitrification process. *Scale bars* indicate 1 mm (**A–H**) and 500 μm (**I–L**)

### 3.6 Analysis of the Cell Morphology and Membrane Integrity During the Vitrification Process Using Cryo-MPLSM

1. Grow the cell colonies and the MEF cells on circular microscopy slides as described in Subheading 3.2.
2. Apply Calcein AM and Hoechst 33324 to the sample culture prior to vitrification experiments at room temperature according to the supplier's application notes.
3. Remove the staining medium and add vitrification medium.
4. Immerse the microscopy slide with cells in the vitrification media according to Subheading 3.3.
5. Put another microscopy slide on top of the sample.
6. Cool the silver block of the cryo-microscopy chamber down to −180 °C.
7. Place the "sample and slide sandwich" onto the precooled silver block in order to vitrify them at the highest possible cooling rate (*see* **Note 5**).
8. Cool the 40×/1.3NA oil immersion objective by the stream of cold, dry gas before attachment.
9. Attach a 40×/1.3NA oil immersion objective carrying a tiny drop of oil to the sample face for observation (*see* **Note 6**).
10. Adjust the focal plane to a favorable position, once the sample is vitrified in the cryo-microcopy chamber and maintained under −140 °C (*see* **Note** 7).
11. Adjust the microscope settings (pixel number, pixel time, replicates, etc.).
12. Perform the image acquisition as for room temperature microscopy.
13. Analyze the acquired images by visual inspection for ice formation events. The green fluorescing areas indicate intracellular sites where the cytosol did not crystallize, i.e., remained amorphous. The blue fluorescing areas indicate the presence of nucleic acids, i.e., the cell nucleus (as long as it is intact). Dark areas inside and between cells mark adverse ice formation. In case of (stem) cell clusters that were cooled with typical slow freezing rates, a lot of frozen water can be found in the intercellular space between shrunken cells due to osmotic processes during the cooling course (*see* Fig. 4c). The interruption of the intercellular cohesion will lead to decomposition of the cell cluster after thawing. Faster cooling will lead to the formation of intracellular ice (*see* Fig. 4d) since the cells still have nearly their natural water content. So far it is the tissue analogy to Mazur's two factor hypothesis. If the cooling rate is finally high enough for vitrification, the entire sample keeps its genuine morphology even at very low temperatures, and no dark ice areas show up at all (*see* Fig. 4a, b) (*see* **Note 8**).

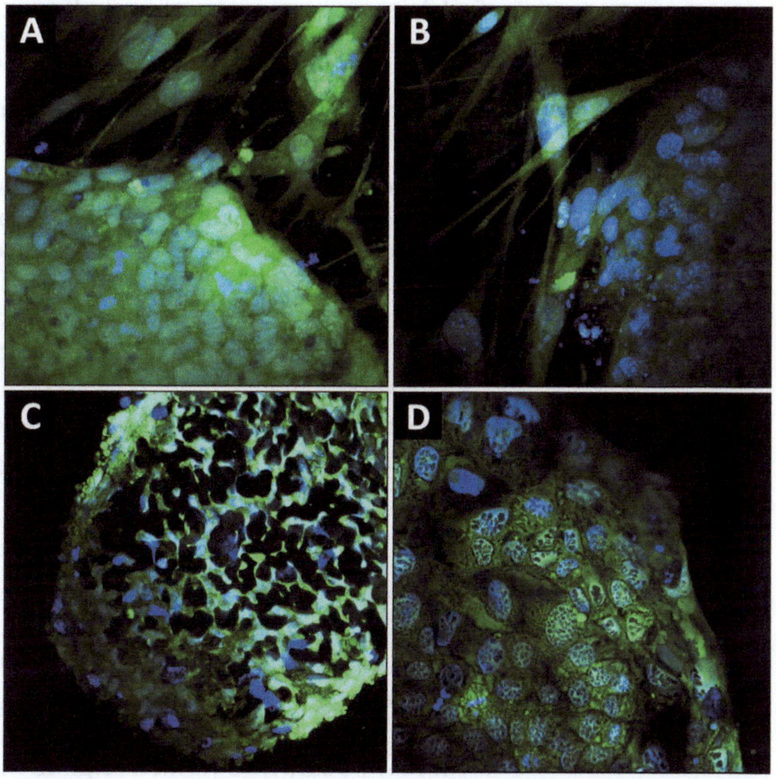


**Fig. 4** MPLSM images of hESCs during vitrification and slow-rate freezing: (**a, b**) hESC colony during successful vitrification at −196 °C; (**c**) frozen water (*black areas*) in the intercellular space of a hESC colony during slow-rate freezing with 1 °C/min at −50 °C; (**d**) formation of intracellular ice in a hESC colony during freezing with 10 °C/min at −50 °C; cells were stained with Calcein AM and Hoechst 33324 as described above

## 4 Notes

1. Avoid cluster formation of the plated hESC fragments in the middle of the dish before the cells attach to the cultivation surface.
2. Remove as much VS2 as possible by manual pipetting to avoid the formation of a meniscus. A medium meniscus results in high colony loss in these areas post-thawing due to a reduced cooling rate not sufficient for vitrification.
3. The vitrification procedure can be performed outside of the laminar flow, when the lid of the cultivation chamber is closed.
4. Few of the vitrified TWIST-substrates show damages after vitrification due to high thermal stress on the material, resulting in damaged hESC colony areas with low viability.

5. It's necessary to perform the vitrification in the cryo-microscopy chamber under a gentle flow of nitrogen (e.g., from the evaporating liquid nitrogen) or precooled dry air to prevent the condensation of air humidity on the sample face and the chamber interior.
6. The oil immersion is needed for sufficient excitation in multi-photon excitation and helps to keep the air humidity away from the sample face. Once the oil is at observation temperature (that means at least −140 °C), it is nearly solid, so further lateral displacements of the sample against the objective are no longer possible. One has to investigate the field of view you choose first.
7. In vitrification protocols usually no particular cooling rates are given. It is an "as fast as possible" approach. Therefore, it is difficult to mimic the cooling rate of a protocol in the microscopy setup. But if the cooling rate of the process in the cryo-microscopy chamber turns out to be too slow for a given sample, this can be easily noticed by the formation of intracellular ice grains.
8. The cryo-microscopy setup presented here allows furthermore a nice insight into the devitrification process. Once the sample is successfully vitrified to the amorphous state below −140 °C, it may be slowly rewarmed (e.g., to mimic the temperature rise due to a withdrawal of a neighbor sample from the sample rack in biobanking). We found the glass transition temperatures of typical vitrification media to be in the range around −120 °C. Above the glass transition temperature ice nucleation and growth is enabled and may be observed by the emergence of dark spots within and between the cells. In order to observe devitrification of the medium, an additional (membrane impermeant) dye has to be added for detection in another channel.

## Acknowledgment

The work with human embryonic stem cells was permitted by the Robert Koch Institute (18. and 44. permission) and carried out according to German law.

### References

1. Odorico JS, Kaufman DS, Thomson JA (2001) Multilineage differentiation from human embryonic stem cell lines. Stem Cells 19:193–204
2. Wright WE, Piatyszek MA, Rainey WE, Byrd W, Shay JW (1996) Telomerase activity in human germline and embryonic tissues and cells. Dev Genet 18:173–179
3. Giri S, Bader A (2015) A low-cost, high-quality new drug discovery process using patient-derived induced pluripotent stem cells. Drug Discov Today 20:37–49

4. Heilker R, Traub S, Reinhardt P, Schöler HR, Sterneckert J (2014) iPS cell derived neuronal cells for drug discovery. Trends Pharmacol Sci 35:510–519
5. Stacey GN, Hunt CJ (2008) Banking stem cell lines for future therapies. In: Polak J, Mantalaris S, Harding SE (eds) Advances in tissue engineering. Imperial College Press, London, pp 229–254
6. Acker JP, Larese A, Yang H, Petrenko A, McGann LE (1999) Intracellular ice formation is affected by cell interactions. Cryobiology 38:363–371
7. Richards M, Fong CY, Tan S, Chan WK, Bongso A (2004) An efficient and safe xeno-free cryopreservation method for the storage of human embryonic stem cells. Stem Cells 22:779–789
8. Wowk B (2010) Thermodynamic aspects of vitrification. Cryobiology 60:11–22
9. Liu JC, Guan X, Ryan JA, Rivera AG, Mock C, Agrawal V et al (2013) High mitochondrial priming sensitizes hESCs to DNA-damage-induced apoptosis. Cell Stem Cell 13:483–491
10. Vajta G, Holm P, Kuwayama M, Booth PJ, Jacobsen H et al (1998) Open pulled straw (OPS) vitrification: a new way to reduce cryoinjuries of bovine ova and embryos. Mol Reprod Dev 51:53–58
11. Beier AF, Schulz JC, Dörr D, Katsen-Globa A, Sachinidis A et al (2011) Effective surface-based cryopreservation of human embryonic stem cells by vitrification. Cryobiology 63:175–185
12. Beier AF, Schulz JC, Zimmermann H (2013) Cryopreservation with a twist – Towards a sterile, serum-free surface-based vitrification of hESCs. Cryobiology 66:8–16
13. www.ibidi.com/xtproducts/de/ibidi-Labware/Open-Slides-Dishes:-ibidi-Standard-Bottom/m-Dish-35mm-high

# Chapter 18

# Vitrification: A Reliable Method for Cryopreservation of Animal Embryos

**B. Singh, G. Mal, and S.K. Singla**

## Abstract

The impact of cryopreservation in assisted reproduction is increasingly appreciated. Cells, oocytes, and embryos preserved under ultralow temperature can endure storage for indefinite time with almost no alteration in their metabolic and genetic components. Advances in cryopreservation and its applications to preserve cells, gametes, and embryos offer opportunities in conservation of biodiversity and dissemination of elite livestock germplasm. This chapter describes the vitrification for cryopreservation of livestock embryos.

**Key words** Vitrification, Cryopreservation, Embryos, Livestock

## 1 Introduction

Consistent supply and preservation of raw biological materials (cells, gametes, and embryos) are the major challenges in the widespread use of assisted reproduction technology (ART) in animals. To deal with the problem of uncertain supply and demand, the researchers often have to store gametes and embryos over extended time period. With the invention of advanced technologies like mammalian cloning, animal transgenesis, and stem cell engineering, the conservation of biological materials through cryopreservation has become imperative (Fig. 1).

Cryopreservation is an established component of ART and biodiversity conservation. Compared with slow freezing, the vitrification is simple, and is the technique of choice in human and mammalian reproductive programs, animal facilities, and genetic manipulation experiments [1]. In view of its simpler technical and equipment requirements, the vitrification appears to be the method of choice [2].

The standard vitrification protocol may vary for oocytes and embryos in different species. For instance, the oocytes are, in general, more susceptible to cooling injuries, due to disassembling

Zsolt Peter Nagy et al. (eds.), *Cryopreservation of Mammalian Gametes and Embryos: Methods and Protocols*, Methods in Molecular Biology, vol. 1568, DOI 10.1007/978-1-4939-6828-2_18, © Springer Science+Business Media LLC 2017

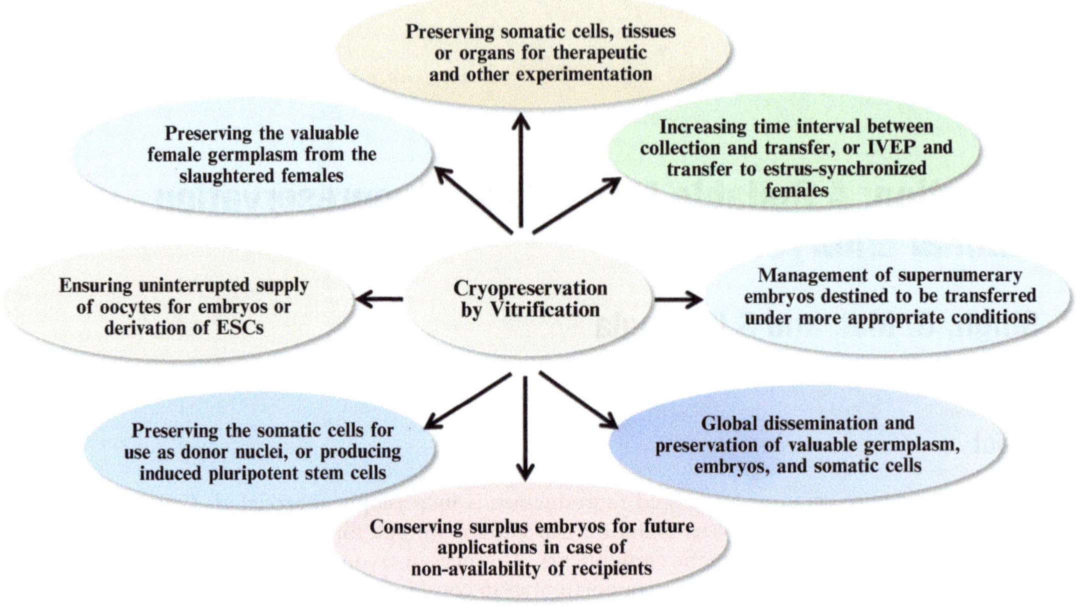


**Fig. 1** A summarized presentation of the applications of cryopreservation in animal assisted reproduction technologies and diversity conservation. The cryopreservation by using vitrification is simpler and economically doable method. *IVEP* in vitro embryo production, *ESCs* embryonic stem cells

of meiotic spindle and subsequent impaired reassembling. Porcine and bubaline oocytes and embryos [3–5] are particularly sensitive to chilling injuries in part due to presence of cytoplasmic fat globules that cause damage to cell membranes during cryopreservation. Cytoplasm degeneration could be the major damage caused by vitrification in bovine oocytes [6, 7].

However, the spindle damage can be minimized significantly by proper methodology and cryoprotective agents (CPAs) [8]. In addition, the cytoskeleton stabilizers such as cytochalasin-B significantly improve developmental competence of vitrified oocytes [9].

Vitrification in open pulled straw (OPS) [10] originally developed for bovine ova and embryos was found to be better than slow freezing for the cryopreservation of cloned bubaline embryos [11, 12]. The studies on bovine IVF blastocysts vitrified for 15 years followed by in vitro culture or embryo transplantation revealed that there were no significant differences in their in vivo developmental competence and pregnancy outcomes [13].

Compared to conventional cryopreservation, the OPS vitrification was reported to be the method of choice for cryopreservation of goat blastocysts [14]. In addition, reduction in lipid contents could improve vitrification cryotolerance of in vitro produced (IVP) ovine embryos [15].

While vitrification has a clear role in livestock assisted reproduction [16–18], we support continued research to establish optimal slow-cooling methods that may assist in alleviating concerns over safety issues, such as storage, transport, and use of high concentration of CPAs in the process.

## 2 Materials and Methods

Prepare all solutions in ultrapure cell culture-grade water (prepared by purifying deionized water to attain sensitivity of 18 MΩ at 25 °C) and high cell culture-grade analytical reagents. Prepare and store the reagents at room temperature (unless indicated otherwise) or diligently following the instructions specified by the manufacturers. Also, follow the waste disposal regulations properly when disposing the waste materials.

### 2.1 Cell Culture Media, Sera, and Supplements

1. Cell culture media.
2. Fetal bovine serum (FBS) and other supplements (L-glutamine and antibiotics).

   Aseptic milieu is required at all stages (*see* **Note 1**). Cell culture media and supplements should be high-grade cell culture tested. Biochemicals, cell culture media, and supplements from Sigma Chemical Company, St. Louis, MO, USA, are suitable for in vitro production and cryopreservation of embryos. In a set of experiment, we strongly advise that culture media, sera, and supplements should be of same batch to avoid variation in results. Rather than preparing media from lyophilized or powdered ingredients, ready-to-use culture media should be preferred. It is obligatory to follow the specifications and instructions recommended by the manufacturer.

### 2.2 Plasticware

1. The 90 × 15 mm cell culture dishes (Nunclon or BD Falcon).
2. Center well organ culture dish (BD Falcon).
3. Unopettes (20 μL capacity, Becton Dickinson and Co. Lincoln Park, NJ, USA).
4. Disposable syringe membrane filters (0.45 μm Nunclon or Millipore Corporation, Bedford MA, USA).
5. French mini straws (0.25 mL, IMV, L'Aigle, France) (*see* **Note 2**).
6. Visotubes 10 mm (IMV, L'Aigle, France).
7. Styrofoam container.
8. Disposable microtips of various sizes and capacity.
9. Autoclaveable micropipettes.
10. Cryoware markers or pens (Nalgene Nunc.) are recommended for manual marking of straws.

### 2.3 Instruments

1. Good quality dissecting or Stereo Zoom microscope (e.g., Olympus or Nikon) with warming stage.
2. Laminar flow hood.
3. Inverted fluorescence microscope (e.g., Olympus, Nikon, or Carl Zeiss).
4. Cryocanes (IMV, L'Aigle, France).
5. Refrigerators.
6. Humidified $CO_2$ incubators.
7. Water baths are required. There should be ample availability of liquid nitrogen. The person involved should be a qualified cryobiologist.

### 2.4 Vitrification of Embryos

1. Due care is recommended in preparation and use of CPAs (*see* **Note 3**). Transfer the blastocysts to an equilibrium solution comprising of 7.5 % EG, 7.5 % DMSO, and 20 % FBS in TCM-199 (T20) medium for 1 min.
2. Take out the cell culture dishes containing blastocyst from humidified $CO_2$ incubator, and transfer them with the help of sterile Unopettes to the next solution consisting of 15 % EG, 15 % DMSO, and 20 % FBS in T20.
3. After incubation for 30 s, transfer the blastocysts to OPS or French mini straw (*see* **Notes 4–6**), and seal them with the help of thermal sealer. Note that embryos enter OPS by capillary action and no suction pressure is needed. Plunge the labeled OPS instantly into liquid nitrogen. Different cryocontainers (*see* **Note** 7) or labeled visotubes should be used to store separate batches of embryos.

### 2.5 Warming

1. Take out the straws from liquid nitrogen and dip in warm water (38 °C). Wipe the warmed straws with sterile blotting paper, and cut one end to allow the embryos gather in cell culture dish containing thawing solution (1.0 M sucrose in T20) for 1 min.
2. Transfer the embryos into dilution solution (0.5 M sucrose in T20) for 3 min.
3. Wash twice or thrice the blastocysts with T20 for 5 min every time.
4. After incubation in T20 for 5–6 h and based on their characteristic morphological attributes, the blastocysts are ready for transfer into estrus-synchronized recipient animals (*see* **Note 8**).

## 3 Notes

1. Contamination should be avoided. Users are recommended to choose a working system that allows for work with cells, oocytes, and embryos. As cloning involves multiple steps

including in vitro maturation (IVM) and micromanipulation of oocytes, and culturing of cells, the labs should be well furnished for handling and maintaining cell lines and embryos.

2. Special care should be taken while selecting carriers. It is necessary to use the types of carrier or vessel materials with rapid heat transfer that support the process of uniform heat exchange to achieve optimal cooling rates. A stepwise exposure of embryos to precooled concentrated CPAs is recommended.
3. Higher concentrations of CPAs allow shorter exposure times to the CPAs. However, this may lead to CPA-induced toxicity to the cells or embryos. It is important to carefully monitor the duration of exposure to the final CPA before plunging the loaded embryos into liquid nitrogen.
4. The loading of blastocysts into OPS (Figs. 2, 3) should be performed under a Stereo Zoom microscope. This ensures proper loading in carrier system.
5. Exposure of the loaded embryos to vapor phase should be minimum. Embryos should be plunged into liquid nitrogen as immediately as possible.
6. While switching the cells between different concentrations of solutions, fill up the pipette with next lower concentration solution before picking up the embryos for transferring to next concentration.

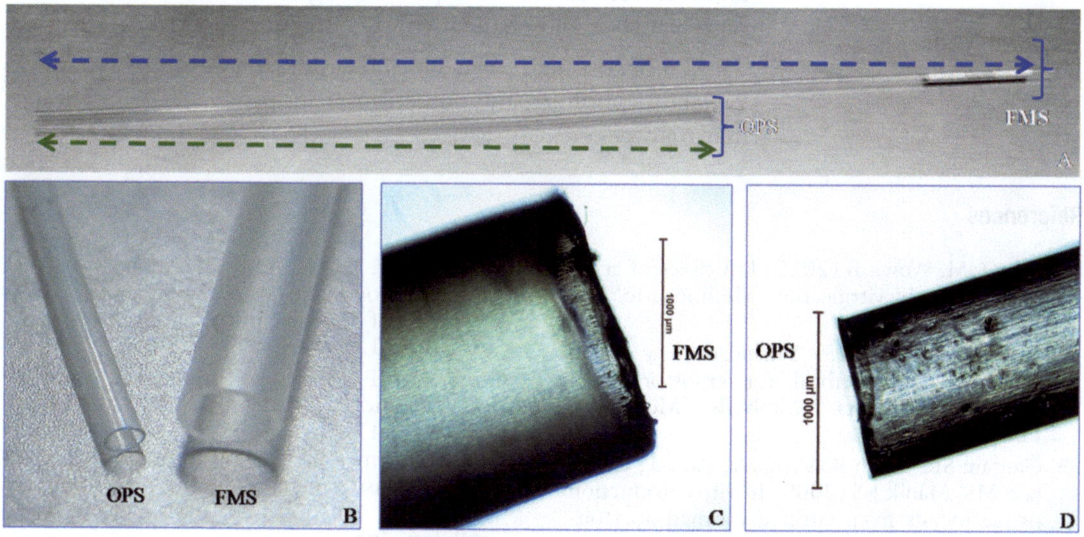


**Fig. 2** Preparing open pulled straw (OPS) for vitrification of embryos. (**A**) A comparative overview of 0.25 mL French mini straw (FMS) and OPS which is much smaller and narrower than FMS. (**B**) Gross appearance and diameters of FMS and OPS, (**C, D**) the microscopic overviews and comparison of the diameters of the FMS and manually prepared OPS

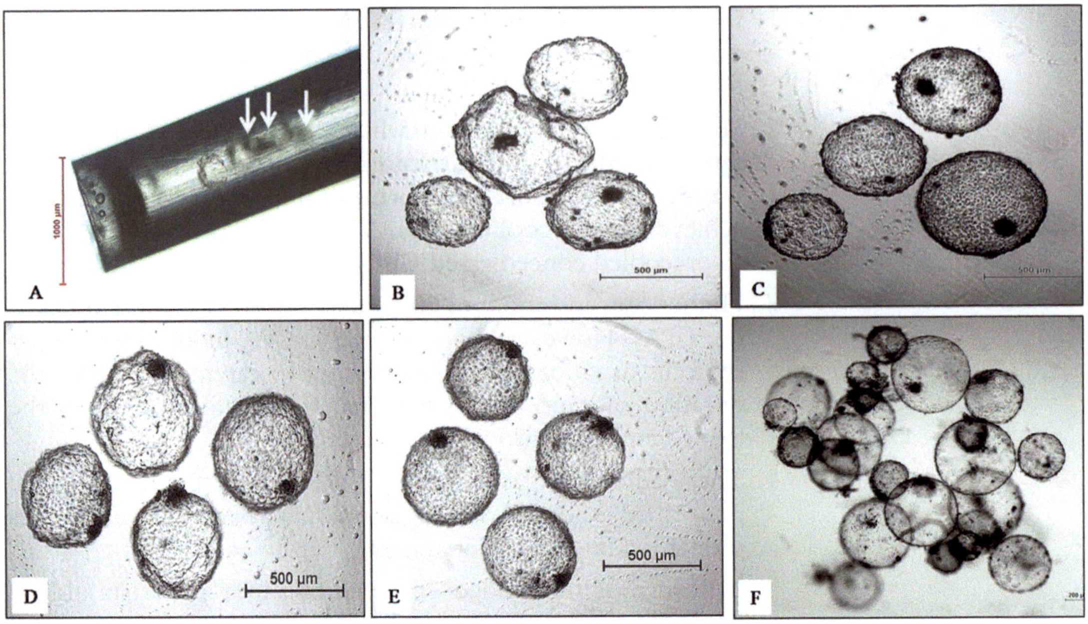


**Fig. 3** (**A**) Cloned bubaline embryos loaded in OPS, (**B**) a group of shrunken zona-free cloned blastocysts immediately after incubation with equilibration solution during vitrification, (**C**) re-expanded cloned blastocysts after incubation for 5–7 min in equilibration solution, (**D**) a group of yet to re-expand cloned blastocysts immediately after warming, (**E**) a group of re-expanded cloned blastocysts after 22–24 h of culture in embryo medium, (**F**) re-expanded cloned blastocyst

7. Cryocontainers should be stored in separate air conditioned laboratories. Levels of liquid nitrogen should be monitored regularly to avoid chances of warming and devitrification.
8. The embryo quality can be determined by TUNEL staining, visual morphology, and detection of apoptosis-inducing molecular markers.

## References

1. Fahy GM, Wowk B (2015) Principles of cryopreservation by vitrification. Methods Mol Biol 1257:21–82
2. Liebermann J (2015) Vitrification: a simple and successful method for cryostorage of human blastocysts. Methods Mol Biol 1257:305–319
3. Gautam SK, Singh B, Verma V, Palta P, Chauhan MS, Manik RS (2009) In vitro production of blastocysts from vitrified-warmed abattoir-derived buffalo (*Bubalus bubalis*) immature oocytes. Indian J Dairy Sci 62:495–502
4. Zhou GB, Li N (2009) Cryopreservation of porcine oocytes: recent advances. Mol Hum Reprod 15:279–285
5. Zhang W, Yi K, Yan H, Zhou X (2012) Advances on in vitro production and cryopreservation of porcine embryos. Anim Reprod Sci 132:115–122
6. Sprícigo JF, Morais K, Ferreira AR, Machado GM, Gomes AC, Rumpf R, Franco MM, Dode MA (2014) Vitrification of bovine oocytes at different meiotic stages using the Cryotop method: assessment of morphological, molecular and functional patterns. Cryobiology 69:256–265
7. Akiyama K, Kobayashi J, Sato Y, Sata R, Ohashi M, Sasaki E, Oda Y, Ogawa Y, Ueda S, Nabenishi H, Matoba S (2010) Calf production

from vitrified bovine sexed embryos following in-straw dilution. Anim Sci J 81:461–466

8. Mandelbaum J, Anastasiou O, Lévy R, Guérin JF, de Larouzière V, Antoine JM (2004) Effects of cryopreservation on the meiotic spindle of human oocytes. Eur J Obstet Gynecol Reprod Biol 113(Suppl 1):S17–S23
9. Wang CL, Xu HY, Xie L, Lu YQ, Yang XG, Lu SS, Lu KH (2016) Stability of the cytoskeleton of matured buffalo oocytes pretreated with cytochalasin B prior to vitrification. Cryobiology 72(3):274–282. doi:10.1016/j.cryobiol.2016.03.005
10. Vajta G, Holm P, Kuwayama M, Booth PJ, Jacobsen H, Greve T, Callesen H (1998) Open Pulled Straw (OPS) vitrification: a new way to reduce cryoinjuries of bovine ova and embryos. Mol Reprod Dev 51:53–58
11. Saha A, Panda SK, Chauhan MS, Manik RS, Palta P, Singla SK (2013) Birth of cloned calves from vitrified-warmed zona-free buffalo (*Bubalus bubalis*) embryos produced by hand-made cloning. Reprod Fertil Dev 25:860–865
12. Sirisha K, Selokar NL, Saini M, Palta P, Manik RS, Chauhan MS, Singla SK (2013) Cryopreservation of zona-free cloned buffalo (*Bubalus Bubalis*) embryos: slow freezing *vs* open-pulled straw vitrification. Reprod Domest Anim 48:538–544
13. Fang Y, Zeng S, Fu X, Jia B, Li S, An X, Chen Y, Zhu S (2014) Developmental competence in vitro and in vivo of bovine IVF blastocyst after 15 years of vitrification. Cryo Letters 35:232–238
14. Al Yacoub AN, Gauly M, Holtz W (2010) Open pulled straw vitrification of goat embryos at various stages of development. Theriogenology 73:1018–1023
15. Romão R, Marques CC, Baptista MC, Barbas JP, Horta AE, Carolino N, Bettencourt E, Pereira RM (2015) Cryopreservation of in vitro-produced sheep embryos: effects of different protocols of lipid reduction. Theriogenology 84:118–126
16. Singh B, Chauhan MS, Singla SK, Gautam SK, Verma V, Manik RS, Singh AK, Sodhi M, Mukesh M (2009) Reproductive biotechniques in buffaloes (*Bubalus bubalis*): status, prospects and challenges. Reprod Fertil Dev 21:499–510
17. Pegg DE (2015) Principles of cryopreservation. Methods Mol Biol 1257:3–19
18. Gupta A, Singh J, Anzar M (2016) Effect of cryopreservation technique and season on the survival of in vitro produced cattle embryos. Anim Reprod Sci 164:162–168

# Chapter 19

# Cryopreservation Effect on Genetic Function: Neonatal Outcomes

**Vanesa Robles, Marta F. Riesco, and David G. Valcarce**

## Abstract

Cryopreservation is a well-established technique commonly used in clinical practice. It is used widely for the conservation of gametes and embryos that will be used later for insemination or in vitro fertilization. However, several studies have shown that this technique can produce changes in messenger RNA levels, in the epigenome and induce DNA damage. Although the embryo has potent mechanisms for DNA repair, and molecular changes in spermatozoa are not necessarily reflected in the embryo, it is important to explore new molecular tests and diagnostic tools to design optimal cryopreservation protocols and avoid undesirable molecular alterations. This chapter describes a protocol to quantify the lesions produced by cryopreservation using a protocol previously published by Rothfuss.

**Key words** Cryopreservation, DNA damage, Quantitative polymerase chain reaction, DNA lesions, DNA extraction

## 1 Introduction

Assisted reproductive technologies (ARTs) have been associated with a higher than normal risk of epigenetic syndromes [1] such as Beckwith-Wiedemann syndrome, Angelman syndrome (AS) and Prader-Willi syndrome (PWS) [2–4]. Moreover, higher than normal rates of low birth weight (LBW) (not attributable to maternal age or multiple births) have also been observed with the use of ARTs [5]. Cryopreservation is a technique used widely in ARTs but it has never been considered as a potential factor in these alterations. However, in humans, it is known that cryopreservation alters the presence of crucial transcripts in the spermatozoa [6], and in model animals such as zebrafish, it modifies the methylation pattern of certain promoters [7]. Additionally, cryopreservation affects DNA in the germline, producing not only fragmentation but also mutations in different genes that cannot be detected with traditional methods for DNA evaluation [8]. Spermatozoan DNA damage has been associated with a significant increase in embryonic loss

Zsolt Peter Nagy et al. (eds.), *Cryopreservation of Mammalian Gametes and Embryos: Methods and Protocols*, Methods in Molecular Biology, vol. 1568, DOI 10.1007/978-1-4939-6828-2_19, © Springer Science+Business Media LLC 2017

after in vitro fertilization and intracytoplasmic sperm injection (ICSI) [9]. Most studies evaluating DNA status in human spermatozoa have been limited to fragmentation analysis. Our group used a quantitative polymerase chain reaction (q-PCR)–based technique published by Rothfuss et al. [10] to quantify lesions in specific genome regions after cryopreservation. First in animal models [7, 8, 11], and then in human sperm samples [12], our results demonstrated that, even when fragmentation is absent, significant DNA damage could be detected in some of the studied genes. This is particularly relevant for those genes considered key players in early embryo development or those genome regions particularly related to certain syndromes. Interestingly, among all the genes and genome regions studied, an elevated number of lesions were found in *SNORD116/PWSAS* (a key region in PWS) and *UBE3A* (key gene in AS) [12].

Thus, the qPCR technique initially described by Rothfuss et al. [10] can be applied successfully for the detection of DNA damage produced by cryopreservation (Fig. 1), and could be crucial in the selection of optimized cryopreservation protocols for human sperm that contribute greatly to the safety and success of ARTs. This chapter provides a detailed description of this technique and its application to the detection of lesions produced by cryopreservation.

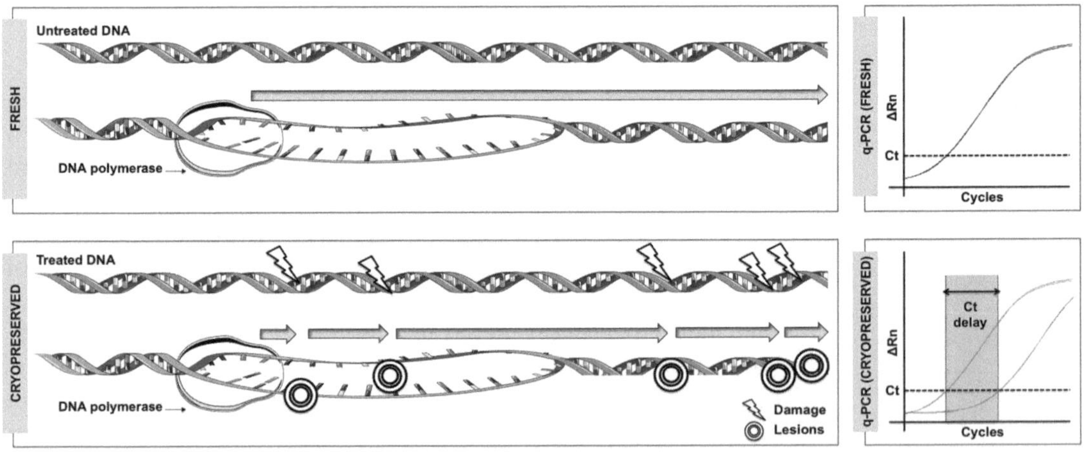


**Fig. 1** Molecular basis of q-PCR-based technique for DNA lesion analysis described by Rothfuss et al. [10]. DNA lesions derived from damage caused after cryopreservation affects DNA polymerase amplification. This disruption in the polymerase complex progression is translated in a threshold cycle (Ct) delay when the analysis is monitored in a real-time PCR system. The graph illustrates non-treated DNA (fresh DNA) and treated DNA (cryopreserved DNA) differences schematically

## 2 Materials

Informed consent should be obtained from donors, and specific national legislation should be followed in order to work with human samples. The approval of the bioethics committee of the working institution is also required.

Perform all protocols at room temperature unless otherwise specified.

### 2.1 Sampling

1. Sterilized sample container.
2. Pasteur pipettes.
3. Calibrated 15 mL tubes.
4. Haemocytometer.
5. Optical microscope.
6. PBS 10×: to a flask, add 2 g KCl (27 mM), 14.2 g $Na_2HPO_4$ (100 mM), 2.4 g $KH_2PO_4$ (17 mM), and 80.8 g NaCl (1.37 M). Add bidistilled water up to 1 L. Adjust pH to 7.4. Prepare a 1:10 dilution to obtain PBS 1×, check the pH, and autoclave.

### 2.2 Cryopreservation and Thawing Protocol

Lesions produced by any cryopreservation protocol can be measured using this technique. In this chapter, one such protocol used for human spermatozoa in fertility clinics is described.

1. Sperm freezing medium (*Origio Medicult Media*).
2. 0.5 mL French straws.
3. Cooling device, consisting of a hermetic box with a lid (made of polystyrene or similar material) filled to half of its volume with liquid nitrogen. A rigid plastic rack is then positioned inside it at the appropriate distance from the surface of the liquid nitrogen where the samples will be placed.

### 2.3 DNA Extraction

1. Lysis buffer. Prepare a solution of 10 mM Tris-Cl, 50 mM NaCl, 1 mM ethylenediamine tetra-acetic acid (EDTA) and 1 % v/v sodium dodecyl sulphate; adjust pH to 8 prior to addition of the sodium dodecyl sulphate. Immediately prior to experimentation, aliquot the required amount of this solution and add 2 mg/mL proteinase K and dithiothreitol to 1 mM.
2. Heater.
3. Phenol/chloroform/isoamyl (25:24:1).
4. Chloroform.
5. Flow hood.
6. Centrifuge.
7. Absolute ethanol.

8. 70 % pre-cooled ethanol (−20 °C).
9. TE buffer. Add 1 mL of 1 M Tris pH 7.5 and 0.2 mL 0.5 M ethylenediamine tetra-acetic acid (EDTA), pH 8, to a flask. Add bidistilled water up to 100 mL, and adjust the pH to 7.4.

#### 2.4 Primer Validation and Establishment of Optimal Melting Temperature

1. Gradient thermocycler.
2. PCR commercial kit (DNA polymerase, dNTPs, $MgCl_2$, specific DNA polymerase buffer).
3. Agarose.
4. Electrophoresis system.

#### 2.5 q-PCR Validation and q-PCR Assays:

1. Real-time PCR system.
2. Fluorescent dye for short fragments.
3. Fluorescent dye for long fragments.
4. Adapted 96-well optical plates for the real-time PCR system.

## 3 Methods

#### 3.1 Sampling

1. Collect the ejaculates in sterilized sample containers previously provided to the patients or donors (*see* **Note 1**). If possible, following WHO guidelines [13], collect sample after 3 days of sexual abstinence.
2. Let the ejaculates liquefy for 20 min (*see* **Note 2**).
3. Evaluate the volume of the sample with a disposable Pasteur pipette (*see* **Note 3**).
4. Analyse sperm concentration using a hemocytometer (*see* **Note 4**). Flick the sample to dilute a homogeneous sample 1:10 in PBS 1× and charge the chambers. Evaluate the agglutination level of the sample, abnormal morphology, and leukocyte population, and take into account any unusual values for interpretation of results (*see* **Note 5**).

#### 3.2 Cryopreservation Protocol and Thawing

Perform the protocol under evaluation at this point. The following is an example of a common cryopreservation protocol.

1. Divide the sample in two aliquots: one will be subjected to the cryopreservation protocol, and the other kept in fresh conditions for comparison. Note that this step ensures cell number count equivalence between cryopreserved and fresh samples.
2. Drop by drop, dilute the semen sample 1:1 in a commercial Sperm Freezing Medium (*Origio Medicult Media*) to a final concentration: $4 \times 10^7$ sptz/mL.
3. Loading of the straws should be performed after exactly 10 min, since this is the equilibration time needed in the cryoprotectant

solution. For loading of 0.5 mL French straws, a P-1000 micropipette can be used (*see* **Note 6**).

4. After loading, expose straws horizontally to liquid nitrogen vapour (2 cm over the surface) in the cooling device (*see* **Note 7**). Close the box to reduce nitrogen evaporation and keep the straws in this atmosphere for 30 min.
5. Immediately following the vapour freezing period, plunge the straws into liquid nitrogen. Then, transfer the straws to the storage liquid nitrogen tank, and keep them until needed (*see* **Note 8**).
6. For thawing, carefully remove the straws from the liquid nitrogen and allow them to thaw at room temperature for 5 min. Thereafter, rub them between your hands for 2 s to ensure they are completely thawed. Hold the straw vertically in an Eppendorf tube with a cotton plug outside. Cut the cotton plug with scissors. Blow gently from above with a micropipette to allow the mixture to collect in the tube.

### 3.3 DNA Extraction

This part of the protocol must be repeated identically with the non-cryopreserved aliquot.

Work in a gas extraction flowhood for **steps 3** and **4**.

1. Take an aliquot containing $10^7$ cells from the thawed cryopreservation sample and centrifuge it at 400 *g* for 10 min.
2. Resuspend the pellet in 0.7 mL lysis buffer. With gently agitation, lyse the samples at 55 °C for 2 h.
3. Add one volume (0.7 mL) phenol/chloroform/isoamyl (25:24:1) and, by hand, agitate the tubes vigorously for 4 min.
4. Centrifuge the tubes at 13,200 × *g* for 5 min. Collect the aqueous phase (upper phase) with a P-1000 micropipette, and recover it to a new tube (*see* **Note 9**). Wash it with a volume (0.7 mL) chloroform by performing exactly the same procedure but this time with 2 min of vigorous agitation and 5 min of centrifugation at 13,200 × *g*. Discard the other phases (*see* **Note 10**).
5. Repeat the previous step with the upper phase after the first wash.
6. Split the volume into two new tubes (0.35 mL). Add 0.875 mL (2.5 volume) absolute ethanol to precipitate the DNA. Maintain it overnight at −20 °C. In many cases, after shaking, DNA is visible.
7. Centrifuge the sample at 13,200 × *g* for 10 min. Carefully discard the ethanol, keeping the precipitated DNA in the tube, then wash the DNA with 0.5 mL of 70 % ethanol cooled to −20 °C.

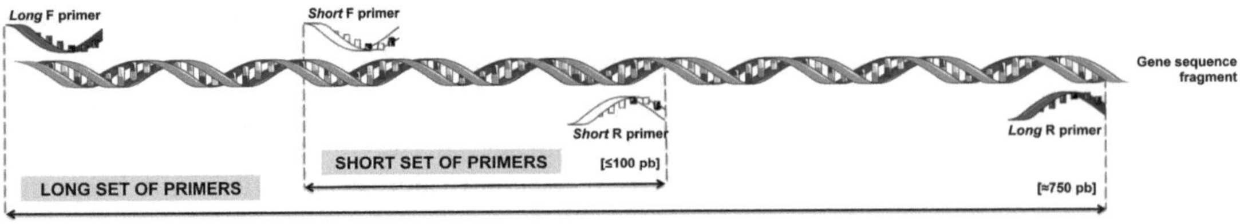


**Fig. 2** Schematic representation of primer design. Two sets of primers are needed for the protocol: (1) *Long primers*: Generated product is ~750 bp; and (2) *Short primers*: generated product is <100 bp. The small amplicon must be contained within the big amplicon

8. Centrifuge the sample for 5 min at 13,200 × *g*. Discard the ethanol with a micropipette and, if needed, use a pulse centrifuge to collect as much ethanol residue as possible, then remove it with a micropipette.
9. Maintain the tubes open for 10–15 min to let the ethanol evaporate.
10. Carefully, resuspend DNA in 30 μL of TE buffer.
11. Evaluate DNA quantity and purity using a Nanodrop spectrometer. Select only high purity ($A_{260}/A_{280} > 1.8$) samples for q-PCR analysis.
12. Dilute DNA samples to the desired concentration for q-PCR, usually 20–25 ng/μL. Keep the DNA samples stored at −80 °C if they are not immediately processed (*see* **Note 11**).

### 3.4 Primer Design

1. Select sequences of the studied genes from a database.
2. Using primer design software, create two pairs of primers for each studied gene. The first set of primers should generate a longer amplicon of around 700–800 bp, and the second should create a shorter one (≤100 bp); this short fragment must be designed to be within the larger amplicon (Fig. 2). Primers should have the following characteristics: 19–25 bases long, G+C content not higher than 70 %, and melting temperature (Tm) between 58 °C and 62 °C (*see* **Note 12**).

### 3.5 Primer Validation and Establishment of Optimal Melting Temperature

1. Check optimal primer melting temperatures (Tm) by conventional PCR in a gradient thermocycler in order to confirm the product size and optimized Tm. Try varying the temperature around the Tm suggested by the primer design software (±1 or 2 °C). Use a commercial kit for this purpose, with a final concentration of around 20 ng/μL in the final reaction mix.
2. Carry out a standard PCR cycle adapted for the expected product length (*see* **Note 13**).
3. Separate the products using 2 % agarose gel electrophoresis to verify their size, and select the optimal Tm (the one yielding the cleanest specific band) for q-PCR assays.

**3.6 q-PCR Validation**

1. Evaluate the efficiency of the designed primers sets for validating q-PCR results. Prepare a dilution series starting from 1 μg/mL DNA sample from 1:10 to 1:100,000, corresponding to approximately 1000–0.01 ng DNA.
2. Amplify each dilution for each gene and set of primers (long and short) using the cycling conditions described below (Subheading 3.7, **steps 1** and **2**).
3. Acceptable efficiencies should range from 80% to 120 %. The relation between Ct and the DNA dilution value must be linear, ideally with regression, $R^2 = 1$.

**3.7 q-PCR**

Perform the q-PCR analysis for long and short fragments for the same gene simultaneously in the same 96 plate in the real-time system (*see* **Note 14**).

1. Prepare the reaction mixture for long fragments (20 μL): 4 μL 5× Fast Start DNA Master plus SyBr Green I (Roche), 1 μL each 10 μM forward and reverse primer, 0.4 μL 50× ROS passive reference dye (Bio-Rad), template DNA (3 ng) and bidistilled water up to 20 μL (*see* **Note 15**).
2. Prepare the reaction mixture for short fragments (20 μL): 10 μL 2× SYBER Green PRC (Applied Biosystems), 1 μL each 10 μM forward and reverse primers, template DNA (3 ng) (*see* **Note 16**).
3. Design the plate conditions for the q-PCR system.
4. Run a q-PCR cycle in a quantitative thermocycler consisting of a pre-incubation phase of 10 min at 95 °C followed by 40 cycles of 15 s at 95 °C, 10 s at the melting temperature of each primer set, and 50 s at 72 °C, with a final extension period of 7 min at 72 °C.
5. Always check product specificity by running a 2 % agarose gel electrophoresis.

**3.8 DNA Lesion Analysis**

Rothfuss and colleagues initially described the following analysis in 2010 (Rothfuss et al., 2010):

1. Calculate the difference in Ct value determined by q-PCR analysis between the cryopreserved (treated) and the fresh samples. Repeat this calculation for each long and short fragment for each gene. In this way, a collection of "Δlong" and "Δshort" variables will be generated.
2. Replace the reported values in this formula described by Rothfuss:

$$\text{Lesion rate [Lesion per 10kb DNA]} = \left(1 - 2^{-(\Delta\text{long}-\Delta\text{short})}\right) \times \frac{1000[\text{bp}]}{\text{size of long fragment [bp]}}$$

## 4 Notes

1. Note that any biological sample that has not been tested for infectious diseases is potentially hazardous, thus compliance with all health and safety measures is paramount in this respect.
2. If evaluation of motility is needed, keep the samples at 37 °C until this has been performed.
3. If motility is going to be analysed, avoid using plastic pipettes as they can negatively influence this parameter.
4. The most common haemocytometer types regularly used for sperm concentration evaluation are: Neubauer chamber and Makler chamber. Computer Assisted Sperm Analysis (CASA) software can also be used for motility and concentration evaluation.
5. If cell type selection is needed for a particular experiment (for example, a study may focus only on motile cells), perform a density gradient adapted to the requirements.
6. Loading French straws require practice. Training may be required before performing the protocol in an experimental setting. Using a P-1000 micropipette is the most common method of loading the straws; however, a system for loading French straws can be created easily with a syringe, a piece of thin plastic tube, and a P-200 tip. Attach the tip to the extreme end of the tube by its narrower end, and then connect this item to the syringe. Place the cotton plug part of the French straw into the tip, and the other end inside the mixture of the cryoprotectants and ejaculate. Aspire 0.5 mL of the mixture until the French straw is loaded (Fig. 3). When working with a lot of different samples, the use of coloured straws facilitates tracing afterwards, and avoids mistakes.
7. To create a cooling device, cut two identical square pieces of polystyrene with a width corresponding to the distance of exposure in your cryopreservation protocol. Then, remove the internal part of the squares to create two perimeters. Cut a plastic rigid net, or similar, to the same dimensions of the polystyrene perimeters. Glue the three parts together (Fig. 4).
8. Working with liquid nitrogen can be dangerous. Extreme caution and specific health and safety guidelines must always be followed.
9. Note that aqueous phases present very dense mucus textures that might hamper their recovery with a micropipette. It is recommended to attempt this step very slowly and carefully, trying to collect as much volume as possible without compromising the sample. It is preferable to work with a smaller volume than to risk contamination with the interphase.If it is not possible to recover the full starting volume, top up with TE to reach the required volume.

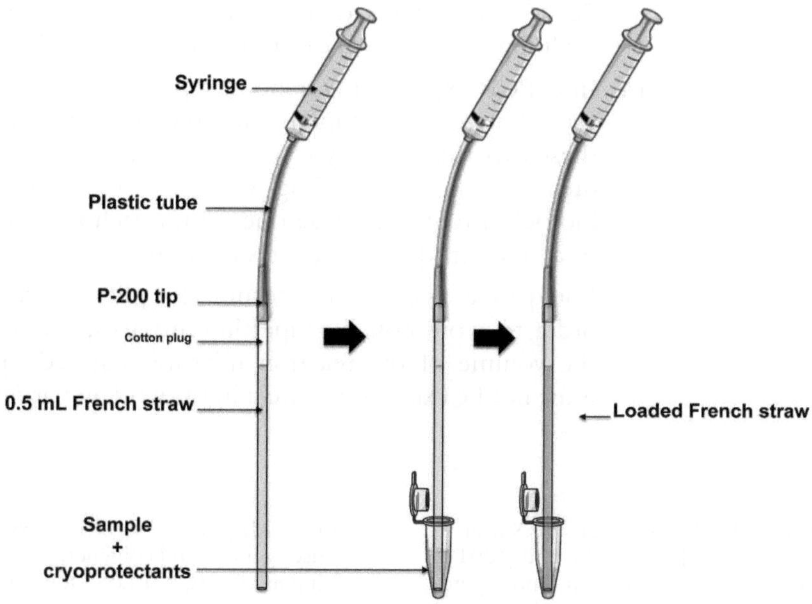


**Fig. 3** Preparing a system for loading French straws. A syringe, a P-200 tip and a plastic tube create a helpful device for loading samples into French straws

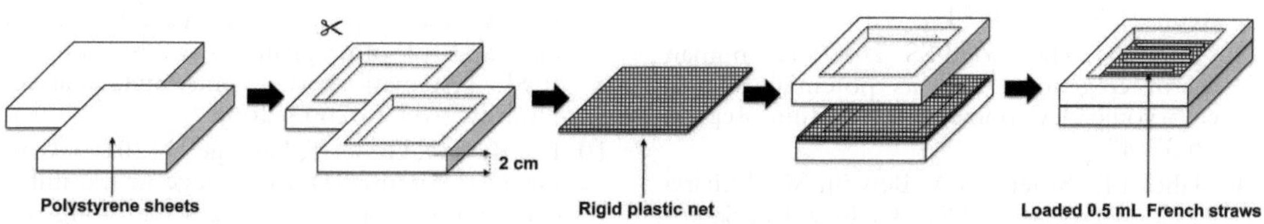


**Fig. 4** Preparing a cooling device. Using polystyrene sheets and a rigid plastic net, a cooling device for French straw cryopreservation can be generated by hand in the laboratory. Adapt sheet width according to the protocol

10. All discarded solutions must be properly processed.
11. If samples are not processed immediately, DNA samples should be stored at −80 °C since, due to the high accuracy of the technique, −20 °C storage could induce detectable alterations in the DNA. For this reason, it is better to process both control and experimental samples instantly processed, or, if this is not possible, they should be kept for the same period of time at −80 °C.
12. It is possible to use the forward or reverse primer from the long amplicon as the forward or reverse primer for the short amplicon. In this way, you can reduce the number of primers you need to design, reducing both the time required and the cost.
13. A conventional PCR cycle usually consists of 5 min at 95 °C followed by 30 cycles of 30 s at 95 °C and 30 s at Tm, and

72 °C at the required amplification time (reference: 60 s for 1000 bp), with a final extension of 7 min at 72 °C.

14. In q-PCR experiments, pipetting precision is crucial. Always use calibrated micropipettes and the same set of materials for these experiments. Try to perform all procedures in a reproducible manner, avoiding variation among technical and biological replicates. Use one tip for each pipetting in order to avoid introducing changes in volumes.
15. For q-PCR assays, three technical replicates are needed. In order to avoid possible pipetting mistakes, prepare 1.5 times the volume of the reaction mixture required. q-PCR dyes must not be exposed to direct light (work in partial darkness).

## References

1. van Montfoort AP, Hanssen LL, de Sutter P, Viville S, Geraedts JP, de Boer P (2012) Assisted reproduction treatment and epigenetic inheritance. Hum Reprod Update 8 (2):171–197
2. Amor DJ, Halliday J (2008) A review of known imprinting syndromes and their association with assisted reproduction technologies. Hum Reprod 23:2826–2834
3. Carrell DT, Hammoud SS (2010) The human sperm epigenome and its potential role in embryonic development. Mol Hum Reprod 16:37–47
4. Maher ER, Brueton LA, Bowdin SC, Luharia A, Cooper W, Cole TR, Macdonald F, Sampson JR, Barratt CL, Reik W et al (2003) Beckwith-Wiedemann syndrome and assisted reproduction technology (ART). Journal of Medical Genetics 40:62–64
5. Maher ER, Afnan M, Barratt CL (2003) Epigenetic risks related to assisted reproductive technologies: epigenetics, imprinting, ART and icebergs? Human Reproduction 18:2508–2511
6. Valcarce DG, Cartón-García F, Herráez MP, Robles V (2013) Effect of cryopreservation on human sperm messenger RNAs crucial for fertilization and early embryo development. Cryobiology 67(1):84–90
7. Riesco MF, Robles V (2013) Cryopreservation causes genetic and epigenetic changes in zebrafish genital ridges. PLoS One 8(6):e67614
8. Riesco MF, Robles V (2012) Quantification of DNA damage by q-PCR in cryopreserved zebrafish primordial germ cells. J Appl Ichthyol 28:925–929
9. Zini A, Boman JM, Belzile E, Ciampi A (2008) Sperm DNA damage is associated with an increased risk of pregnancy loss after IVF and ICSI: systematic review and meta-analysis. Hum Reprod 23:2663–2668
10. Rothfuss O, Gasser T, Patenge N (2010) Analysis of differential DNA damage in the mitochondrial genome employing a semi-long run real-time PCR approach. Nucleic Acids Research 38:e24
11. Cartón-García F, Riesco MF, Cabrita E, Martínez-Pastor F, Herráez MP, Robles V (2013) Quantification of lesions in nuclear and mitochondrial genes of *Sparus aurata* cryopreserved sperm. Aquaculture 402–403: 106–112
12. Valcarce DG, Cartón-García F, Riesco MF, Herráez MP, Robles V (2013) Analysis of DNA damage after human sperm cryopreservation in genes crucial for fertilization and early embryo development. Andrology 1(5):723–730
13. World Health Organization (2010) WHO laboratory manual for the examination and processing of human semen

# Chapter 20

# Gavi-Automated Vitrification Instrument


## Tammie K. Roy, Susanna Brandi, and Teija T. Peura



## Abstract

Gavi is intended for use in a laboratory or clinic environment for the preparation and vitrification of oocytes, cleavage stage embryos and blastocysts. Gavi is designed to automate the equilibration steps in the vitrification process to minimize the variability that occurs during cryopreservation. This automated process reduces the potential for errors and ensures a standardized, repeatable procedure for vitrification in a controlled, closed-system environment.



**Key words** Automation, Cryopreservation, Embryo, Gavi, IVF, Standardization, Vitrification


## 1 Introduction

Cryopreserving oocytes and embryos is essential to the operation of in vitro fertilization (IVF) clinics around the world. Cryopreservation allows clinics to maximize their pregnancy rates from a single oocyte collection procedure as well as to ensure the safety of patients through a reduction in ovarian hyperstimulation syndrome (OHSS). There are two main methods routinely used for the cryopreservation of human oocytes and embryos: slow freezing and vitrification. Vitrification is often described as the most effective method to cryopreserve oocytes and embryos [1]. Vitrification utilizes high concentrations of cryoprotectants along with ultra-rapid cooling to preserve cells in a glass-like (vitreous) state preventing detrimental ice crystal formation [1]. Clinics that have mastered vitrification report extremely high embryo recovery rates, in excess of 90 %, and pregnancy, live birth, and neonatal outcomes equivalent (or better) than those from fresh transfers [2–4].

**Electronic supplementary material:** Supplementary material is available in the online version of this chapter at 10.1007/978-1-4939-6828-2_20. Videos can also be accessed at http://www.springerimages.com/videos/978-1-4939-6826-8.


Zsolt Peter Nagy et al. (eds.), *Cryopreservation of Mammalian Gametes and Embryos: Methods and Protocols*, Methods in Molecular Biology, vol. 1568, DOI 10.1007/978-1-4939-6828-2_20, © Springer Science+Business Media LLC 2017

While these outcomes make vitrification very appealing, the actual process involves a number of manual steps, which can make it a time-consuming procedure that requires highly skilled operators. These can result in outcomes that may vary between operators and ART clinics [5], leading some clinics to train a limited number of staff that specialize in the technique to ensure consistency and outcomes.

There are currently a large variety of vitrification devices and protocols available to IVF clinics. For all vitrification devices and processes, there are numerous variables that need to be addressed. These include the type and concentration of cryoprotectants, temperature and timing of oocyte and embryo exposure, the rate of cooling and subsequent warming, and whether or not the embryo comes in direct contact with liquid nitrogen [6, 7]. All of these variables make it difficult to standardize the procedure when performed manually.

The high number of variables, as well as the high-labor and high-skill demands of vitrification, make it a good candidate for automation. Automation has the potential to provide a more standardized and repeatable vitrification process that will allow all users to achieve the best outcomes from their cryopreservation program. The Gavi system has been developed to automate the equilibration of oocytes and embryos prior to vitrification using a novel closed-system device [8]. The system has been designed to allow the control of a number of the variables of vitrification, including the concentration of cryoprotectants, exposure rates of the oocyte and embryo to the cryoprotectant, and the temperature and timing of exposure to these cryoprotectants, variables that can be difficult to control manually. The automated Gavi system has been shown to produce equivalent in vitro outcomes to that of the gold-standard Cryotop open system [8, 9]. The following method details the steps required for vitrification and warming of oocytes, cleavage stage and blastocyst stage embryos using the Gavi system.

## 2 Materials

The Gavi instrument has been developed as part of a system. The use of specific Gavi-related consumables will ensure that the best outcomes are achieved. The use of non-Gavi consumables may impact the embryo recovery and survival outcomes.

The Gavi Pod, Gavi Tip and Seal Cartridge, and Gavi Medium Cartridge are designed for single use only. Do not attempt to refill or reuse items.

Installation and setup of the Gavi instrument should be completed as per the Gavi user manual supplied with the Gavi instrument.

#### 2.1 Equipment and Consumables Required for Vitrification

The following is a list of the general equipment needed for the preparation of the Gavi consumables and accessories:

- Pipettes with sterile tips suitable for moving embryos.
- Pipette with flexible tip capable of dispensing 2 μL.
- Patient identification labels or xylene-free permanent marker.
- PPE required for handling liquid nitrogen.
- Liquid nitrogen.
- Liquid nitrogen storage Dewar's.
- VitBase solution (*see* **Note 1**).
- Two 4-well culture dishes.
- Microscope with a non-heated stage.
- Gavi Pods.
- Gavi cassettes.
- Gavi Tip and Seal Cartridges.
- Gavi Medium Cartridges (*see* **Note 2**).
- Gavi tweezers.
- Gavi operating tray.
- Gavi storage dividers already in liquid nitrogen storage Dewar's.
- 37 °C incubator with gas off.
- Timer with count-up function.

#### 2.2 Equipment and Consumables Required for Warming

1. The following is a list of the general equipment needed for the warming of pods vitrified on the Gavi:
   - Microscope with heated stage turned OFF.
   - Water bath set at 37 °C.
   - Suitable 4-well IVF dishes.
   - Pipettes—P1000 and P10/20 with sterile tips.
   - PPE required for handling liquid nitrogen.
   - $LN_2$ bucket.
   - Gavi working station.
   - Sealed Gavi Pods from Gavi vitrification.
   - Gems Warming Set (WarmSol1, WarmSol2, and WarmSol3) (*see* **Note 3**).
   - MilliQ/deionized/tap water.
   - Calibrated timer.
   - Tweezers.
   - Kimwipes/paper towels.

## 3 Methods

All steps should be completed at room temperature unless otherwise specified.

### 3.1 Preparation of VitBase Dishes

1. VitBase dish for oocyte and embryo equilibration: Prepare and label a 4-well culture dish. Add 500 μL of VitBase in each well required. For example, if three oocytes/embryos are to be vitrified, add 500 μL of VitBase in each of three wells. Place the dish in a 37 °C non-gassed incubator. Allow sufficient time for the VitBase to equilibrate to 37 °C.
2. VitBase dish for pod loading: Prepare and label another 4-well culture dish with 500 μL of VitBase in a single well. Place the dish onto the bench top and allow sufficient time to warm to room temperature.

### 3.2 Switching on Gavi

1. Switch on Gavi using the power switch located on the left-hand side of the instrument.
2. Select the oocyte, cleavage or blastocyst protocol icon from the Protocol Selection Screen. Gavi will begin an internal warm-up process (*see* **Note 4**).

### 3.3 Preparation of the Gavi Operating Tray

1. Load each tip and seal cartridge into the operating tray's tip and seal cartridge dock so that the cartridge's loading tab sits just below the operating tray's top and the tip and seal cartridge handle clicks gently into place (Fig. 1).

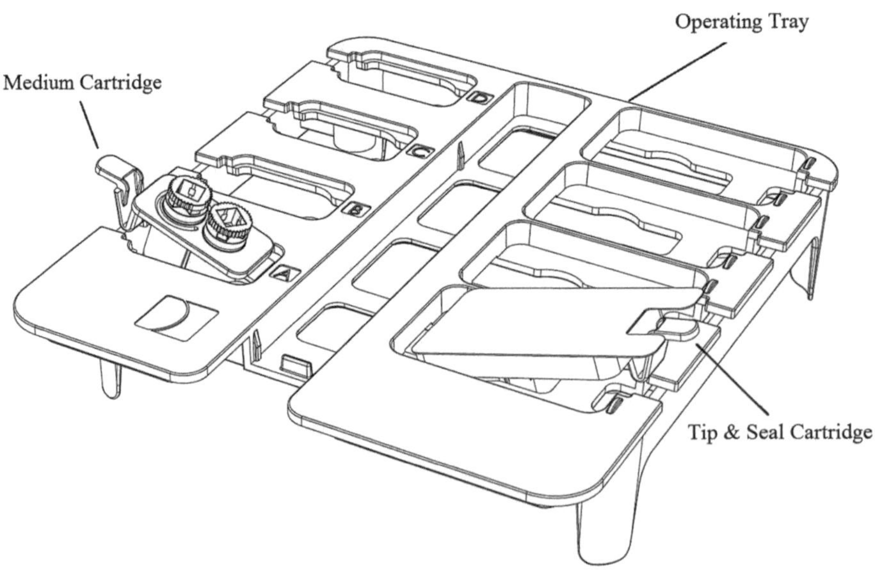


**Fig. 1** Loading of medium and tip and seal cartridges into operating tray. Load one consumable for each pod to be processed in the positioned as shown. Ensure that each consumable cartridge clicks into place. Always start loading from the A position

2. Load each medium cartridge into the operating tray's medium cartridge dock so that the cartridge's loading tab sits just below the operating tray's top and the medium cartridge handle gently clicks into place (Fig. 1).
3. Load the operating tray into Gavi. Open the Gavi access door and gently place the operating tray over the Gavi operating tray dock (*see* **Note 5**).
4. Do not remove the covers from the tip and seal cartridge or the tops from the medium cartridge vials at this stage.

### 3.4 Preparation of the Gavi Pods and Cassette

1. Place an identification label [or mark] onto both label areas of the cassette (*see* **Note 6**).
2. Remove each pod that is to be used from its packaging.
3. Place each pod into the cassette so that the pod's loading tab sits inside the cassette's pod bracket and the pod handle sits over the magnetic location holder on the cassette (Fig. 2) (*see* **Note 7**).
4. Place an identification label [or mark] onto the label area of each pod that is to be used.
5. Place the cassette with loaded but empty pods onto the bench top. To minimize the chance of debris falling into the empty pods, the cassette with loaded pods may be placed upside down on the bench top.

### 3.5 Preparation of the $LN_2$ Bucket

1. Fill the $LN_2$ bucket with liquid nitrogen. Ensure the $LN_2$ bucket is filled to the fill line and returned to its position on the Gavi instrument.
2. Place the lid on the $LN_2$ bucket to reduce liquid nitrogen evaporation.

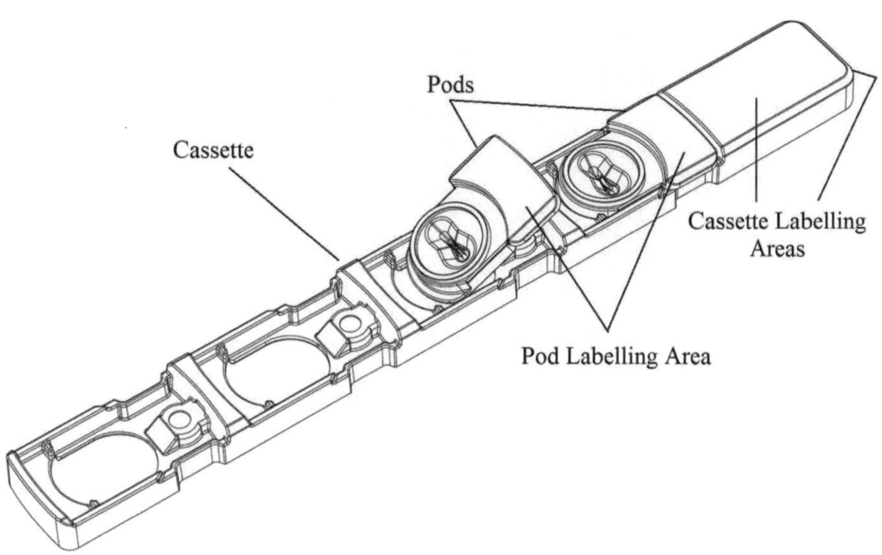


**Fig. 2** Insertion of pod/s into cassette. Ensure that pods are inserted as shown and fit securely into place before use. Label all areas of the pod and cassette as shown

### 3.6 Equilibration of Oocytes/Embryos in VitBase

1. Retrieve the oocytes/embryos to be vitrified, and the 4-well culture dish in the 37 °C un-gassed incubator prepared earlier. Place both onto the microscope stage (*see* **Note 8**).
2. Transfer the oocytes/embryos to be vitrified to the 4-well dish. Place the number of oocytes, cleavage embryos or blastocysts that will be vitrified in each pod into each well. Note: For blastocysts only 1 blastocyst should be vitrified in each pod. For oocytes and cleavage stage embryos 2 can be vitrified in a single pod.
3. Return both dishes to their respective incubators.
4. The 4-well dish with the oocytes/embryos to be vitrified should remain in the 37 °C un-gassed incubator for an additional 5 min. The 5 min timer should start once the dishes have been placed within the incubator.
5. The following two steps (final instrument preparation and preparation of pods with VitBase) are to be completed within the 5 min period in which the oocytes/embryos are equilibrating in VitBase (*see* **Note 9**).

### 3.7 Final Instrument Preparation

1. Remove the tops from the medium cartridge vials (Fig. 3).
2. Remove the covers from the tip and seal covers (Fig. 4).
3. Check there is sufficient $LN_2$ in the $LN_2$ bucket.
4. After Gavi has warmed up, the following warning message appears on the instrument screen: "Check $LN_2$ Levels." After checking sufficient $LN_2$ in $LN_2$ bucket, select the tick icon.
5. Use the + and – buttons to input the number of pods to be vitrified into the instrument screen. Based on the input, Gavi will highlight the items on the operating tray it expects to receive (*see* **Note 10**).

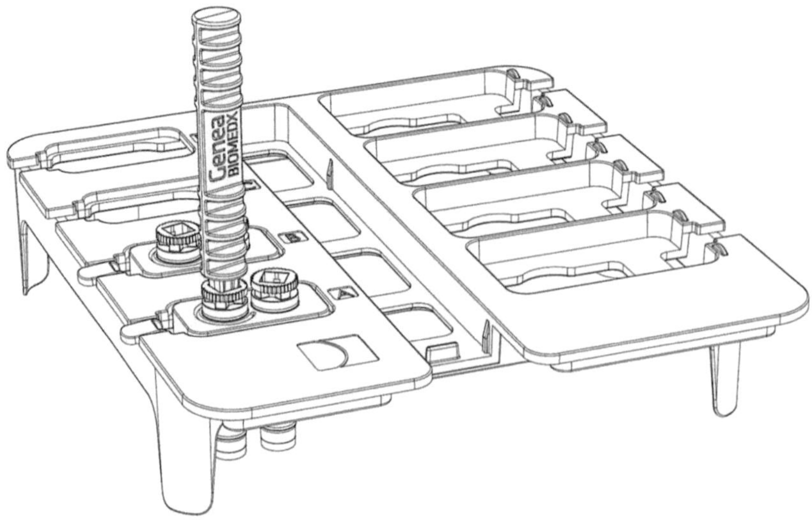


**Fig. 3** Removal of medium cartridge caps. Use the vial decapper as shown to carefully remove the caps from the medium cartridge vials

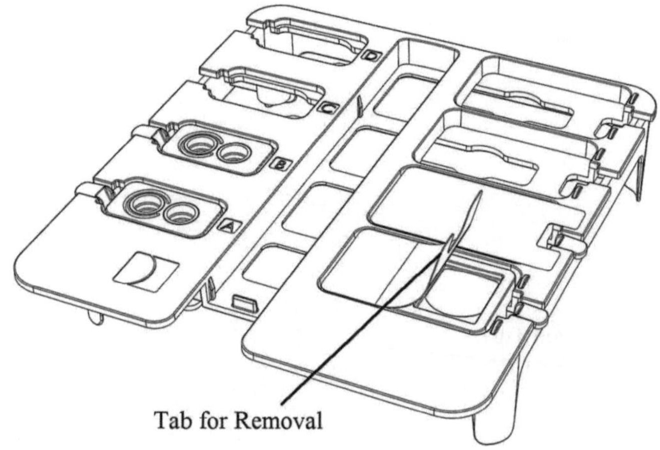


**Fig. 4.** Removal of covers from tip and seal cartridge. Carefully peel off the cover from the cartridge using the tabs. Be careful not to dislodge the position of the foil seal in the cartridge

### 3.8 Preparation of Gavi Pods with VitBase

1. Retrieve the previously prepared 4-well culture dish with 500 μL of VitBase in a single well equilibrating at room temperature.
2. Retrieve the cassette and pods prepared earlier and place under a microscope. Microscope stage heater should be OFF.
3. Set the pipette with flexible tip to 2 μL. Aspirate 2 μL of VitBase from the 4-well dish using the second stop of the pipette (*see* **Note 11**).
4. Complete the following steps under the microscope.
5. Place the pipette tip into the pod divot (Fig. 5).
6. Fill slowly the divot area first ensuring no bubbles are created.
7. Continue to dispense the remaining 2 μL (to the first pipette stop), by dragging the pipette tip to pipette well and then up to cover the entire pod channel. Ensure to spread the VitBase around the pod to fill area (Fig. 5) by gently dragging the pipette tip around the edge of the pod channel (*see* **Note 12**).
8. Using the same pipette with flexible tip, repeat the above steps for all pods that are to be run in the current cassette.
9. To minimize the risk of evaporation, the completion of this step should coincide with the end of the 5 min period that the oocytes/embryos are equilibrating in VitBase in the 37 °C un-gassed incubator. (*see* **Note 13**).

### 3.9 Loading of Oocytes/Embryos into the Pods

1. Retrieve the 4-well culture dish with embryos to be vitrified from the un-gassed 37 °C incubator. Place the dish on the microscope stage. Note: Microscope stage heater should be OFF (*see* **Note 14**).

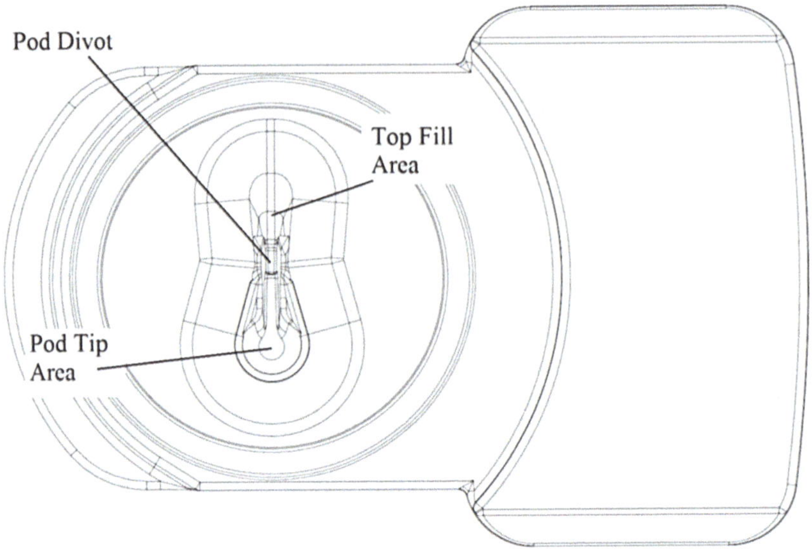


**Fig. 5** Pod loading with VitBase. In preparation for placing the oocyte/embryo in the pod, ensure that VitBase is loaded and spread evenly from the pipette tip area to the *top* fill area with no bubbles present in the pod

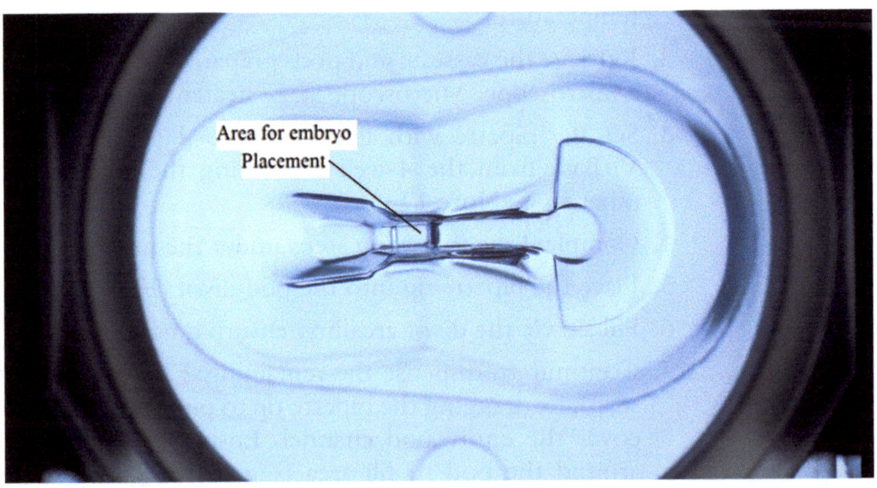


**Fig. 6** Oocyte and embryoloading into the pod. Oocytes/embryos should be placed in the area indicated. All pods should be checked for correct oocyte/embryo placement prior to placing the cassette on the instrument

2. Transfer the oocytes, cleavage embryos or blastocyst from the 4-well culture dish to the pod in pod dock 1. Place the oocytes/embryos such that they are positioned within the pod divot (Fig. 6) (*see* **Note 15**).
3. Repeat this step and load any remaining oocytes/embryos into each of the pre-prepared pods.
4. After placing all oocytes/embryos into their pods, perform a final check to ensure each oocyte/embryo has remained in the

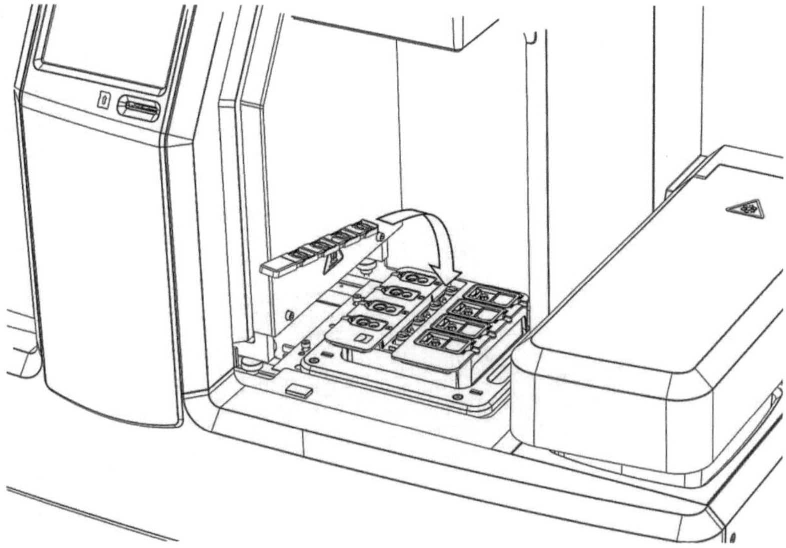

**Fig. 7** Placement of cassette onto Gavi. Carefully place the loaded cassette onto the Gavi instrument as shown in the figure above

pod divot. If they have moved, return them to the specified pod divot position.

### 3.10 Loading of Cassette into Gavi

1. Gently place the cassette into the operating tray cassette dock so that it sits flush against the back of the tray. Start by placing the distal end of the cassette onto the tray and then lowering the cassette handle onto the tray. The magnets in the cassette and pods will "clip" into place ensuring correct positioning (Fig. 7) (*see* **Note 16**).
2. Close the access door.

### 3.11 Running the Gavi Protocol

1. Select the "Play" icon on the Gavi screen (*see* **Note 17**).
2. As part of the protocol, Gavi automatically detects and displays error alerts if it has been loaded and prepared incorrectly (*see* **Note 18**).
3. The user interface will display a countdown timer showing the time remaining until protocol completion.
4. The first WARNING ALARM sounds with approximately 30 s remaining until protocol completion. At the first WARNING ALARM, immediately return to the Gavi instrument (*see* **Note 19**).
5. Open the $LN_2$ bucket cover. Remove the $LN_2$ bucket lid.
6. The second WARNING ALARM sounds at completion of the protocol. At the second WARNING ALARM, open the Gavi access door.

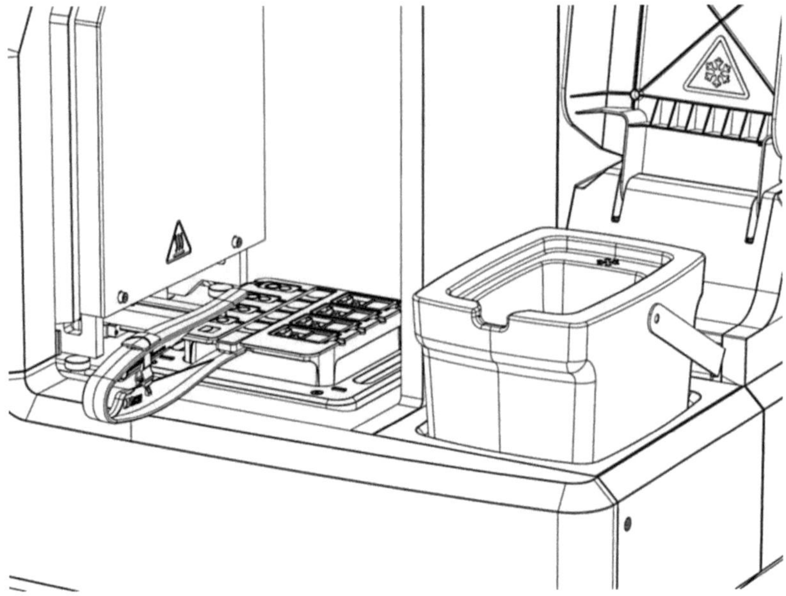

**Fig. 8** Removal of cassette from Gavi. On completion of the protocol, use the Gavi tweezers to grip the cassette at the handle as shown to ensure secure and fast removal of the cassette

7. The operating tray should be moving back to its original position.
8. When the operating tray has come to a complete stop, use the tweezers to grip the cassette handle (Fig. 8) (*see* **Note 20**).
9. Remove the cassette from the operating tray and immediately dunk the cassette into the $LN_2$ bucket. Ensure all pods are completely submerged. The cassette should be moved around within the liquid nitrogen in a swirling motion for a minimum of 5 s (Fig. 9) (*see* **Note 21**).
10. Release the cassette from the tweezers. Replace the $LN_2$ bucket lid to minimize liquid nitrogen evaporation.
11. Return to the Gavi screen. Select the "tick" icon to confirm completion of the protocol run.
12. After completion of the protocol run, remove the operating tray and dispose of the tip and seal cartridge and the medium cartridge.
13. When ready, remove the $LN_2$ bucket and transfer the cassettes to long-term storage Dewars (*see* **Note 22**).
14. The instrument can now be loaded with another operating tray and used again.

### 3.12 Preparation of Warming Dishes

1. Dishes for culture of warmed oocytes/embryos should be prepared as per laboratory protocol.

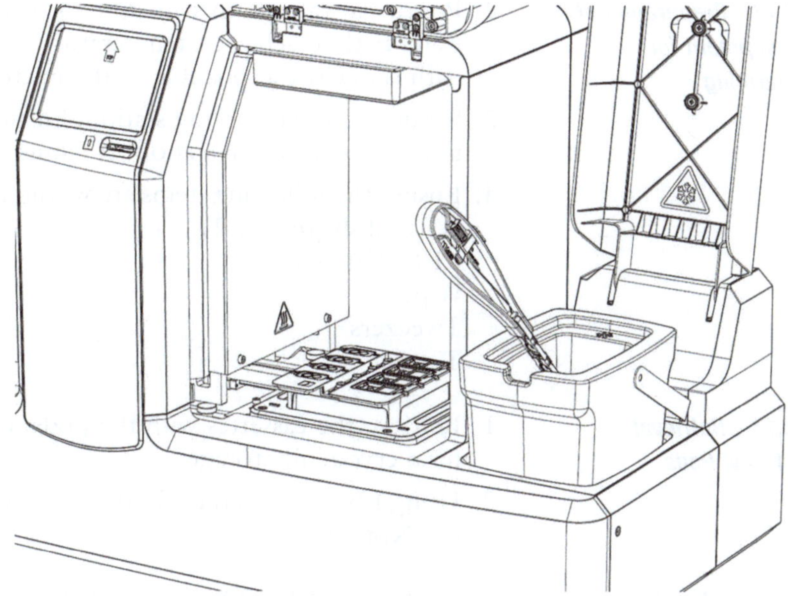

**Fig. 9** Dunking of cassette into $LN_2$ Bucket. Quickly dunk the cassette into the $LN_2$ bucket and agitate for at least 5 s

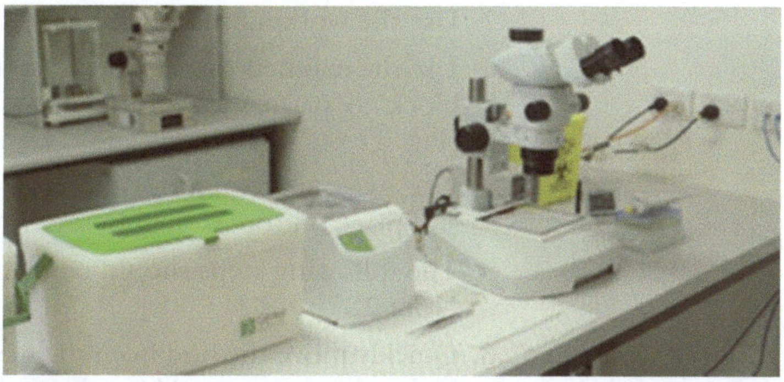

**Fig. 10** Warming setup. From *left* to *right*, the Gavi working station, water bath, and microscope

2. Retrieve a Gems Warming Set (*see* **Note 23**).
3. Using aseptic techniques, prepare a 4-well dish as follows:

   Aliquot 500 μL of WarmSol1 into well 1.
   Aliquot 500 μL of WarmSol2 into well 2.
   Aliquot 500 μL of WarmSol3 into well 3 and well 4.
4. Allow the dish to equilibrate to room temperature.
5. Prepare sufficient dishes for the number of oocytes/embryos to be warmed. Note: each dish can be used for a maximum of two rounds of warming (i.e., no more than two pods from the same patient) (*see* **Note 24**).

1. Set up a stereo microscope with the heated stage OFF.

### 3.13 Preparation of Equipment for Warming

2. Prepare a water bath with MilliQ or deionized water. Set the water bath to achieve a temperature of 37 °C. Ensure the water bath is as close as possible to the microscope (Fig. 10).
3. Set up the Gavi working station. Ensure there is sufficient $LN_2$ in the Gavi working station. If required, top up with $LN_2$.
4. Ensure the following items are within reach of the microscope:
   Warm dish (refer 4.2)
   P10/P20 pipette
   Wipes
   Tweezers
   Timer

### 3.14 Retrieval of Gavi Pods

1. Retrieve the cassettes with the pods to be warmed from long-term cryogenic storage.
2. Using $LN_2$ bucket, transfer the cassettes to the warming station (*see* **Note 25**).

### 3.15 Warming of Gavi Pods

1. Using the P10/P20 pipette, pre-load a sterile tip with 10 μL of WarmSol1 from well 1 of the warm dish. Keep the pipette within reaching distance to the microscope. Proceed quickly to the next step (*see* **Note 26**).
2. Use the working station to access the pod to be warmed.
3. Use the magnets on the bottom of the internal working station block to place the cassette on its back against the magnets (Fig. 11).
4. Use tweezers to remove the pods required from the cassette and place on top of the block (*see* **Note 27**).
5. The following steps should be completed in less than 20 s (*see* **Note 28**).
6. Quickly move the pod from the working station into the water bath using tweezers. Move the pod around the water for 2–3 s (*see* **Note 29**).

**Fig. 11** Working station. Place cassette on internal metal block using the magnets to locate and secure the cassette in place

7. As the pod is removed from the water bath, wipe the pod using a wipe/paper towel to remove water from the external surfaces of the pod.
8. Place the pod under the microscope.
9. Remove the lid seal from the pod. If there is difficulty in removing the lid seal, this step can be done on the bench before moving the pod under the microscope.
10. Carefully dispense the 10 μL of WarmSol1 from the pre-loaded pipette into the pod
11. For oocytes, cleavage embryos and blastocysts, the oocytes/embryos should remain in WarmSol1 in the pod for 1 min. During this time, locate the oocytes/embryos in the pod (*see* **Note 30**).
12. For oocytes and cleavage stage embryos *only*, transfer the oocytes/cleavage embryos to the WarmSol1 in well 1 of the warming dish prepared earlier and release the oocyte/embryo at the bottom of the dish. Wash the oocyte/embryo around in the solution three times before leaving it in the solution for a further 1 min.
13. Transfer the oocyte/embryo from the pod (blastocyst) or well 1 (oocyte/cleavage embryo) to WarmSol2 in well 2 of the warm dish. Release the oocyte/embryo at the bottom of well 2. Although the oocyte/embryo may float, do not move oocytes/embryos during this time. Leave the oocyte/embryo to re-equilibrate for three minutes. The oocyte/embryo should start to show signs of re-expansion.
14. After 3 min, transfer the oocyte/embryo from well 2 to WarmSol3 in well 3. Release the oocyte/embryo at the bottom of well 3. Although the oocyte/embryo may float, do not move the oocytes/embryos in this time. Leave the oocyte/embryo to re-equilibrate for 5 minutes. The oocyte/embryo should continue to re-expand.
15. After 5 min, transfer the oocyte/embryo from well 3 to WarmSol3 in well 4. Leave the oocyte/embryo in well 4 for 1 min. During this time gently move the oocyte/embryo around the well.
16. After 1 min, transfer the oocyte/embryo from well 4 into the standard embryo culture dish.
17. Repeat for each pod to be warmed (*see* **Note 31**).

## 4 Notes

1. VitBase is a HEPES buffered medium that is used to maintain oocytes/embryos for a short period of time in a non-gassed environment.
2. Gavi Medium Cartridge contains two solutions. The first is an equilibration solution that contains 8 % ethylene glycol (EG) and 8 % DMSO. The second is the vitrification solution that contains 16 % EG, 16 % DMSO, and 0.68 M trehalose. Only these media should be used in conjunction with the Gavi instrument.
3. Gems Warming Set is used for the warming of embryos that have undergone vitrification using the Gavi instrument. The set contains three solutions. WarmSol1 is a 1 M trehalose solution, WarmSol2 is a 0.5 M trehalose solution, and WarmSol3 is VitBase. Only these media should be used in conjunction with the Gavi instrument.
4. It is important to allow the Gavi instrument time to warm up prior to use. Completing this step at this time allows the instrument to be ready for use once the oocytes/embryos are loaded into the pods.
5. Gavi is equipped with heating and cooling elements beneath the operating tray dock to equilibrate the Gavi Medium Cartridge to the set temperature. Therefore, it is important to load the operating tray with the consumables at this point to allow equilibration of the medium cartridge to the correct temperature.
6. Cassettes and pods can be labeled with either specially designed labels or with a marker that is suitable to use with $LN_2$. If using pre-purchased labels, a pod label can be placed on the top of the cassette in order to help identification when stored vertically.
7. When less than four pods are to be vitrified, the pods should be arranged sequentially starting at the proximal end (closer to the label area) of the cassette. For example, if two pods are to be vitrified, the pods should be placed in pod location A and pod location B only and not into any other location.
8. VitBase is the initial holding solution for oocytes/embryos being processed in Gavi. Oocytes/embryos require equilibration in VitBase before being loaded into pods. This ensures the oocytes/embryos remain in place in the pod during processing on the instrument.
9. This timing is critical and should be practiced to ensure it can be adhered to prior to clinical use of instrument.

10. Inputting the correct number of pods is vital as the instrument will complete checks of consumables and process only the positions selected on this screen.
11. Loading the correct amount of VitBase into the pod is crucial to ensuring the success of the instrument. The instrument is programmed to remove a specified amount of VitBase prior to dispensing the equilibration solution from the medium cartridge. If too much VitBase is loaded at this step, it can result in dilution of the equilibration solution. If too little VitBase is loaded into the pod, this may result in over-draining of the pod which could lead to drying out of the divot area where the oocyte/embryo resides. Reverse pipetting is also recommended in order to prevent any bubbles being created in the pod during loading.
12. It is important at this stage to ensure the divot is full of VitBase and has no bubbles. If a bubble forms in the divot of the pod please discard the pod and prepare a new one. Bubbles have the potential to either dislodge the oocyte/embryo from its correct position or attach to the oocyte/embryo and cause floating. Both of these have the potential to impact the recovery of the embryo. The pipette well must also be filled properly with VitBase. This ensures that the pipette tip will have contact with the VitBase to ensure its correct removal by the instrument. Staff should be competent in this process prior to clinical use of instrument.
13. As the pods have been loaded with a small volume, it is very important that these steps are completed to coincide with the end of the 5 min period that the oocytes/embryos have been equilibrating in the VitBase at 37 °C in an un-gassed environment. If there is too long between loading the pods and placing the oocytes/embryos in them you risk evaporation reducing the volume of VitBase which may impact the osmolality of the solution and also the ability of the instrument to remove the correct amount of VitBase. Staff should be competent in this process prior to clinical use of instrument.
14. Microscope stage needs to be off to limit evaporation from the VitBase that is loaded into the pods.
15. The pods are designed to hold 2 oocytes, 2 cleavage embryos but only one blastocyst. Increasing the number of each of these types loaded into each pod may impact recovery and survival. It is important to ensure the oocyte/embryo is placed and remains within the pod divot. Incorrect positioning of the oocyte/embryo may result in incorrect processing of the oocyte/embryo by Gavi.
16. While oocytes/embryos are very stable when loaded into the pods in the correct position, some care should be taken to not drop the cassette too roughly into position.

17. You can abort a run at any time by pressing the "red cross" icon.
18. Gavi automatically detects and displays error alerts if it has been loaded and prepared incorrectly. If an error is detected it is important to ensure the problem is fixed prior to pressing play again as the instrument will only process the pods that are selected and assessed to be correct during the instrument check. More details of these errors can be found in the Gavi user manual.
19. While the instrument is running the protocol, the embryologist is free to leave. If there is more than one patient that has oocytes/embryos to be vitrified, the time can be used to start to set up the next operating tray and prepare the oocytes/embryos and pods.
20. Using the supplied tweezers ensures that the cassette is held in a safe and secure position while dunking the cassette and pods into the $LN_2$ bucket. This aids in ensuring the vitrification occurs as quickly as possible.
21. The cassette should be placed into the $LN_2$ as quickly as possible to ensure that vitrification takes place as soon as possible after protocol completion. This ensures that the time of exposure to the vitrification solution is as consistent as possible to reduce any chances of impact on the survival of the oocyte/embryo from cryoprotectant toxicity.
22. Take care to limit the exposure of the vitrified pods to room temperature during moving and storage. Exposure time should be less than 2 s.
23. Instrument development and testing has been completed using Gems Warming Set. Use of another warming set may impact embryo survival.
24. Limiting the use of the dishes will ensure that the solutions stay as pure as possible. With each oocyte/embryo movement, transfer of some of the previous solution occurs. This can change the concentration of the solutions which may impact how the oocyte/embryo is equilibrated and the cryoprotectants removed.
25. While not absolutely required, the working station allows the user to access the pods in a horizontal position while still remaining under $LN_2$. This ensures that the pod will remain in the vitrified state until the actual warming step. This prevents exposure of the oocyte/embryo to temperatures and conditions that may encourage ice crystal formation which can damage the cells and impact survival.
26. It is important to pre-load the solution to ensure that it can be added to the pod as quickly as possible. Care should be taken

not to leave this pre-loaded tip for too long as evaporation may occur which may impact oocyte/embryo survival when it is introduced into the pod during the subsequent steps.

27. At this stage the cassette could be returned to long-term storage or placed down the side of the working station if more than one pod is intended to be warmed.
28. This timing is critical to ensure the warming process is completed as quickly as possible and to ensure that oocytes/embryos are exposed to the first warming solution as quickly as possible after warming. This timing should be practiced to ensure it can be adhered to prior to clinical use of instrument.
29. This timing has been tested during development and has been shown to give the fastest warming rates. This timing is critical and should be practiced to ensure it can be adhered to prior to clinical use of instrument.
30. Oocytes and embryos tend to float in this solution; so if the oocyte/embryo is not immediately obvious, focusing of the pod area up and down through different focal planes will usually help in locating the oocyte/embryo. The oocyte/embryo is quite often found on the surface of the warming solution.
31. Gavi pods are single use only and should not be reused.

## References


1. Kuwayama M (2007) Highly efficient vitrification for cryopreservation of human oocytes and embryos: the Cryotop method. Theriogenology 67:73–80
2. Roy TK, Bradley CK, Bowman MC et al (2014) Single-embryo transfer of vitrified-warmed blastocysts yields equivalent live-birth rates and improved neonatal outcomes compared with fresh transfers. Fertil Steril 101:1294–1301
3. Kato O, Kawasaki N, Bodri D et al (2012) Neonatal outcome and birth defects in 6623 singletons born following minimal ovarian stimulation and vitrified versus fresh single embryo transfer. Eur J Obstet Gynecol Reprod Biol 161:46–50
4. Takahashi K, Mukaida T, Goto T et al (2005) Perinatal outcome of blastocyst transfer with vitrification using cryoloop: a 4-year follow-up study. Fertil Steril 84:88–92
5. Gosden R (2011) Cryopreservation: a cold look at technology for fertility preservation. Fertil Steril 96:264–268
6. Alpha Scientists in Reproductive Medicine (2012) The Alpha consensus meeting on cryopreservation key performance indicators and benchmarks: proceedings of an expert meeting. Reprod Biomed Online 25:146–167
7. Kader AA, Choi A, Orief Y et al (2009) Factors affecting the outcome of human blastocyst vitrification. Reprod Biol Endocrinol 7:99
8. Roy TK, Brandi S, Bradley CK et al (2016) Automatic vitrification: development of the Gavi system. In: Tucker MJ, Liebermann J (eds) Vitrification in Assisted Reproduction, 2nd edn. CRC Press, Florida
9. Roy TK, Brandi S, Tappe NM et al (2014) Embryo vitrification using a novel semi-automated closed system yields in vitro outcomes equivalent to manual Cryotop method. Hum Reprod 29(11):2431–2438

# Part V

## Appendices

# Appendix A: Cryotech® Vitrification Thawing

## Goral Gandhi, Masashige Kuwayama, Sakina Kagalwala, and Priyanka Pangerkar

### Abstract

In the last 15–20 years, many centers are using vitrification as a method of choice for cryopreservation of human oocytes and embryos. As vitrification technologies have improved their success profiles, new applications seem to have emerged, making IVF treatments more successful and more flexible.

This appendix describes the Cryotech® method, which is the latest "minimal volume approach" method, suitable for cryopreservation of oocytes and embryos of any developmental stage, including blastocysts. Dr. Masashige Kuwayama, who has introduced major advances in oocyte and embryo cryopreservation, has developed this method. A detailed protocol has been described with finer tips for the accurate use of the method for perfect survival and safety.

**Key words** Cryotech vitrification, Cryotech, Cryotech, Embryos, Blastocysts, Thawing

## 1 Introduction

Vitrification refers to the physical process by which a highly concentrated cryoprotective solution supercools to very low temperatures and finally solidifies into a metastable glass, without undergoing crystallization at a practical cooling rate. Samples are thus cryopreserved without detrimental intracellular ice formation.

Vitrification is now accepted as a very reliable and reproducible technique to cryopreserve human gametes. Its physical and biological principles are now much better understood. Laboratories the world over are reporting very high survival rates [1–3]. This is leading to more numerous and more successful applications of vitrification.

There are many areas where vitrification is playing a key role today.

1. The ultimate aim of any ART procedure is the birth of a single, live, healthy baby that can be achieved through an elective

Zsolt Peter Nagy et al. (eds.), *Cryopreservation of Mammalian Gametes and Embryos: Methods and Protocols*, Methods in Molecular Biology, vol. 1568, DOI 10.1007/978-1-4939-6828-2_21, © Springer Science+Business Media LLC 2017

single-embryo transfer. Consequently, if we want to reduce the number of embryos that we transfer, optimizing the embryo cryopreservation program is essential.

2. Ovarian hyperstimulation syndrome (OHSS) is a potentially life-threatening, iatrogenic complication of IVF resulting from the extensive use of gonadotropins [4]. A variety of preventive strategies have been adopted to reduce the risk of OHSS, but none of these methods have consistently demonstrated efficacy in preventing the syndrome except for cancellation of the cycle before hCG administration [5]. Although cycle cancellation is the most effective and safest approach, it is obviously frustrating and costly for the patients, and a second attempt of ovarian stimulation may lead to the same risk. Vitrification has enabled us to cryopreserve all eggs or embryos and avoid a fresh embryo transfer.
3. Minimal stimulation protocols are thought to produce few eggs, but eggs of better quality. Women with a poor ovarian reserve, who commonly do not respond to heavy stimulation, are left with few options when planning a family. One such option is embryo banking with a minimal stimulation IVF protocol. This approach allows women to have consecutive cycles before the follicular reserve is depleted. It will maximize the ovaries' already limited life span, allowing patients the opportunity to store embryos while oocyte production is still active. Results suggest the efficacy of this approach for patients who would normally be counselled for donor egg IVF [6, 7]. Minimal stimulation IVF is unthinkable without a strong vitrification program.
4. Oocyte vitrification is giving a chance to women to preserve their fertility and plan their families. A woman's ovarian reserve quantity decreases with advancing age, as does the quality of her oocytes (e.g., there is a higher incidence of chromosomal anomalies) that, in turn, reduces the chances of successful fertilization, implantation, and early embryo development [8]. By age 35, a woman has lost ~95 % of her eggs, and the rest are aging rapidly. Fertility preservation with egg banking is a great way for women to plan their families.

Today, we can make any IVF program much more efficient with the strong support of an excellent vitrification program.

During the development of vitrification methods, scientists have made great efforts to find innovative methods using minimal volume techniques to increase the thermal change during the process. A higher cooling rate can facilitate vitrification, and a higher warming rate can prevent devitrification at lower concentration of cryoprotectants. This can reduce toxicity from vitrification solutions. Minimum volume methods can prevent ice crystal formation.

The high rates of cooling and warming resulting from the minimum volume methods can prevent the harmful chilling injuries that happen between 15 and −15 °C [9]. Minimum volume methods can also avoid fracture injury.

Dr. Masashige Kuwayama has more than 20 years of laboratory and clinical experience with both animal and human oocytes and embryos. He has introduced several novel procedures to vitrify oocytes and embryos, including various dilution methods, as well as the Cryotop method and the Cryotip method [10, 11]. His minimal volume vitrification techniques are used worldwide producing excellent results. Cryotech vitrification, a minimal volume approach, is the latest method of oocyte and embryo cryopreservation introduced by Dr. Kuwayama. Using Cryotech method of vitrification, oocyte survival rate of 97.1 % [12] and embryo survival rate of 97.34 % have been reported [6].

## 2 Product Description

### 2.1 Solutions

Cryoprotectant toxicity is a fundamental limiting factor for the successful cryopreservation of living systems by both freezing and vitrification. Mixtures of cryoprotectants have less toxicity and are more effective than single-agent cryoprotectants. With two cryoprotectants, the concentration of each can be lower than that needed when either is used separately, thereby making the solution less toxic [13].

Extracellular disaccharides and macromolecules, such as sucrose and trehalose, are commonly added to vitrification solutions to lower the concentration of cryoprotectants. This helps to draw water out of the cell to attain better dehydration and reduce osmotic shock. Sucrose that is routinely used has been replaced with trehalose in Cryotech solutions, thus reducing endotoxicity due to sucrose [14].

There is no added serum and synthetic serum supplements in Cryotech vitrification solution. Hence, risk of serum-derived virus contamination is avoided. It contains hydroxypropyl cellulose (HPC), which has optimal high viscosity providing better resistance to damage during cooling, storage, and warming [15]. Vitrification solution is completely chemically defined and is stable for one year at 4–8 °C and 3 months at room temperature. Each component of vitrification and warming solution is specifically selected to get optimum results (Fig. 1).

### 2.2 Loading Device: "Cryotech"

The carrier device is a polypropylene strip attached to a plastic handle accompanied with a protective covering for safe handling and storage (Fig. 2). Longer and wider handle and sheet make it easy for writing details and loading samples. It can be used as "closed cooling" after sealing or "open cooling" before sealing (Fig. 3).

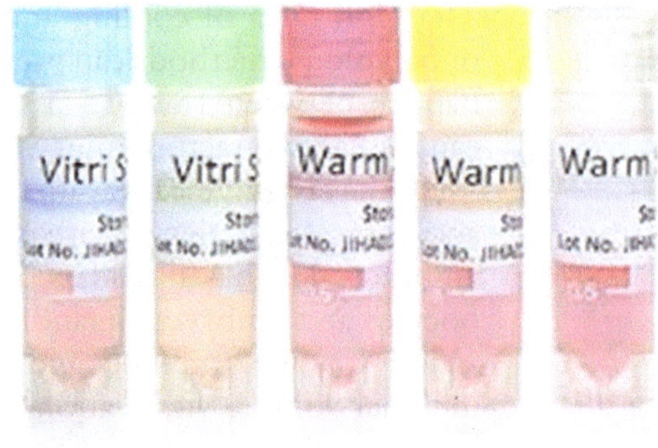


**Fig. 1** Cryotech vitrification and warm solutions

**Fig. 2** Cryotech carrier device

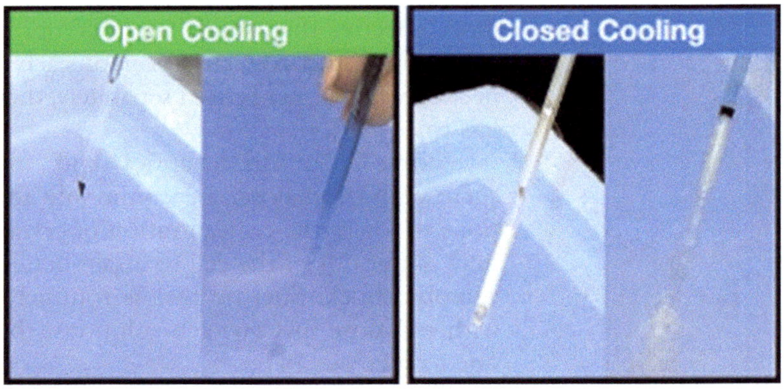


**Fig. 3** Cryotech carrier device can be used as both open and closed cooling system

### 2.3 Vitri Plate and Warm Plate

The Vitri plate (Fig. 4a) used for vitrification has a special holder for the Cryotech. Thus, the focus remains the same while washing the oocytes in vitrification solution and placing them on the Cryotech device. One hand of the embryologist is completely free while loading the device. It gives tremendous ease of handling to the embryologist, thus greatly increasing the overall efficiency of the method.

The warm plate (Fig. 4b) has a well for the first solution, called the "thawing solution (TS)" of the warming process followed by three other wells for diluent solution (DS) and washing solution (WS). This increases the speed and efficiency of the warming

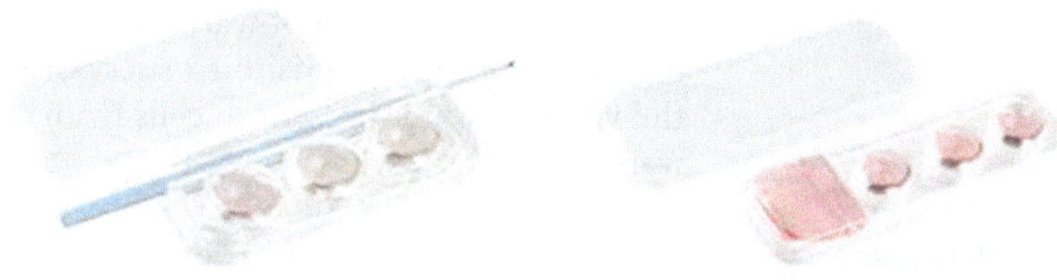

**Fig. 4** (**a**) *Left image*—Vitri plate, (**b**) *right image*—warm plate

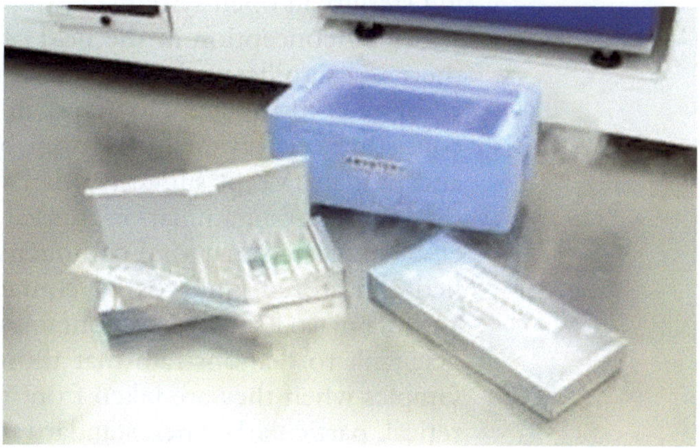

**Fig. 5** Cryotec Vitrification Kit and $LN_2$

process. High warming rates are extremely crucial to survival of the gametes, and the design of the warm plate facilitates very high warming rate, aiding in the overall efficiency of the process.

### 2.4 Protocol

The toxicity of a cryoprotectant is correlated to its concentration, the time of exposure, and the temperature.

*Time*: All cryoprotectants are toxic, and it is important to monitor the duration of exposure to the final cryoprotectant very carefully before plunging into liquid $N_2$. Survival rates after vitrification improved with the evolution of two-step protocols. The Cryotech method involves 10–15 min equilibration in the 50 % diluted solution of the final permeable cryoprotectant medium at room temperature and short exposure (<90 s) to the final concentrated cryoprotectant. This approach is relatively harmless for sensitive structures including human oocytes, while it increases considerably their survival rate and developmental competence [16].
*Temperature*: The fast entrance and the degree of toxicity of the cryoprotectant can be influenced by temperature. Equilibration at 37 °C avoids the re-expansion of the cell especially the first step of warming. Using CPAs at room temperature (22–25 °C) or lower rather than 37 °C may decrease their toxicity. The Cryotech method uses room temperature equilibration for significant improvement in survival rate.

Only the first step of warming is done at 37 °C as rapid warming rates are a prerequisite to successful vitrification as it prevents the vitreous water present in cells from crystallization at the time of warming.

### 2.5 Theoretical Risk of Cross Contamination vs. Successful Outcome

A high freezing rate and even higher warming rate are crucial to successful vitrification and survival [17]. This can be best achieved via direct contact between the sample and liquid nitrogen (open system vitrification). However, there are technical concerns regarding sterility in open systems.

A misconception in the field of reproductive medicine is that there is a significant risk of cross contamination during gamete or embryo cryostorage. However, in a recent article by Pomeroy consisting of a review of the available literature on animal models and human IVF, it suggests otherwise [18]. There is a negligible risk of cross contamination in IVF working conditions. Bielanski et al. suggested that viral pathogens could be transmitted to vitrified and stored embryos by contaminated nitrogen [19]. These authors placed small virus-contaminated particles of ice in a nitrogen tank to demonstrate that those particles could attach to the samples when they are taken from the tank. The risk is only theoretical, particularly when standard universal precautions are used.

In ART centers cross contamination by these agents and transmission between samples have not occurred, and no such disease transfer has yet been reported in an estimated of 600,000–1,000,000 vitrified embryos. At present, most embryos and oocytes are vitrified with open systems worldwide, indicating a higher overall efficiency and consistency.

## 3 Cryotech Vitrification Protocol

### 3.1 Materials Required

- Cryotech Vitrification Kit.

  Equilibration solution (ES).
  Vitrification solution (VS).
  Cryotech.
  Vitri plate.
- Microscope (turn off the heating plate).
- Stopwatch (with count up function).
- Tweezers.
- Micropipette for 300 μL.
- Cooling rack.
- Liquid nitrogen.

*Note*: Solutions are to be used within 30 days of opening the vials. Sterile aseptic conditions of dispensing should be maintained.

### 3.2 Preparation for Vitrification

1. Bring ES and VS vials to room temperature (25–27 °C) at least 1 h before vitrification.
2. Fill 90 % of the cooling rack with fresh liquid nitrogen.
3. Take the culture dish containing oocyte/embryo out from the incubator. Check the quality of the oocyte, embryo, or blastocyst.

*Note*: Use a right size Pasteur pipette: 140–170 μm for oocyte and cleavage-stage embryo and 170–300 μm for blastocyst. For oocyte vitrification take the cumulus cells off. Best timing of vitrification of blastocyst: the size (diameter) should be between 160 and 220 μm for perfect survival after vitrification.

### 3.3 Equilibration (12–15 min)

1. Write ES, VS1, and VS2 on the lid of the Vitri plate. Fill the wells of Vitri plate with 300 μL ES, 300 μL VS1, and VS2, respectively. Put the lid on the Vitri plate immediately (Fig. 6).
2. Aspirate the oocyte/embryo at the middle of the fine part of a Pasteur pipette.
3. Put the oocyte/embryo with small amount of medium on the surface of ES well (Fig. 7). Start the stopwatch. As oocyte/embryo sinks to the bottom, it will shrink (Fig. 7b).
4. Put the lid and wait for the recovery of the shrinkage. When the oocyte/embryo volume is completely recovered, it is the end of

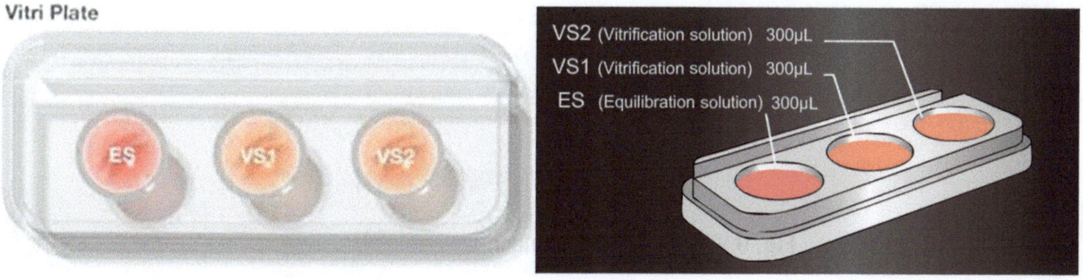


**Fig. 6** Vitri plate with vitrification solutions

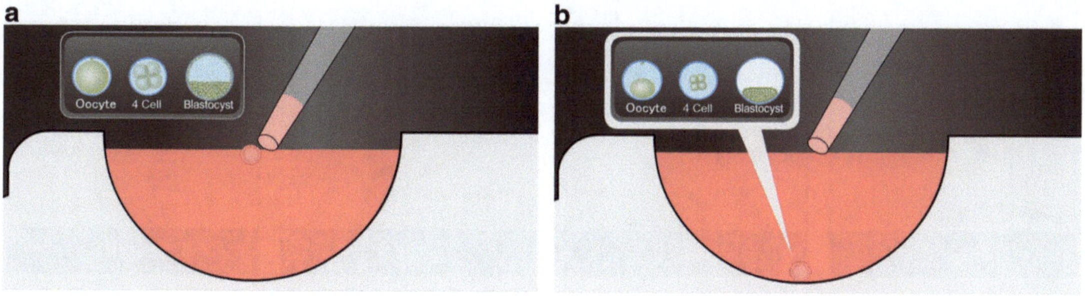


**Fig. 7** (**a**) Time 0 (**b**) After 90 seconds. Equilibration step—shrinking of the oocyte/embryo

this step. If you can't confirm the complete recovery, the limit time of this step is 15 min for oocyte and blastocyst and 12 min for 4–8 cell embryo (Fig. 8).

5. Write information of oocyte/embryo on the handle of Cryotech and set on Vitri plate (Fig. 9).

*Note*: Oocyte vitrification is complete when the width of perivitelline space becomes equal to the width before immersing in ES.

### *3.4 Vitrification Step in VS1 (30–40 s)*

1. Aspirate the oocyte/embryo with *ES* till the middle of the pipette. Transfer the oocyte/embryo at the middle depth of *VS1* with *ES* (Fig. 10, **step 1**).

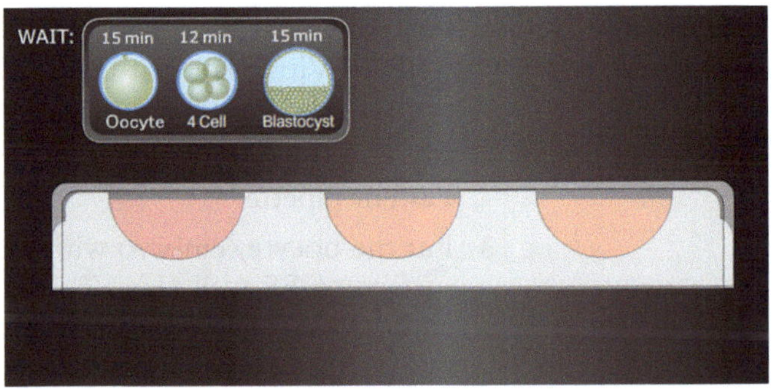


**Fig. 8** Equilibration step—recovery of the oocyte/embryo

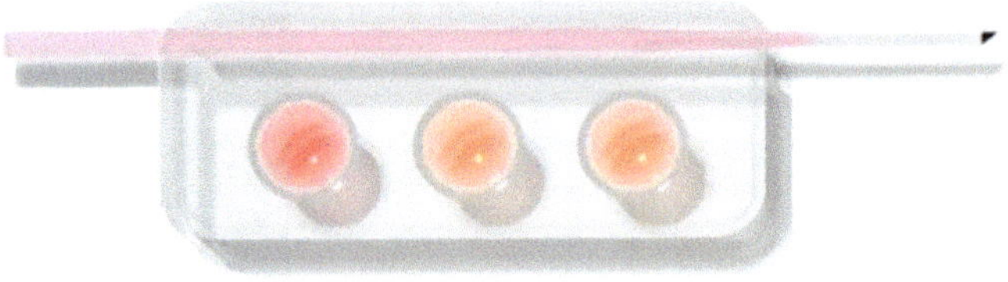

**Fig. 9** Vitri plate with Cryotech

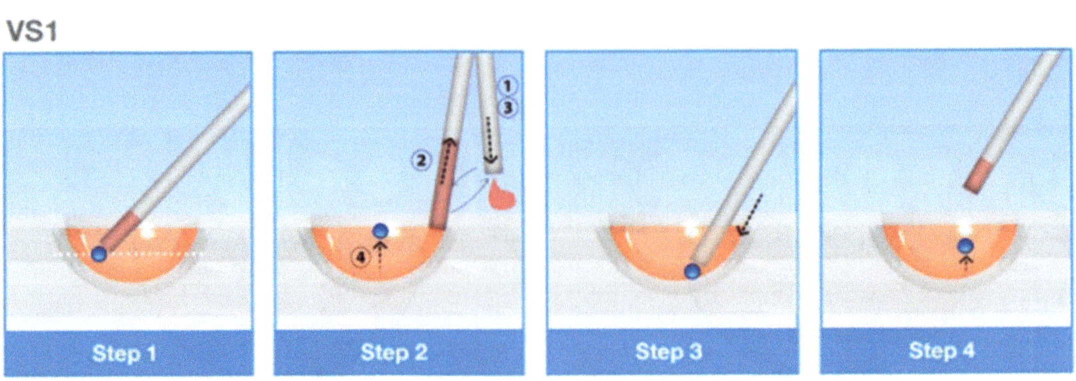


**Fig. 10** VS1 washing of oocyte/embryo (*steps 1–4*)

2. Expel the remaining *ES* outside the well (Fig. 10, **step 2/1**) and aspirate fresh *VS1* from the edge of the wall (Fig. 10, **step 2/2**). Oocyte/embryo floats immediately to the surface of *VS1* (Fig. 10, **step 2/3**).
3. Aspirate oocyte/embryo at the top inside of pipette. Transfer it again to the bottom of *VS1* (Fig. 10, **step 3**).
4. The oocyte/embryo floats slowly to the middle depth and stops (Fig. 10, **step 4**).
5. Expel *the* remaining *VS1* outside the well and aspirate fresh *VS2*. Aspirate oocyte/embryo at the top inside of the pipette.

### *3.5 Vitrification Step in VS2 (10–20 s)*

1. Transfer the oocyte/embryo to the middle depth of VS2 (Fig. 11, **step 5**).
2. Expel the remaining VS outside the well (Fig. 11, **step 6/1**) and aspirate fresh VS2 from the edge of the wall (Fig. 11, **step 6/2**).
3. Expel VS2 around the oocyte/embryo and mix the solution around oocyte/embryo to exchange the remaining previous solution (Fig. 11, **step** 7). Expel and wash inside of the pipette with fresh VS2.
4. Take oocyte/embryo at the top of pipette (Fig. 11, **step 8**).

### *3.6 Loading Cryotech*

1. Place the oocyte/embryo near the end of Cryotech sheet with minimal volume of VS2 (one oocyte/embryo per droplet is recommended) (Fig. 12). Immediately submerge the Cryotech into fresh liquid nitrogen.
2. Put the straw cap on Cryotech in the liquid nitrogen.

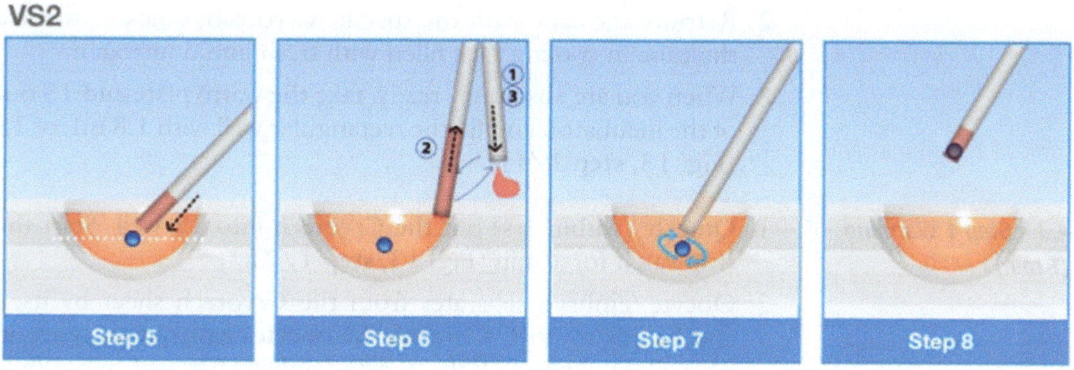


**Fig. 11** VS2 washing of oocyte/embryo (*steps 5–8*)

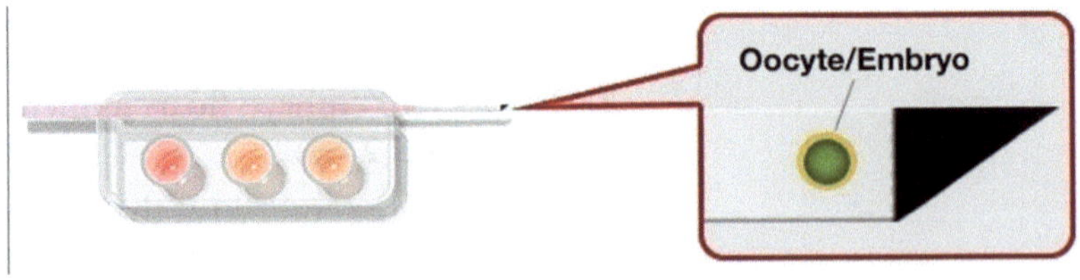


**Fig. 12** Loading of Cryotech

## 4 Cryotech Warming Protocol

### 4.1 Materials Required

- Cryotech Warming Kit.
  - Warming solution (TS).
  - Diluent solution (DS).
  - Washing solution (WS).
  - Warm plate.
- Microscope (turn off the heating plate).
- Stopwatch (with count up function).
- Tweezers.
- Micropipette for 300 μL.
- Cooling rack.
- Liquid nitrogen.

### 4.2 Preparation for Warming

1. Place the warm plate and TS vial (with cap) in the incubator at 37 °C >3 h before warming (overnight storage is recommended). Bring DS and WS vials to room temperature (25–27 °C) at least 1 h before warming.
2. Retrieve the cane with the specific Cryotech; quickly immerse the cane in cooling rack filled with fresh liquid nitrogen.
3. When you are absolutely ready, take the warm plate and TS out of the incubator, and fill the rectangular well with 1.8 mL of TS (Fig. 13, **step 1/1**).

### 4.3 Step 1 Warming (1 min)

1. Quickly (within 1 s) put the Cryotech into TS well. Start the stopwatch for 1 min (Fig. 13, **step 1/2**).
2. Oocyte/embryo separates from the Cryotech sheet by itself and begins to float. Confirm the oocyte/embryo existence in TS well. Do not touch the oocyte/embryo before 1 min. While waiting, fill the DS well with 300 μL of dilution solution (Fig. 13, **step 2/1**).

### 4.4 Step 2 Dilution (3 min)

1. At the end of 1 min, aspirate the oocyte/embryo and 3 mm long of TS into the pipette (Fig. 14, **step 1**).

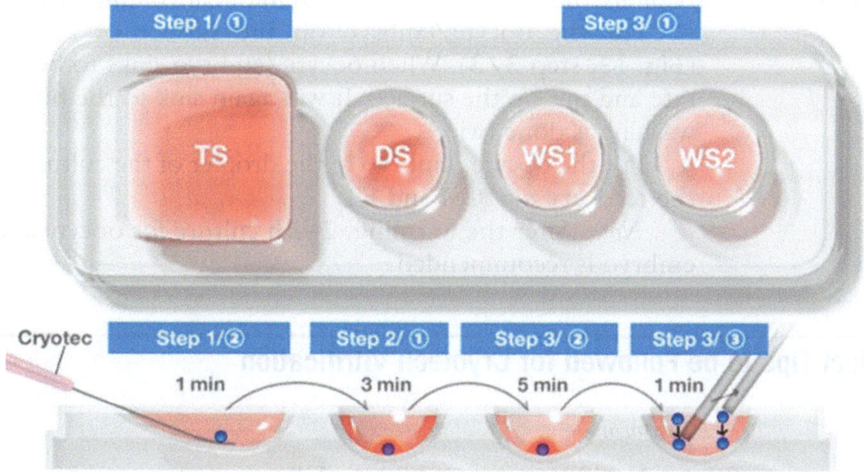


**Fig. 13** Procedure for warming (*steps 1–3*)

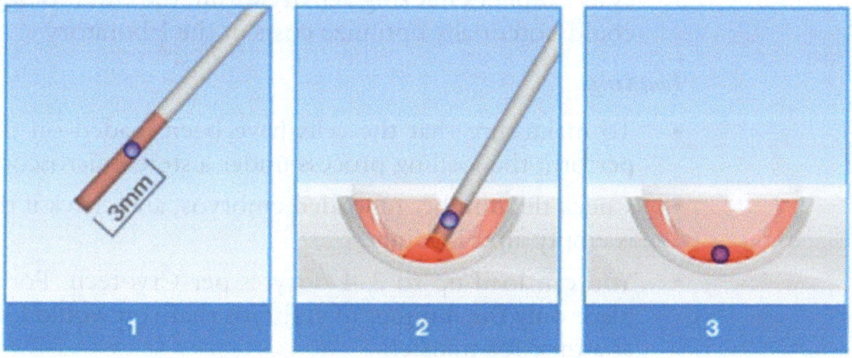


**Fig. 14** Gradual replacement of solutions (TS → DS, DS → WS1)

2. Transfer TS to the bottom center of DS (Fig. 14, **step 2**), and expel the oocyte/embryo slowly at the bottom of TS layer in DS well (Fig. 14, **step 3**). This is for most gradual displacement from TS to DS. Wait for 3 min. While waiting, fill the WS1 and WS2 well with 300 μL of washing solution.

### 4.5 Step 3 Washing

#### 4.5.1 Washing 1 (5 min)

1. Aspirate the oocyte/embryo and 3 mm long of DS into the pipette (Fig. 14, **step 1**).
2. Transfer DS to the bottom center of WS1 (Fig. 14, **step 2**), and expel the oocyte/embryo slowly at the bottom of DS layer in WS1 well (Fig. 14, **step 3**). This is for most gradual displacement from DS to WS1. Wait for 5 min.
3. Give a survival judgment at the end of this step depending on the recovery of the shrunken oocyte/embryo.

#### 4.5.2 Washing 2 (1 min)

Aspirate the oocyte/embryo with minimal volume of WS1.

Put the oocyte/embryo on the surface of the WS2 well (Fig. 13, **step 3/3**). When oocyte/embryo sinks to bottom, aspirate and put on the surface. It will again sink to the bottom; thus, washing is done twice.

Put the oocyte/embryo in the droplet of the culture media for the recovery for ICSI and ET.

*Note*: After the warming, 2–4 h culture for oocyte and 3 h for embryo is recommended.

## 5 Practical Tips to be Followed for Cryotech Vitrification

*Protocol*

- Vitrification is a highly skill-oriented technique, and close adherence to the recommended protocol of the manufacturer is absolutely essential to achieve optimal results.
- A single Cryotech protocol for oocytes, embryos, and blastocysts simplifies freezing activity within the embryology labs and could potentially optimize costs in the laboratory.

*Loading*

- To make sure that the cells have been loaded on the carrier, perform the loading process under a stereomicroscope.
- Check the number of loaded embryos, and check if the pipette is empty after loading.
- You can load up to 3–4 oocytes per Cryotech. For embryos, place only the number of embryos that you would like to thaw at a time for transfer.
- While using minimal volume techniques, each oocyte/embryo has to be loaded in separate drops while loading multiple oocyte/embryo on a single Cryotech. Don't load all in a single blob of solution. Smaller volumes allow better heat transfer, thus facilitating higher cooling and warming rates.

*Immersion in liquid nitrogen*

- During vitrification, higher cooling rates can be achieved by quickly passing through the damaging vapor phase above the liquid nitrogen.
- During vitrification after immersing in liquid nitrogen, do rapid front-back movement of Cryotech to disperse the layer of gaseous nitrogen around the surface of Cryotech for uniform cooling.

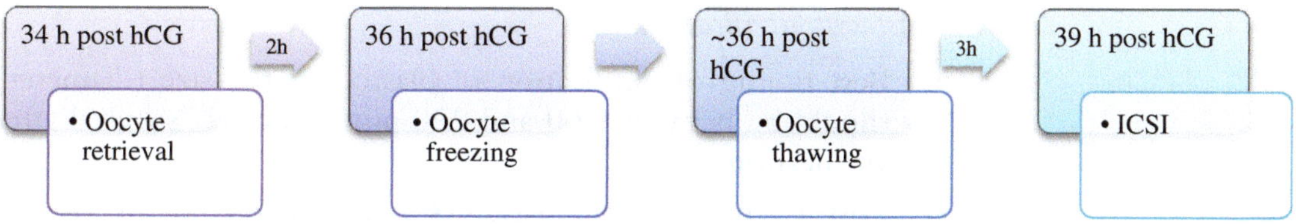


**Fig. 15** Timing of oocyte vitrification

*Warming*

- The first step of the procedure of immersing the Cryotech from liquid nitrogen into warm TS at 37 °C is most critical. This step must be performed with utmost efficiency to achieve very high warming rate.
- Move the sample from one solution to the other by keeping it in a small quantity of the previous solution and releasing it into the next solution. In this way the passage will be less shocking.

*Oocyte vitrification*

- Removing cumulus cells is better for vitrification of mature oocytes.
- Immature oocytes should be vitrified with the cumulus cells, as the cumulus cells may help the in vitro maturation of oocytes after warming.
- Timing of oocyte vitrification (Fig. 15)—the meiotic spindle of the oocyte is very sensitive to cryoprotectants and low temperature. Spindle undergoes transient disappearance or disorganization immediately after thawing. The spindle apparatus recovers after thawing and seems to do so without alterations in >80 % of the cases after 1–3 h of in vitro culture at 37–38 °C. Inseminating oocytes soon after thawing, when there is serious spindle disorganization, adversely affects fertilization outcome and growth of embryos. The optimum time for intracytoplasmic sperm injection (ICSI) for human oocytes is between 37 and 41 h post human chorionic gonadotropin (hCG). The time interval between oocyte thawing and ICSI is crucial for normal fertilization and subsequent development. Therefore, considering competing aspects of oocyte aging and spindle recovery, timing of events is essential for a successful oocyte cryopreservation program. Hence, oocyte recovery can be performed at 34 h post hCG, vitrification of oocytes being performed at 2 h after oocyte recovery, and ICSI being performed at 3 h post thaw (post hCG 39 h).

*Embryo vitrification*

- Freeze the number and grades of embryos per Cryotech as per the requirement for the frozen embryo transfer.

*Blastocyst vitrification*

- Best timing of vitrification of blastocyst: the size (diameter) should be between 160 and 220 μm for perfect survival after vitrification.
- Post-warming, viable blastocysts re-expand and are usually allowed 3 h of incubation to regain their vitality before being transferred. Re-expansion is the sign of viability.

*Storage and expiry*

- Heat dissipation occurs much faster in vitrified specimens as they have smaller surface area. As a consequence, anyone handling vitrified cells should be extremely cautious when transferring vitrified specimens to and from the storage dewar.
- Failure to check and top up dewar can have disastrous consequences for cryopreserved material, and top-up can be managed via a staff that is strictly adhered to.

## References


1. Chian RC, Kuwayama M, Tan L, Tan J, Kato O, Nagai T (2004) High survival rate of bovine oocytes matured in vitro following vitrification. J Reprod Dev 50(6):685–696
2. Katayama KP, Stehlik J, Kuwayama M, Kato O, Stehlik E (2003) High survival rate of vitrified human oocytes results in clinical pregnancy. Fertil Steril 80(1):223–224
3. Loutradi KE, Kolibianakis EM, Venetis CA, Papanikolaou EG, Pados G, Bontis I, Tarlatzis BC (2008) Cryopreservation of human embryos by vitrification or slow freezing: a systematic review and meta-analysis. Fertil Steril 90(1):186–193 Epub 2007 Nov 5
4. Delvigne A, Rozenberg S (2002) Epidemiology and prevention of ovarian hyperstimulation syndrome (OHSS): a review. Hum Reprod Update 8:559–577
5. Amso NN, Ahuja KK, Morris N, Shaw RW (1990) The management of predicted ovarian hyperstimulation involving gonadotropin-releasing hormone analog with elective cryopreservation of all pre-embryos. Fertil Steril 53 (6):1087–1090
6. Gandhi GN, Allahbadia GN, Kagalwala S, Khatoon A, Hinduja R, Allahbadia A (2014) IVF Lite—a new strategy for managing poor ovarian responders. IVF Lite 1:22–28
7. Tal L, Shir D, Ekaterina S et al (2014) Comparison of IVF outcomes between minimal stimulation and high-dose stimulation for patients with poor ovarian reserve. Int J Reprod Med 2014:581451. doi:10.1155/2014/581451
8. Baker TG (1963) A quantitative and cytological study of germ cells in human ovaries. Proc R Soc Lond B Biol Sci 158:417–433
9. Martino A, Songsasen N, Leibo SP (1996) Development into blastocysts of bovine oocytes cryopreserved by ultra-rapid cooling. Biol Reprod 54(5):1059–1569
10. Kuwayama M (2007) Highly efficient vitrification for cryopreservation of human oocytes and embryos: the Cryotop method. Theriogenology 67(1):73–80 Epub 2006 Oct 20
11. Kuwayama M, Ieda S, Zhang J, Kato O (2005) The CryoTip method: aseptic vitrification of oocytes and embryos. Fertil Steril 84(Suppl 1):S187 (abstract 98)
12. Kagalwala S, Gandhi G, Allahbadia G, Kuwayama M, Allahbadia A, Chipkar V et al (2013) A comparison of two different vitrification methods for cryopreservation of mature human oocytes. Hum Reprod 28(1): i149–i206
13. Li R, Trounson A (1991) Rapid freezing of the mouse blastocyst: effects of cryoprotectants and of time and temperature of exposure to cryoprotectant before direct plunging into liquid nitrogen. Reprod Fertil Dev 3:175–183
14. Kolibianakis EM, Venetis CA, Tarlatzis BC (2009) Cryopreservation of human embryos by vitrification or slow freezing: which one is better? Curr Opin Obstet Gynecol 21:270–274
15. Mori C, Yabuuchi A, Ezoe K, Murata N, Takayama Y, Okimura T, Uchiyama K, Takakura K, Abe H, Wada K, Okuno T,

Kobayashi T, Kato K (2015) Hydroxypropyl cellulose as an option for supplementation of cryoprotectant solutions for embryo vitrification in human assisted reproductive technologies. Reprod Biomed Online 30 (6):613–621

16. Kuwayama M, Vajta G, Kato O, Leibo SP (2005) Highly efficient vitrification method for cryopreservation of human oocytes. Reprod Biomed Online 11(3):300–308

17. Seki S, Mazur P (2009) The dominance of warming rate over cooling rate in the survival of mouse oocytes subjected to a vitrification procedure. Cryobiology 59(1):75–82

18. Pomeroy KO, Harris S, Conaghan J, Papadakis M, Centola G, Basuray R, Battaglia D (2010) Storage of cryopreserved reproductive tissues: evidence that cross-contamination of infectious agents is a negligible risk. Fertil Steril 94 (4):1181–1188

19. Bielanski A, Vajta G (2009) Risk of contamination of germplasm during cryopreservation and cryobanking in IVF units. Hum Reprod 24 (10):2457–2467

# Appendix B: Solid Surface Vitrification

## Mohan S. Kamath and K. Muthukumar

### Abstract

Solid surface vitrification method involves direct contact of carrier loaded with droplet containing gametes or embryos with precooled metal surface. Over the years, following certain modifications, solid surface vitrification has emerged as an efficient method for vitrifying human gametes and embryos. Here, we describe the principle and methodology of solid surface vitrification.

**Key words** Solid surface vitrification, Blastocyst vitrification

## 1 Introduction

Vitrification is now an established method of cryopreservation for human gametes and embryos. It has the advantage of lack of intracellular ice formation in comparison with the traditional slow freeze technique. Disadvantages include embryo exposure to relatively higher concentration of toxic cryoprotectants and theoretical concerns regarding risk of transmission of viral infections due to contact with liquid nitrogen [1, 2].

Various combinations of carriers and loading devices are now currently being used in both open and closed vitrification methods [3]. Several carrier systems have been further modified and protocols refined to improve post-thaw outcomes [3–6].

Solid surface vitrification (SSV) methodology involves direct contact of carrier (fiber plug) loaded with media droplet containing the gametes or embryos with precooled metal block. In 1990, *Drosophila* embryos were successfully preserved by keeping them in a metal grid on a cold metal surface [7]. This resulted in more efficient heat transfer, and similar principal was used to develop solid surface vitrification. The solid surface vitrification as described by Dinnyés et al. [8] consists of metal cube covered by aluminum foil partially submerged in liquid nitrogen (Fig. 1). In the original study, the microdrops containing the bovine oocytes were dropped onto the metal surface (precooled to −150 °C to −180 °C by

Zsolt Peter Nagy et al. (eds.), *Cryopreservation of Mammalian Gametes and Embryos: Methods and Protocols*, Methods in Molecular Biology, vol. 1568, DOI 10.1007/978-1-4939-6828-2_22, © Springer Science+Business Media LLC 2017

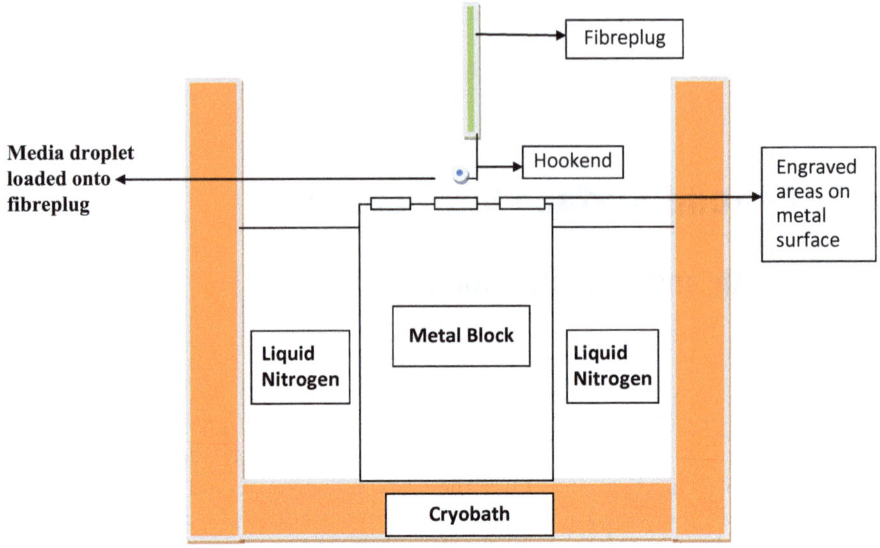


**Fig. 1** Solid surface vitrification device consisting of sterile metal block partially submerged in liquid nitrogen in a container (Cryobath). The media drops containing the gametes or embryos are brought in contact with metal surface, leading to instantaneous vitrification of the drops

partial immersion into liquid nitrogen) leading to instantaneous vitrification of the droplets. This containerless method resulted in sterile handling of the droplets due to clean metal surface along with higher cooling rates [8]. Subsequent study compared SSV with cryoloop vitrification method for cryopreserving goat oocytes and embryos and found similar survival and cleavage rates [9]. The survival and development of porcine oocyte improved with SSV method when it was combined with pretreatment with cytochalasin B [10].

There has been refinement of the SSV protocol over the years. A standard kit is now available commercially (CryoLogic Vitrification Method—CVM). However the survival rate following CVM was found to be lower than open Cryotop method. In a subsequent study, modification of original protocol was introduced to enhance the survival. Firstly, a thin glass pipette was used to reduce the vitrification volume to less than 0.2 μL compared to 1 μL per fiber plug media in the original protocol. Secondly, the fiber plug was kept in contact with metal wall within the hole for approximately 10 s before it was pushed down into the sleeve to minimize temperature changes. The modified CVM protocol yielded similar cryosurvival rates in the mice model compared to Cryotop method [11].

## 2 Materials for Vitrification

1. Sterile hollow metal block covered with lid (weighs around 1.2 kg) with six pairs of holes(*see* **Note 1**).

2. Carrier device preferably sterile fiber plug (nylon make) with sleeve.
3. Liquid nitrogen (3–4 L).
4. Vessel to hold the liquid nitrogen around the metal block (Cryobath).
5. Four-well sterile culture dish.
6. MOPS (3-(N-morpholino)propanesulfonic acid) or HEPES (4-(2-hydroxyethyl)-1-piperazineethanesulfonic acid) buffer.
7. Equilibration and vitrification solution.
8. Heated stage with stereomicroscope.
9. Stripper handle with strippers.

## 3 Methods

### 3.1 Preparation of Metal Block for Solid Surface Vitrification

The metal block is placed in the Cryobath (CryoLogic, Victoria, Australia). The metal block is covered by a lid to prevent the liquid nitrogen or vapors from coming in contact with media drop which is being vitrified. The metal block has a provision to hold the sleeves facilitating the placement of carrier (Figs. 2 and 3).

The liquid nitrogen is slowly poured around the hollow metal block. Liquid nitrogen initially boils, and when the temperatures of metal block and liquid nitrogen are similar, the boiling stops. The level of liquid nitrogen is raised steadily until it reaches the level of top hole on the sides of the block.

We have observed sedimentation of liquid nitrogen vapors on the surface of the block if the lid is not closed and metal surface is exposed for long duration (*see* **Note 2**). Due to exposure of media

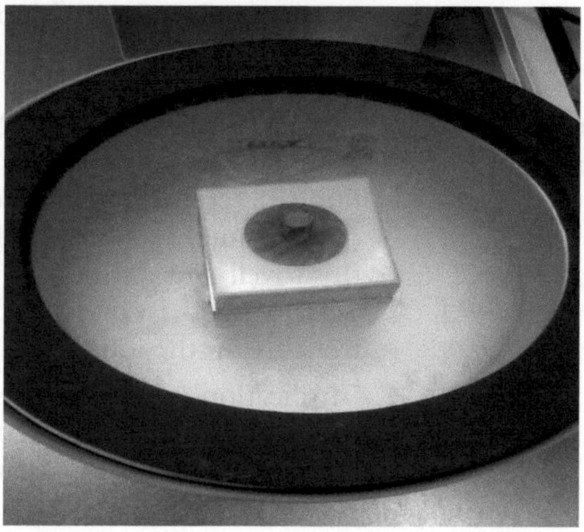

**Fig. 2** The hollow metal block placed inside the Cryobath surrounded by liquid nitrogen and the surface of the block shown closed by the lid

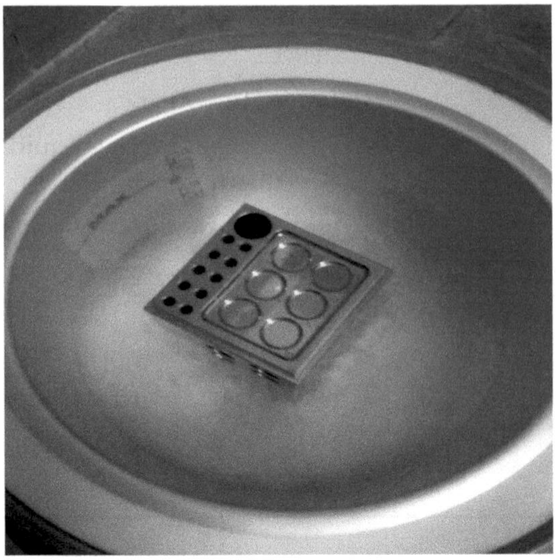

**Fig. 3** The engraved slots on solid metal surface and holes to hold the sleeves

droplets to liquid nitrogen vapors, SSV is classified as a semi-closed system. The lid should be placed immediately over the block to minimize the vapor sedimentation. If the lid is not closed properly, sedimentation can take place or liquid nitrogen can percolate through the gap onto the metal surface. Sedimentation of vapors or liquid nitrogen on the surface can reduce the efficiency of heat transfer resulting in ice crystal formation. Such areas of sedimentation can be removed using sterile plastic to ensure that a clean metal surface is available for vitrifying the droplets.

There are multiple engraved slots on the metal block. Each slot can be used for individual patient. In practice, the individual slot surface area can be used to vitrify the embryos/gametes derived from patient infected with blood borne viruses. The adjoining slot can be used for the next patient.

### 3.2 Preparation of the Solutions

1. Aliquot the buffer (HEPES or MOPS) in the first and second well of four-well dish. Subsequently aliquot equilibration solution (ES) consisting of 7.5 or 8 % ethylene glycol (EG) and 7.5 or 8 % dimethyl sulfoxide (DMSO) in HEPES or MOPS in the third well and vitrification solution (VS) consisting of 15 or 16 % EG and 15 or 16 % DMSO in 0.5 M sucrose or 0.68 M trehalose in the fourth well, respectively.
2. Vitrification solution consisting of sucrose is left at room temperature, and the one with trehalose is kept at 37 °C for 10 min before starting the vitrification.

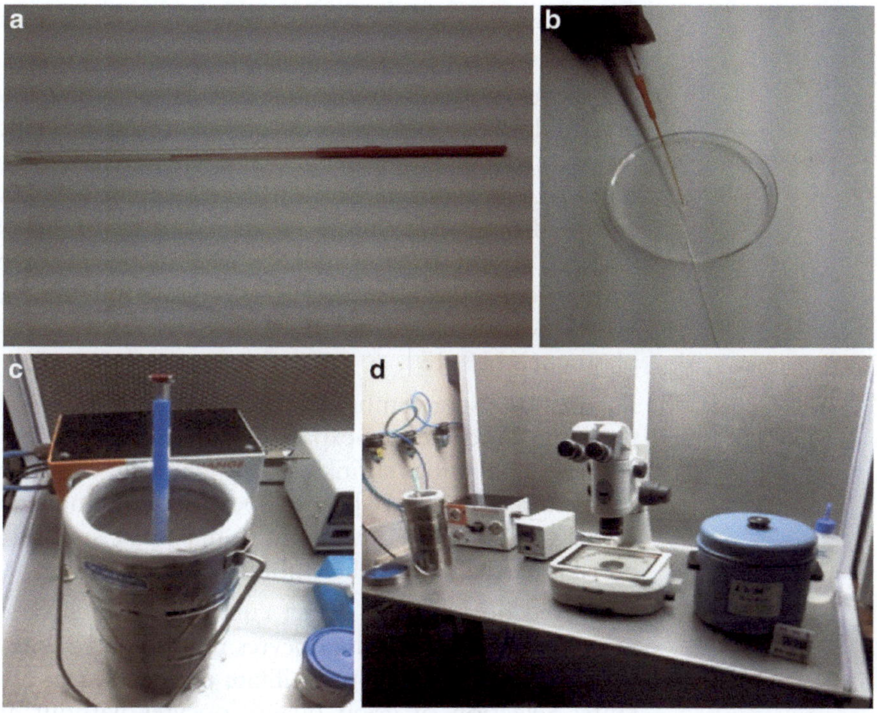

**Fig. 4** Fiber plug labeled with patient's identification details (**a** and **b**). Dewar flask filled with liquid nitrogen holding aluminum cane and goblet (**c**). Work station with stereomicroscope and heated stage, Cryobath, and stopwatch arranged before starting vitrification procedure (**d**)

### 3.3 Preparation of the Carrier and Holding Material

1. For patient identification, we can either use label with name and hospital number using thermal printer (Brady labels, Milwaukee, United States) or manually write using cryomarker near the holding end of the fiber plug (Fig. 4a, b). The sleeve is placed in the hole of metal block after cooling in liquid nitrogen with open end on the top for facilitating the placement of fiber plug.
2. The aluminum cane with goblet (Cryo Bio System, Irvine Scientific) with patient's name and identity number is kept submerged in liquid nitrogen in the 1 L Dewar flask (Fig. 4c, d). Sterile handling of the materials remains essential throughout the procedure.
3. Records of the patient in the embryology worksheet along with details of number of embryos to be vitrified in each fiber plug need to be maintained.

### 3.4 Vitrification Method for Oocytes and Embryos

1. The oocyte and embryos are rinsed in buffer and then moved into the ES and incubated for 5–10 min till they reach osmotic equilibrium. Subsequently they are placed in the VS for less than 30 s and then loaded onto the fireplug (carrier) near the hook end. The volume is between 1–3 μL to achieve higher

cooling rates when the drop is brought in contact with metal surface.

2. Blastocoel collapse using mechanical pipetting method is practiced for rapid permeation of cryoprotectants which increases the post warming blastocyst survival rates [12](*see* **Note 3**).The exposure time in ES (containing 8 % DMSO) for collapsed blastocyst should be approximately 1 min 50 s and between 5 and 10 min in ES (7.5 % DMSO) for blastocyst without collapsed blastocoel. The cryosurvival rate was around 80–85 % with the use of ES with 8 % DMSO and collapsed blastocoels [13–15].
3. For cleavage stage embryos, the exposure time depends upon the time taken for re-expansion of the blastomeres of the embryo after initial shrinkage and generally considered to be between 5–10 min [16]. The exposure of 3 min is suggested if 8 % DMSO in ES is being used to maximize survival rates and prevent partial lysis of the embryos [17].
4. Collected oocytes with intact cumulus are incubated for 2 h prior to vitrification. The oocytes require more gradual exposure using microdrops to facilitate permeation of cryoprotectants. Selection of good quality oocytes without any major intracytoplasmic dysmorphism is advised prior to vitrification.
5. For zygote stage vitrification, the protocol is similar to cleavage stage embryo vitrification.
6. After exposing to the VS for less than 30 s, the media droplet containing the oocytes or embryos (1–3 per droplet) are loaded onto the hook end of the fiber plug and brought in contact with precooled metal surface. The drop turns into a glassy bead (Fig. 5). The loaded fiber plug is then pushed into the sleeve (Fig. 6).

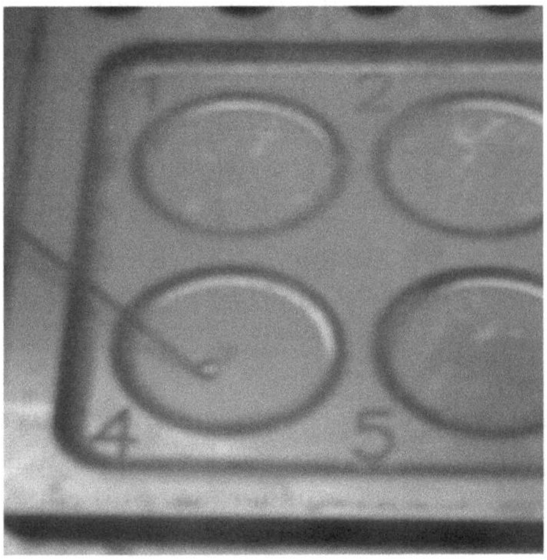


**Fig. 5** The loaded carrier is brought in contact with engraved slot on the precooled metal surface leading to formation of glassy bead vitrified drop

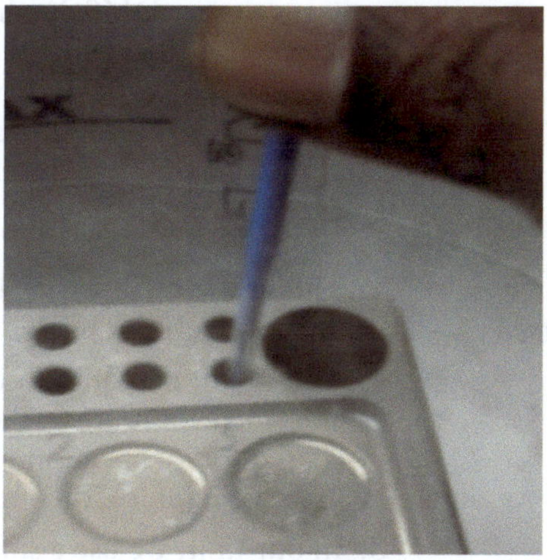

Fig. 6 The carrier with vitrified droplet inserted into the sleeve and sealed

### 3.5 Warming Protocol

1. Warming method following SSV is similar to warming protocol following other commonly used vitrification protocols for embryo/oocyte cryopreservation.
2. The carrier is removed from the sleeve and placed in 0.33 M trehalose to identify the blastocysts. Subsequently the blastocysts are transferred into well containing 0.33 M trehalose solution for rinsing and removing the toxic cryoprotectants (duration 5 min). They are shifted and incubated in the next well containing 0.22 M trehalose and finally in buffer solution for 5 minutes each. Finally they are placed in blastocyst medium till transfer.
3. In the warming protocol for cleavage stage, 1.0 and 0.5 M sucrose replaces the trehalose solution. The exposure time also is 1 and 3 min for the first (1.0 M) and second (0.5 M) sucrose well and 5 minutes in the buffer solution. Subsequently they are placed in cleavage (for day 2 embryos) or blastocyst medium (for day 3 embryo) till transfer. Warmed zygotes are placed in cleavage stage medium as well.
4. Oocytes have similar exposure time as cleavage stage protocol, and subsequently they are transferred to fertilization medium and incubated 2 h for spindle to reorganize before ICSI.
5. Ensure that the warm stage temperature is maintained at 37 °C during the entire warming procedure.

### 3.6 Cryopreservation of Spermatozoa by SSV

The main principle of spermatozoa vitrification involves vitrifying low volume of sample to maintain higher cooling rate which minimizes cryopreservation-induced damages. The technique has been refined by utilization of only nontoxic, non-permeable cryoprotectants (glucose, sucrose, and trehalose) and doing away with toxic cryoprotectants (DMSO), diminishing the risk of cryoinjury to the spermatozoa [18, 19]. The protocol has been standardized and involves simple carrier system [20]. The SSV protocol has been compared with standard vapor freezing method and has been found to comparable in terms of yielding fewer numbers of DNA-damaged sperms [20].

### 3.7 Sperm Vitrification Method

1. The sterile metal block for SSV is prepared as described in the previous subheading.
2. The sperm preparation is done using density gradient method and small volume of pellet is collected. The reconstituted pellet is used for vitrification. The SSV method is efficient for vitrifying severe oligoasthenoteratozoospermic samples.
3. In our unit, we use a simple stripper-based carrier to vitrify the spermatozoa using low-volume droplet for facilitating vitrification.
4. We use 4 μL sperm suspension, diluted with 0.5 M sucrose (dilution ratio 1:1) at 37 °C (Fig. 7). We load it into MEA tested stripper (275 μL capacity) (Fig. 8) and the sample pushed out onto the precooled metal solid surface to vitrify as droplet (Fig. 9). The vitrified drop is attached to the end of the stripper. The stripper is then placed into sperm straw and sealed near the holding end to finally close the sperm straws (CBS high security straw, Irvine Scientific, USA) for storage in liquid nitrogen (Fig. 10).

### 3.8 Warming of Vitrified Spermatozoa

1. In our unit, we use the following protocol for warming the vitrified spermatozoa as described below.
2. The warm stage temperature should be maintained at 37 °C. We need to aliquot 0.150 mL of buffer solution in a sterile test tube.

**Fig. 7** Sperm suspension mixed with sucrose in HEPES or MOPS (1:1) ratio at 37 °C in heated stage

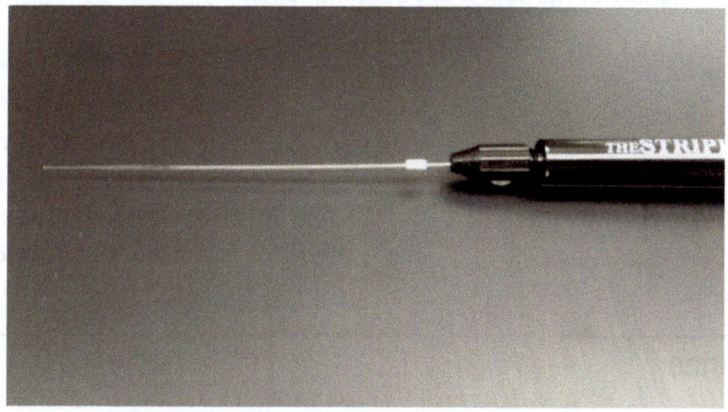

**Fig. 8** A stripper (capacity 275 μL) with silicon seal connected to handle used to pipette less than 3 μL sperm suspension with sucrose buffer

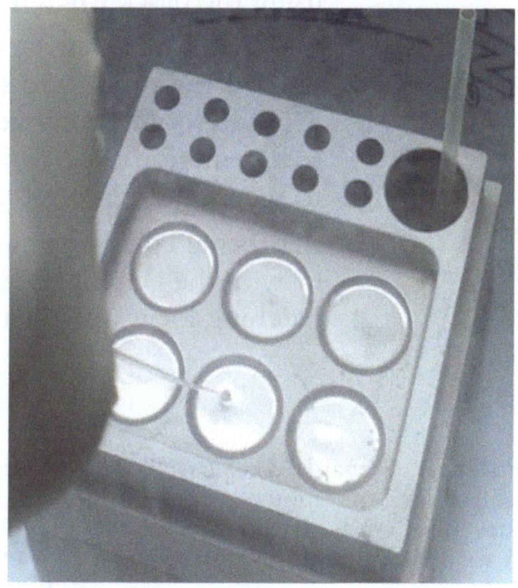

**Fig. 9** The sperm suspension droplet brought in contact with metal surface to vitrify as a droplet

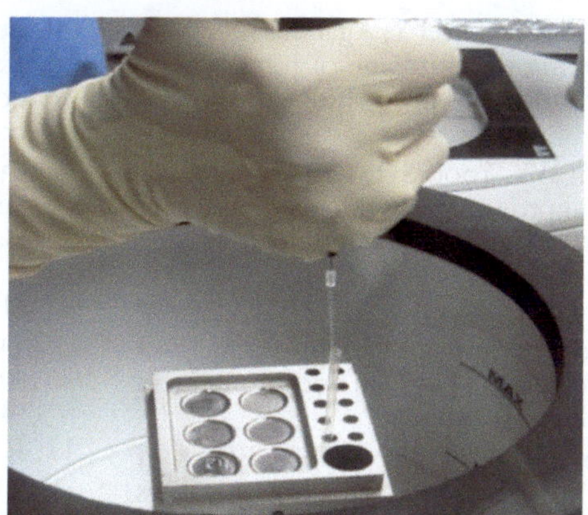

**Fig. 10** The vitrified droplet in the stripper end is carefully inserted into the straw and the stripper is detached and left inside the straw secured by silicon seal and later transferred into the cryobank

3. The stripper containing the vitrified spermatozoa is placed into the buffer solution to warm and release the spermatozoa. The warmed spermatozoa can be directly used for ICSI since they are not exposed to permeable cryoprotectants. However we use a separate drop in the ICSI dish, and selected spermatozoa are washed before use to remove any sucrose.

## 4 Notes

1. The solid surface metal block can be sterilized prior to use by autoclaving. After its usage, the block should be left for drying and the deposited dried sediments should be removed. Avoid using the block when it is hot.
2. The liquid nitrogen should be filled in Cryobath just prior to start of the procedure since too early filling results in longer exposure of metal surface to liquid nitrogen vapors leading to sedimentation. Alternatively, we can use a sterile aluminum foil to secure the gap between the block and the lid.
3. Blastocoel collapsing can be done using mechanical pipetting or with the help of laser. In our unit, we perform laser-assisted hatching for warmed day 5/6 blastocysts prior to transfer.

### References

1. Shaw PW, Fuller BJ, Bernard AG, Shaw RW (1991) Vitrification of mouse oocytes: improved rates of survival, fertilization and development to blastocyst. Mol Reprod Dev 33:210–214
2. Beilanski A, Vajta G (2009) Risk of contamination of germplasm during cryopreservation and cryobanking in IVF units. Hum Reprod 24:2457–2467
3. Kader AA, Choi A, Orief Y, Agarwal A (2009) Factors affecting the outcome of human blastocyst vitrification. Reprod Biol Endocrinol 7:99
4. Cho HJ, Son WY, Yoon SH, Lee SW, Lim JH (2002) An improved protocol for dilution of cryoprotectants from vitrified human blastocysts. Hum Reprod 17:2419–2422
5. Kuwayama M, Vajta G, Kato O, Leibo SP (2005) Highly efficient vitrification method for cryopreservation of human oocytes. Reprod Biomed Online 11:300–308
6. Liebermann J, Tucker MJ (2002) Effect of carrier system on the yield of human oocytes and embryos as assessed by survival and developmental potential after vitrification. Reproduction 124:148–152
7. Steponkus PL, Myers SP, Lynch DV, Gardner L, Bronshteyn V, Leibo SP, Rall WF, Pitt RE, Lin TT (1990) Cryopreservation of Drosophila melanogaster embryos. Nature 345:170–172
8. Dinnyés A, Dai Y, Jiang S, Yang X (2000) High developmental rates of vitrified bovine oocytes following parthenogenetic activation, in vitro fertilization, and somatic cell nuclear transfer. Biol Reprod 63:513–518
9. Begin I, Bhatia B, Baldassarre H, Dinnyés A, Keefer CL (2003) Cryopreservation of goat oocytes and in-vivo derived embryos using the cryoloop (CLV) and solid surface vitrification (SSV) methods. Theriogenology 59:1839–1850
10. Somfai T, Dinnyés A, Sage D, Marosan M, Carnwath JW, Kikuchi K, Kaneko H, Noguchi J, Niemann H (2006) Development to the blastocyst stage of parthenogenetically activated in vitro matured porcine oocytes after solid surface vitrification (SSV). Theriogenology 66(2):415–422
11. Jo J, Jee BC, Suh CS, Kim SH (2013) Closed vitrification of mouse oocytes using the CryoLogic vitrification method: a modification that improves developmental competence. Clin Exp Reprod Med 40(4):148–154

12. Hiraoka K, Hiraoka K, Kinutani M, Kinutani K (2004) Blastocoele collapse by micropipetting prior to vitrification gives excellent survival and pregnancy outcomes for human day 5 and 6 expanded blastocysts. Hum Reprod 19 (12):2884–2888
13. Muthukumar K, Mangalaraj AM, Kamath MS, George K (2008) Blastocyst cryopreservation: vitrification or slow freeze? Fertil Steril 90: S426–S427
14. Kamath MS, Mangalaraj AM, Muthukumar K, Cullinan R, Aleyamma TK, George K (2011) Blastocyst cryopreservation using solid surface vitrification: a preliminary study. J Hum Reprod Sci 4(3):114–120
15. Muthukumar K, Kamath MS, Mangalaraj AM, Aleyamma TK, Chandy A, George K (2013) Comparison of clinical outcomes following vitrified day5/6 blastocyst transfers using solid surface methodology with fresh blastocyst transfers. J Hum Reprod Sci 6(1):59–64
16. Kuwayama M, Vajta G, Ieda S, Kato O (2005) Vitrification of human embryos using Cryotip method. Reprod Biomed Online 11:608–614
17. Wilding MG, Capobianco C, Montanaro N, Kabili G, Di Matteo L, Fusco E, Dale B (2010) Human cleavage-stage embryo vitrification is comparable to slow-rate cryopreservation in cycles of assisted reproduction. J Assist Reprod Genet 27(9–10):549–554
18. Gilmore JA, Liu J, Gao DY, Critser JK (1997) Determination of optimal cryoprotectants and procedures for their addition and removal from human spermatozoa. Hum Reprod 12:112–118
19. Schulz M, Muñoz M, Risopatrón J, Sánchez R (2006) Cryopreservation of human spermatozoa by vitrification. Int J Morphol 24:31
20. Satirapod C, Treetampinich C, Weerakiet S, Wongkularb A, Rattanasir S, Choktanasiri W (2012) Comparison of cryopreserved human sperm from solid surface vitrification and standard vapor freezing method: on motility, morphology, vitality and DNA integrity. Andrologia 44(1):786–790

# Appendix C: Automated Vitrification of Mammalian Embryos on a Digital Microfluidic Device


## Jun Liu*, Derek G. Pyne*, Mohamed Abdelgawad, and Yu Sun



## Abstract

This chapter introduces a digital microfluidic device that automates sample preparation for mammalian embryo vitrification. Individual microdroplets manipulated on the microfluidic device were used as microvessels to transport a single mouse embryo through a complete vitrification procedure. Advantages of this approach, compared to manual operation and channel-based microfluidic vitrification, include automated operation, cryoprotectant concentration gradient generation, and feasibility of loading and retrieval of embryos.



**Key words** Automated vitrification, Digital microfluidics, Embryo vitrification


## 1 Introduction

This Chapter is modified from our previous work in digital microfluidic processing for vitrification of embryos [1]. Two commonly used cryopreservation techniques for freezing embryos are the slow-freezing method and the vitrification method. Both techniques aim to minimize cell damage caused by freezing that is largely due to the formation of intracellular ice crystals [2]. Conventionally, cells are frozen through the slow-freezing method where cells are placed in a large freezer that can accurately control the freezing rate down to liquid nitrogen temperatures, with low concentrations of cryoprotectants [3]. During slow freezing, extracellular water freezes away from the embryo, using a seeding technique, which creates an osmotic gradient that draws water out of the cell until it finally freezes without the formation of intracellular ice crystals [4]. This procedure requires sophisticated equipment to control the

*These authors contribute equally to this work


Zsolt Peter Nagy et al. (eds.), *Cryopreservation of Mammalian Gametes and Embryos: Methods and Protocols*, Methods in Molecular Biology, vol. 1568, DOI 10.1007/978-1-4939-6828-2_23, © Springer Science+Business Media LLC 2017

freezing rate, which ranges between 0.3 and 1.0 °C/min, and produces a relatively poor survivability rate [5].

On the other hand, vitrification offers an alternative approach in which cells are frozen at extremely high rates, usually by directly plunging the sample into liquid nitrogen, after bathing them in a sequence of high concentration cryoprotectants [6]. Vitrification reduces intracellular ice formation, which is the primary cause of cell death, by freezing the sample in a glass-like state before the molecules have a chance to form crystal structures. This results in a higher cell survival rate after thawing compared to conventional slow freezing without the need for a seeding procedure or a programmable freezer [5, 7]. However, vitrification requires precise washing sequences and timings in each cryoprotectant medium since higher concentrations are used and there are risks of toxicity if overexposed. The process is expensive in terms of technical skills required. In IVF clinics, processing an embryo/oocyte in cryoprotectant medium typically costs a highly skilled embryologist 10–15 min.

Digital microfluidics, which enables the manipulation of liquid droplets over an array of electrodes [8], is a useful tool for sequential sample processing and has been used in many biological applications such as PCR, cell culture, and immunoassays [9–11]. It was also used for testing cryoprotectant concentrations in slow-freezing cryopreservation [12]. Given its capabilities, digital microfluidics is well poised to automate embryo processing for embryo vitrification. The key to automating the vitrification process is to replicate the washing and timing steps of a given protocol while also keeping complete control of the embryo. Droplets on the digital microfluidic platform can act as microvessels to move an embryo and subject it to a series of cryoprotectants of different concentrations, as required by IVF vitrification protocols (Fig. 1).

Here we introduce a digital microfluidic device to automate embryo preparation for the vitrification procedure. Although vitrification was demonstrated inside microchannels [13–15], the digital microfluidic device does not undesirably "park" the embryo to a confined area, which could expose the embryo less uniformly to cryoprotectant medium and has less intricacy of embryo introduction and retrieval onto and from the device. By keeping the embryo in a single droplet, as opposed to microchannels or wells, users are also able to constantly track and control the embryo's locations within the field of view of the imaging system throughout the procedure to avoid cell loss.

## 2 Materials

1. Mouse embryos were produced by natural mating female mice and were gathered 2.5 days past conception, which corresponds to most embryos being in the eight-cell stage.

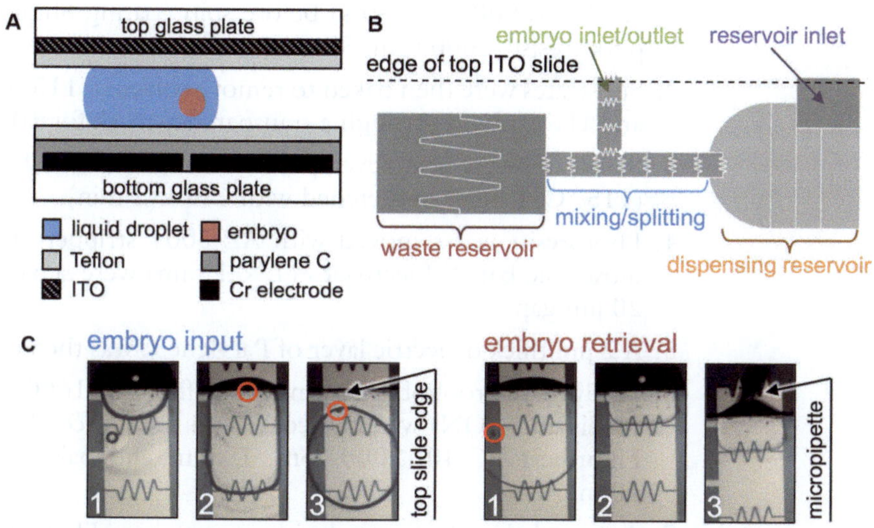


**Fig. 1** Key device design elements and fabrication layers. **(a)** Device structure and fabrication. Electrodes (1 × 1 mm) were separated by a 20 μm gap. **(b)** Regions for vitrification medium dispensing and for embryo loading/extraction. The top ITO slide is placed on the device in a manner that exposes portions of the top electrodes in the dispensing reservoir and exposes portions of the leg of the T-shaped electrodes, for medium and embryo loading, respectively. **(c)** Embryo is input and extracted by actuating electrodes at edge of top glass slide

2. Vitrification solution (VS) typically contains antifreezing agents or cryoprotectants, such as dimethyl sulfoxide (DMSO), some small molecular sized glycols (e.g., ethylene glycol), or sucrose [16]. A combination of DMSO and sucrose was used in this work.

3. The vitrification solution (VS) was made by diluting DMSO in serum-free KSOM medium (EMD Millipore, Billerica, USA) at 33 % concentration, with 1.0 M sucrose. The equilibrium solution (ES) was at half concentration of VS (i.e., 16.5 % DMSO + 0.5 M sucrose).

4. VS was preloaded on the DMF chip before each experiment. The first mixing step, which mixes VS with embryo culture medium (i.e., serum-free KSOM), generates the equilibrium solution ES.

## 3 Methods

### 3.1 Device Fabrication

The devices were constructed via microfabrication.

1. Pre-coated chromium glass slides (Deposition Research Labs Inc., MO) were first primed with hexamethyldisilazane (HMDS) P-20 (Shin-Etsu MicroSi, Phoenix, AZ) by spin

coating (3000 rpm, 30 s) before spin coating Shipley S1811 photoresist (3000 rpm, 30 s).

2. Substrates were then baked to remove solvents (115 °C, 2 min) and UV exposed through a transparency mask for 10 s.
3. Substrates were next developed in MF-321 (2 min), hard baked (115 °C, 1 min), and etched with CR-4 (2 min).
4. Photoresist was removed with AZ-300T stripper (15 min in ultrasonic bath). Electrodes (1 × 1 mm) were separated by a 20 μm gap.
5. A 2 μm thick dielectric layer of Parylene C was then deposited.
6. Lastly, a hydrophobic coating of Teflon AF 1600 (Dupont, Mississauga, ON) was spin coated on the device (1 % w/w in Fluorinert FC-40, 2000 rpm, 1 min) and baked (160 °C, 10 min).
7. A second glass slide coated with un-patterned ITO (Deposition Research Labs Inc., MO) and Teflon AF was used as the ground electrode.
8. Finally, the devices (Fig. 1a) were assembled by placing the top ITO slide over the patterned device using two pieces of double-sided tape as a spacer. This produces a gap of approximately 100 μm between the top and bottom structures.

### 3.2 Device Design

1. Voltages applied to actuate droplets were 55–75 Vrms at 15 kHz.
2. Cryoprotectant droplets were actuated inside silicone oil (2.0 cSt, Gelest Inc., Morrisville, PA) to reduce friction and evaporation.
3. Different regions of the device were designed to achieve the general digital microfluidic functions of transporting, mixing, dispensing, and splitting droplets.
4. A large reservoir was used to hold and dispense the high concentration cryoprotectant medium, as shown in Fig. 1b. This reservoir was split up into many sections to handle variations in liquid volume in the reservoirs as droplets are dispensed during the vitrification protocol.
5. The second reservoir was used as a waste reservoir and, thus, was split into two large electrodes only.
6. A central inverted T-shaped array of electrodes was used for droplet transport, mixing, and splitting. Electrodes in this array were interdigitated to allow droplet overlap with adjacent electrode and increase electrodynamic forces applied on droplets.
7. Top electrode in the leg of the T-shape was an input/output region where half of the electrode was exposed out of the ITO slide to enable embryo loading.

### 3.3 Embryo Loading and Retrieval

As shown in Fig. 1c, to input an embryo, a small droplet (200 nL) containing the embryo is pipetted onto the loading electrode and then actuated into the device, as described in [17]. This technique minimizes exposure of the embryo to outside air. Extraction of the embryo is completed in the opposite manner by transporting the embryo-containing droplet to the edge of the device and retrieving it by a micropipette. Once the embryo is extracted from the device, it is directly frozen in the micropipette inside liquid nitrogen.

### 3.4 Cryoprotectant Mixing

An embryo is input into the device in a small droplet of embryo culture medium, and 100 % VS is input into the device in larger volumes ("reservoir inlet" in Fig. 1b). The cryoprotectant bathing procedure is then performed through a serial mixing/splitting process (*see* Fig. 2). This is accomplished by mixing the embryo-containing droplet with a vitrification solution droplet (VS), thus increasing the concentration of cryoprotectant around the embryo. The resulting droplet is then split into two daughter droplets with the one containing the embryo identified and kept, while the other droplet is moved to the waste reservoir.

After the first mixing step, the droplet reaches 50 % cryoprotectant concentration (i.e., equilibrium solution or ES); the embryo is kept in the ES droplet as long as the specific protocol requires. Then the cryoprotectant concentration is increased again by droplet mixing and splitting. Contrary to its state in ES medium, embryo volume sharply decreases in VS medium and does not recover. The overall mixing profile generated with a single

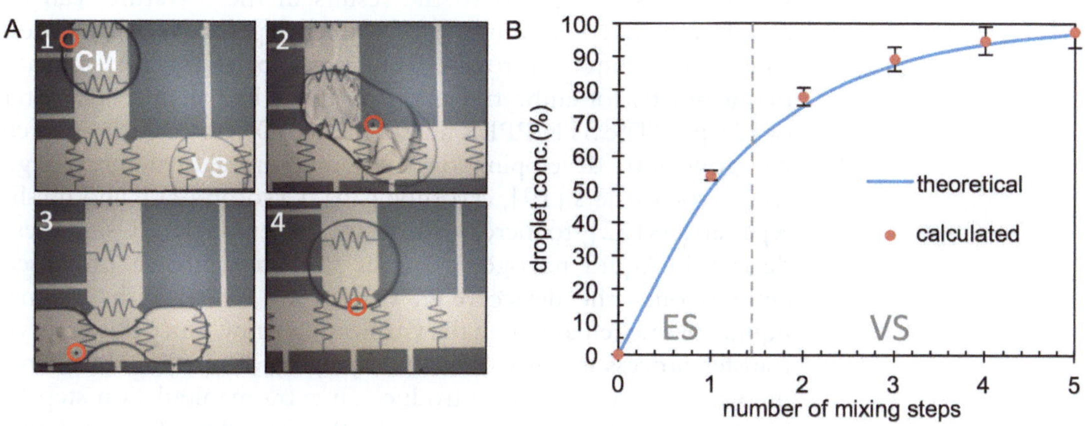


**Fig. 2** Droplet mixing protocol and resulting concentration profile. (**a**) 1: Embryo (*red circle*) contained in culture medium (CM) droplet. 2: Embryo droplet mixed with VS droplet. 3: Droplet split into two droplets (*left* contains embryo). 4: Droplet containing embryo is kept and other droplet is sent to waste. Process is repeated to increase VS concentration. (**b**) Mixing profile showing the generation of ES medium and VS medium. Droplet volumes were calculated by imaging droplet boundaries and modeling as cylinders. Concentrations were then calculated using these volumes before and after each mixing step

dispensing reservoir is $C(n) = 100(1 - \frac{1}{2^n})\%$, where $C(n)$ is the concentration of the droplet and $n$ is the number of mixing steps.

This protocol mimics a typical two-step human embryo/oocyte protocol. However, mouse embryos are typically frozen with only a single-step protocol where the embryo is directly transferred to VS medium and then frozen. On our device, this corresponds to simply removing the waiting time in the ES medium step. Both protocols were performed; however, the results presented were done with the mouse embryo timings to follow the protocol provided by Canadian Mouse Mutant Repository (CMMR).

After complete transfer of the embryo into the VS medium, the droplet containing the embryo is moved toward the edge to be collected by a micropipette (Fig. 1c) and then plunged into liquid nitrogen. Contrary to conventional vitrification protocols and manual operation, which subject embryos to sudden changes in medium concentration, the digital microfluidic approach gradually increases the VS medium concentration, (Fig. 2), which is generally accepted by IVF practitioners to be more benign to embryos due to lower osmotic stress [18].

## 4 Notes

Previous studies of embryo vitrification (4–16-cell stages) reported survival rates in the range of 80–100 %. Based on the limited sample size we have tested, vitrification using digital microfluidic device in our work produced a survival rate of 77 %, and our own manual vitrification trials resulted in a survival rate of 73 %. These lower survival rates, compared to the results in the literature, can be mainly attributed to our use of a micropipette (vs. vitrification straw) inside liquid nitrogen. The micropipette tip is a standard plastic pipette for embryo manipulation and had an inner diameter of 125 μm (The STRIPPER micropipetter, Origio). Much research has gone into developing different mechanical structures (e.g., straw-type carriers [19], cryotube [20], Cryotop [21], and mesh-type carriers [22]) to increase the heat transfer rate [3]. Using these devices in liquid nitrogen requires the transfer of a processed embryo onto the device (e.g., vitrification straw) with minimal liquid volume remained on the vitrification straw. Since this embryo transfer process is not within the capability of our present microfluidic device and would introduce an extra manipulation step by hand, this work focused on proving the feasibility of using digital microfluidics for embryo processing. Therefore, the microfluidically processed embryos were directly frozen inside the micropipette tip in our experiments.

Due to the programmability of the digital microfluidic devices, one key benefit is the ability to implement/test a number of vitrification protocols for efficacy comparisons. Although different human and mouse vitrification protocols use different cryoprotectants, number of mediums, and timings, all of these protocols, at their core, involve the controlled increase in cryoprotectant concentration over a given period of time. This is convenient for DMF device design as it always produces a 50 % concentration after the first mixing (neglecting the time required to homogenize concentration of the resulting droplet which is in the order of 2~3 s when the droplet is rolled over a few electrodes to enhance mixing). This means that a typically two-step protocol involving an initial 50 % concentration ES step, followed by a short 100 % VS step, can be easily realized by following the mixing curve with a pause after the first mixing step to allow the embryo to reach equilibrium.

One limitation of the digital microfluidic platform was handling culture mediums containing high serum concentrations. Serum contains a mixture of proteins that can adsorb on the Teflon-coated surfaces of the device, eventually accumulating to the point that the surface becomes hydrophilic, making droplets immovable [23]. Some strategies have been developed to help overcome this problem, such as the use of Pluronic additives [24], silicone oil baths [25], and superhydrophobic surfaces [26]. Pluronic additives were avoided in our work as embryos are highly sensitive to additives. Superhydrophobic surfaces were also avoided as they require significant additional fabrication efforts. A silicone oil bath was used which did increase droplet movability; however, this approach did not work with sufficient effectiveness for conventional embryo culturing mediums that contain high serum concentrations. Therefore, in this work, a serum-free culture medium was chosen for this proof-of-principle study. In further work, we will explore droplet movability improvement for serum-containing culture medium and the automation of transferring embryos onto vitrification devices (e.g., straw or Cryotop).

## Acknowledgment

This work was supported by the Grants from NSERC (Natural Sciences and Engineering Research Council of Canada) via a Discovery Grant and the Canada Research Chairs Program.

### References

1. Pyne DG, Liu J, Abdelgawad M, Sun Y (2014) Digital microfluidic processing of mammalian embryos for vitrification. PLoS One 9(9): e108128
2. Pegg DE (2010) The relevance of ice crystal formation for the cryopreservation of tissues and organs. Cryobiology 60:S36–S44
3. Whittingham D (1971) Survival of mouse embryos after freezing and thawing. Nature 233:125–126
4. Saragusty J, Arav A (2011) Current progress in oocyte and embryo cryopreservation by slow freezing and vitrification. Reproduction 141:1–19

5. Vajta G, Nagy ZP (2006) Are programmable freezers still needed in the embryo laboratory? Review on vitrification. Reprod Biomed Online 12:779–796
6. Rall WF, Fahy GM (1985) Ice-free cryopreservation of mouse embryos at −196 degrees C by vitrification. Nature 313:573–575
7. AbdelHafez FF, Desai N, Abou-Setta AM, Falcone T, Goldfarb J (2010) Slow freezing, vitrification and ultra-rapid freezing of human embryos: a systematic review and meta-analysis. Reprod Biomed Online 20:209–222
8. Pollack MG, Fair RB (2000) Electrowetting-based actuation of liquid droplets for microfluidic applications. Appl Phys Lett 77:1725–1726
9. Barbulovic-Nad I, Au SH, Wheeler AR (2010) A microfluidic platform for complete mammalian cell culture. Lab Chip 10:1536–1542
10. Chang Y-H, Lee G-B, Huang F-C, Chen Y-Y, Lin J-L (2006) Integrated polymerase chain reaction chips utilizing digital microfluidics. Biomed Microdevices 8:215–225
11. Sista RS, Eckhardt AE, Srinivasan V, Pollack MG, Palanki S et al (2008) Heterogeneous immunoassays using magnetic beads on a digital microfluidic platform. Lab Chip 8:2188–2196
12. Park S, Wijethunga PAL, Moon H, Han B (2011) On-chip characterization of cryoprotective agent mixtures using an EWOD-based digital microfluidic device. Lab Chip 11:2212–2221
13. Heo YS, Lee H-J, Hassell BA, Irimia D, Toth TL et al (2011) Controlled loading of cryoprotectants (CPAs) to oocyte with linear and complex CPA profiles on a microfluidic platform. Lab Chip 11:3530–3537
14. Lai D, Ding J, Smith GD, Takayama S (2013) Automated microfluidic gradient cryoprotectant exchange platform for murine oocyte and zygote vitrification reduces osmotic stress and improves embryo developmental competence. Fertil Steril 100:S107
15. Song YS, Moon S, Hulli L, Hasan SK, Kayaalp E et al (2009) Microfluidics for cryopreservation. Lab Chip 9:1874–1881
16. Ali J, Shelton JN (1993) Design of vitrification solutions for the cryopreservation of embryos. J Reprod Fertil 99:471–477
17. Berthier J, Clementz P, Roux J, Fouillet Y, Peponnet C (2006) Modeling microdrop motion between covered and open regions of EWOD microsystems. NSTI Nanotechnology Conference and Trade Show Boston, USA, Vol. 1. pp. 685–688.
18. Swain JE, Lai D, Takayama S, Smith GD (2013) Thinking big by thinking small: application of microfluidic technology to improve ART. Lab Chip 13:1213–1224
19. Otoi T, Yamamoto K, Koyama N, Tachikawa S, Suzuki T (1998) Cryopreservation of mature bovine oocytes by vitrification in straws. Cryobiology 37:77–85
20. Nakao K, Nakagata N, Katsuki M (1997) Simple and efficient vitrification procedure for cryopreservation of mouse embryos. Exp Anim 46(3):231–234
21. Kuwayama M, Vajta G, Ieda S, Kato O (2005) Comparison of open and closed methods for vitrification of human embryos and the elimination of potential contamination. Reprod Biomed Online 11:608–614
22. Martino A, Songsasen N, Leibo SP (1996) Development into blastocysts of bovine oocytes cryopreserved by ultra-rapid cooling. Biol Reprod 54:1059–1069
23. Au SH, Kumar P, Wheeler AR (2011) A new angle on pluronic additives: advancing droplets and understanding in digital microfluidics. Langmuir 27:8586–8594
24. Luk VN, Mo GC, Wheeler AR (2008) Pluronic additives: a solution to sticky problems in digital microfluidics. Langmuir 24:6382–6389
25. Srinivasan V, Pamula VK, Fair RB (2004) An integrated digital microfluidic lab-on-a-chip for clinical diagnostics on human physiological fluids. Lab Chip 4:310–315
26. Jönsson-Niedziółka M, Lapierre F, Coffinier Y, Parry SJ, Zoueshtiagh F et al (2011) EWOD driven cleaning of bioparticles on hydrophobic and superhydrophobic surfaces. Lab Chip 11:490–496

# Appendix D: Irvine Scientific® Vitrification System


## Matthew VerMilyea and Amber Brewer



## Abstract

This chapter will describe the use of the Irvine Scientific® vitrification and warming solutions (Vit Kits®) along with a detailed protocol for the correct use of the CryoTip® vitrification device for human embryos and oocytes. Successful pregnancies have been reported after carrying out rapid vitrification methods of oocytes, cleavage stage embryos, and blastocysts using the CryoTip® (Popwell et al. Fertil Steril, 101:e20, 2014; Kuwayama et al. Reprod Biomed 11:608–614, 2005; Kuwayama et al. Fertil Steril 84:S187, 2005; Kuwayama et al. Vitrification of human embryos using the CryoTip TM method. Reprod Biomed 11:608–614). Compared to other vitrification carrier devices, the CryoTip® is considered a *closed* carrier for vitrification, thereby eliminating the theoretical risk of disease transmission through contaminated liquid nitrogen during cooling and storage (Bielanski et al. Cryobiology, 40:110–116, 2000). The CryoTip® is cleared by the FDA and has CE mark approval as a closed device to carry gametes or embryos in a specialized medium during cryopreservation procedures and subsequent long-term storage in liquid nitrogen. The CryoTip® has been shown to be a safe and reliable vitrification device, and when compared to other open system vitrification devices, it has provided similar results (Martino et al. Reprod Biol Endocrinol 11:27, 2013, Valbuena et al. Fertil Steril 97:209–217, 2012; Kuwayama et al. Reprod Biomed 11:608–614, 2005). The CryoTip® has also been shown to be suitable for use as a vitrification device for the cryopreservation of small volumes of sperm (Tanaka et al. Fertil Steril 90: S292, 2008).



**Key words** Irvine Scientific, CryoTip®, Vit Kits®, embryo vitrification


## 1 Introduction

The CryoTip® vitrification device has been optimized as a closed system for cryopreservation procedures. It consists of a straw with an ultrafine tip (250 μm inner diameter, 20 μm wall thickness, and 3 cm in length) connected to a wide end (2000 μm inner diameter, 150 μm wall thickness, and 4.5 cm length) and equipped with an adjustable protective metal cover sleeve (Fig. 1). The devices can also be equipped with a classical PETG sperm straw for easy labeling and specimen identification (Fig. 2). Embryos are loaded into the ultrafine tip of the device in approximately 1 μL of vitrification


Zsolt Peter Nagy et al. (eds.), *Cryopreservation of Mammalian Gametes and Embryos: Methods and Protocols*, Methods in Molecular Biology, vol. 1568, DOI 10.1007/978-1-4939-6828-2_24, © Springer Science+Business Media LLC 2017

**Fig. 1** The CryoTip® vitrification device with ultrafine tip, loading marks, and metal protective sleeve

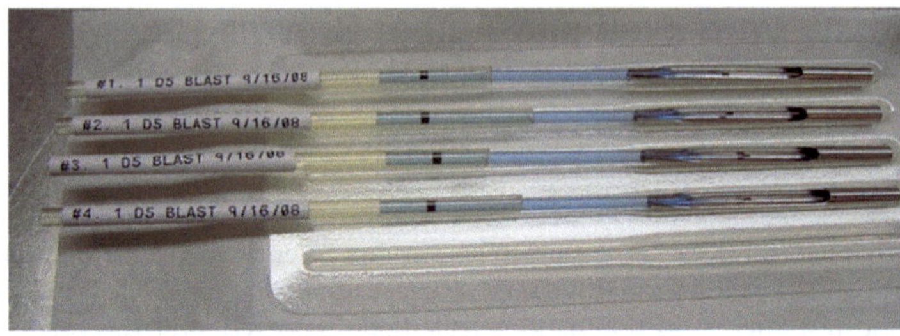


**Fig. 2** CryoTip®(s) with sterile classical PETG sperm straw and label for sample identification

solution. The gamete/embryo and vitrification solution is aspirated into the device by a reusable connector and syringe. The straw is then hermetically heat-sealed at both ends, and the protective metal sleeve is pulled over the ultrafine tip for added protection during handling. The entire device is then plunged into liquid nitrogen and stored for future use.

Since the device has been optimized as a closed system for cryopreservation procedures, it is important to validate that the impulse heat sealer is working optimally. A validation protocol should be followed to assure consistency of the individual sealer settings and confirm a sufficient heat seal. It is recommended that the heat sealer validation procedure be performed annually to ensure seals are made properly.

For warming, the CryoTip® is removed from liquid nitrogen and immediately plunged into a 37 °C water bath for 3 s. The device is wiped dry with a sterile tissue, and both sealed ends are cut with sterile scissors after sliding the protective metal sleeve away from the ultrafine tip. A syringe and connector are attached to the wide end of the straw, and the contents are carefully expelled onto a sterile dish.

When the CryoTip® is used in conjunction with the Irvine Scientific® Vit Kits® for freezing and thawing, exceptional embryo recovery and survivability rates can be achieved (Table 1) along with clinical outcomes (Table 2). Additionally, the CryoTip® and Vit Kits® have been optimized to be used together at room temperature (22–27 °C), providing an easy and reliable protocol for closed system vitrification.

**Table 1**
**Warming results: laboratory**

| Cycles | 471 |
|---|---|
| Embryos warmed (mean) | 852 (1.81) |
| Embryos recovered | 833 (98 %) |
| Embryos survived | 807 (97 %) |
| Embryos transferred (mean) | 807 (1.71) |

Data collated from 1/2012 to 12/2014 from Fertility Institute of Hawaii.

**Table 2**
**Warming results: clinical**

| Cycles | 471 |
|---|---|
| Embryos transferred (mean) | 807 (1.71) |
| Clinical pregnancies | 290 (62 %) |
| Sacs | 400 (50 %) |
| Singleton pregnancies | 183 (63 %) |
| Twin pregnancies | 104 (36 %) |
| Triplet pregnancies | 3 (1 %) |

Data collated from 1/2012 to 12/2014 from Fertility Institute of Hawaii.

## 2 Materials (Fig. 3)

- Impulse heat sealer: with heating element width of 2 or 5 mm, to seal both ends of the CryoTip®.
- Vit Kit®-Freeze: Irvine Scientific® P/N #90133-DSO.
- Vit Kit®-Thaw: Irvine Scientific® P/N #90137-DSO.
- CryoTip®(s): Irvine Scientific® P/N# 40709.
- Connector: Irvine Scientific® P/N# 40736.
- Cryogenic labels.
- 25 μL Hamilton Gastight Luer Tip Syringe or equivalent aspiration tool (using a small volume Hamilton syringe provides greater control during loading and unloading of the CryoTip®).
- 250 μL Hamilton Gastight Luer Tip Syringe (Recommended to be used for warming protocol).
- Small, sharp surgical grade scissors.
- Tweezers or forceps.
- Vitrification solution (VS): Irvine Scientific® P/N #90132.

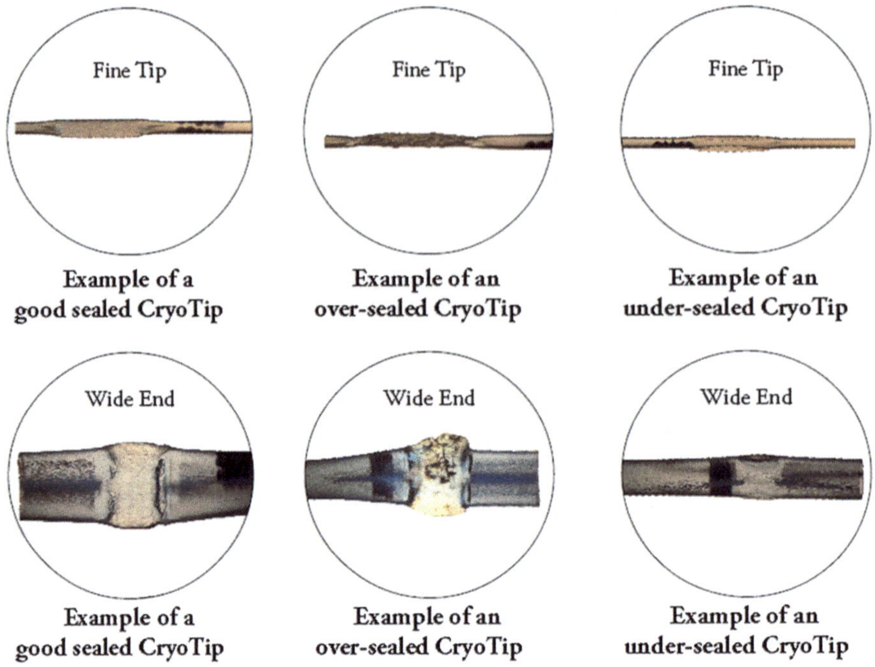


**Fig. 3** Examples of heat-sealed CryoTip®(s)

- Liquid nitrogen (LN2).
- Liquid nitrogen reservoir or Styrofoam box.
- 37 °C water bath.
- Minimum 500 mL beaker.
- Thermometer.
- Glass pipettes.
- Sterile embryo culture dishes.
- Stereo microscope.
- Timer.

## 3 Methods

### *3.1 Impulse Heat Sealer Validation Protocol*

1. Fill the liquid nitrogen reservoir with liquid nitrogen (LN2).
2. Attach the syringe to the wide end of the connector, and then attach the wide end of the CryoTip® to the other (narrow) end of the connector; pull the plunger of the syringe slightly to ensure that there is no capillary action when placing the CryoTip® in the VS solution.
3. Slide the metal cover sleeve up along the CryoTip® to expose the ultrafine tip end.

4. Load the CryoTip® with VS solution up to the 3rd mark by gently pulling on the plunger of the syringe.
5. Set the impulse sealer to scale #3, holding the CryoTip® perpendicular to the sealer, place the ultrafine tip on the middle of the heating element, and brace your hand to ensure a steady seal. Gently push the handle of the heat sealer down to heat-seal the ultrafine tip end at or below the 1st mark. You will hear a beep and the light will turn off to indicate the seal is complete.
6. Observe the seal under the microscope to ensure you have achieved a complete seal. Heat-seal the tip one more time just above the first seal, if the seal is not complete.
7. Slide down the cover sleeve to protect the ultrafine tip.
8. Disconnect the CryoTip® from the connector and syringe; set the impulse sealer to scale #6, holding the CryoTip® perpendicular to the sealer; place the wide end on the middle of the heating element; and brace your hand to ensure a steady seal. Gently push the handle of the heat sealer down to heat-seal the wide end at or above the 4th mark. You will hear a beep and the light will turn off to indicate the seal is complete.
9. Observe the seal under the microscope to ensure you have achieved a complete seal. Heat-seal the tip one more time just below the first seal if the seal is not complete.
10. Plunge the CryoTip® (metal covered ultrafine tip side down) into the LN2 and gently swirl for 3 s.
11. Prepare a 500 mL beaker with 37 °C water bath, and place the thermometer in the water bath to monitor the temperature.
12. Use the tweezers/forceps to gently grab the wide end of the CryoTip® (submerged in LN2), and then immediately plunge it into the 37 °C water bath.
13. Wipe the CryoTip® dry with a Kim wipe, slide up the metal cover sleeve, and carefully observe for leakage of the VS solution along the entire ultrafine tip end. The same volume of VS solution (up to the 3rd mark on the CryoTip®) should be present if no leakage occurred.
14. Withdraw the plunger of the syringe (with connector attached) to the halfway position.
15. Cut the wide end of the CryoTip® on the 4th mark (to remove the heat seal), and then attach it to the connector and syringe.
16. Apply light pressure to the CryoTip® by lightly pushing down on the plunger of the syringe.
17. Carefully observe for leakage of the VS solution and note the location of any leaks. In rare cases, solution may leak from locations not associated with the heat seal as a result of:

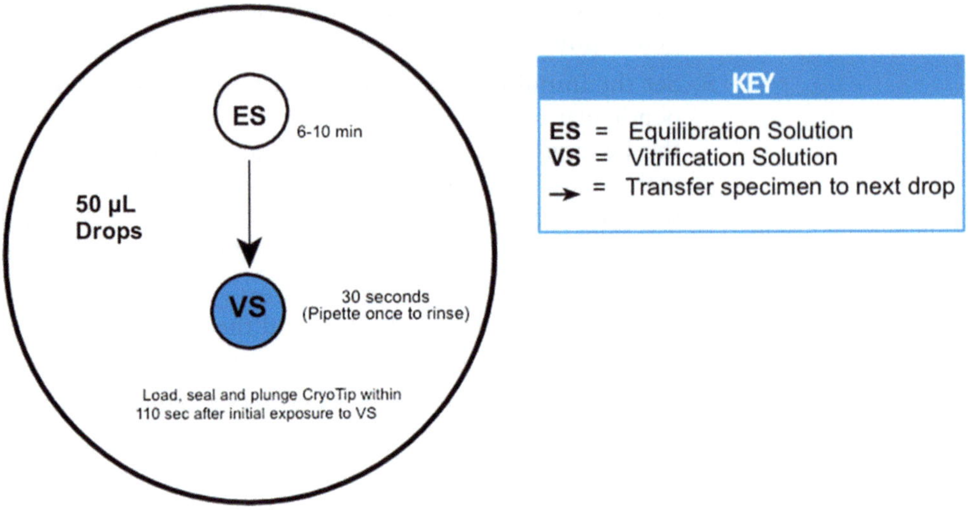


**Fig. 4** Schematic of solution distribution and method for embryo vitrification

- Exceeding the pressure capacity of the sealed CryoTip®.
- Cracks in the CryoTip® (may occur from damage or compromise of the tip integrity prior to or during use).
- The heat sealer interface is not sufficiently flexible and may damage the CryoTip® with creases or cracks during heat sealing; such heat sealers would FAIL the validation.

18. Repeat the entire procedure for a total of 15 CryoTip®(s) per validation, to assure consistency (sealer settings and sufficient heat seal) of the heat sealer (Fig. 4). *Note*: Sealer settings may vary between heat sealers; you should adjust your settings accordingly to achieve a complete seal. We suggest writing the sealer settings on the side of the sealer.
19. It is also strongly encouraged to validate both ends (ultrafine tip and wide end) of the CryoTip® to ensure the sealer settings are sufficient for an adequate seal at the bottom and top of the device.

*3.1.1 Heat Sealer Validation Protocol Acceptance Criteria*

PASS:

When properly heat-sealed, there should be no apparent leaks at either heat seal location and no crimp-induced cracks on the CryoTip®(s). If a leak occurs at the point of the seal, the heat sealer settings may be too high or too low and should be re-tested at an adjusted scale setting.

FAIL:

1. Heat sealer is consistently too hot and does not form a secure seal.
2. Heat sealer is not sufficiently flexible at the point of the heating element which could induce crimps and cracks in the CryoTip® during sealing.

*NOTE*: If frequent incomplete seal is observed due to wearing off of the heat sealer, the heating element should be replaced and then heat sealer should be re-validated.

The recommended heat settings of the impulse sealer may vary in different heat sealers.

### 3.2 Simplified Embryo Vitrification Protocol (2PN to Blastocyst)

All procedures must be performed at room temperature (22–27 °C).

Artificial collapse of the blastocoel cavity is highly recommended prior to vitrification.

Have all necessary materials, tools, and equipment ready and handy before starting procedure.

1. Aseptically dispense one (1) 50 μL drop of ES in a sterile embryo culture dish (Fig. 5).
2. Transfer embryo(s) (2 maximum) to the ES drop and expose undisturbed for 6–10 min.
   *NOTE*: In pre-blastocyst stage embryos, the embryo(s) will shrink and then gradually return to original size, indicating that equilibration is complete.
3. During above equilibration in ES, aseptically dispense one (1) 50 μL drop of VS 2 minutes prior to complete equilibration.
4. Transfer embryo(s) with minimal volume of medium from ES to the VS drop for 30 s before loading.
5. Gently pipette embryo(s) once within VS drop to ensure complete rinse with VS.
   *NOTE*: To minimize floating, after 10 s pipette the specimen(s) to the bottom center of the VS drop.

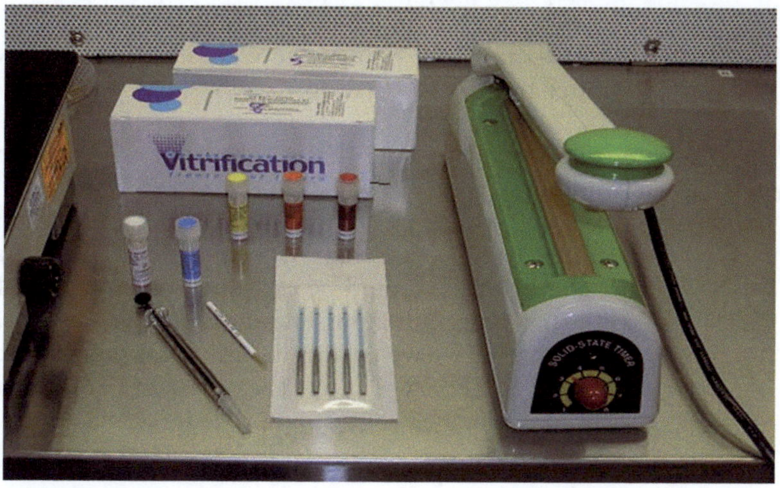


**Fig. 5** Materials required at workstation: Vit Kit®, CryoTip®, labeled straws, impulse heat sealer, 25 μL Hamilton Gastight Luer Tip Syringe, and connector

6. Load, seal, and plunge CryoTip® into LN2 within 80 s. Total maximum exposure to VS is 110 s.
7. Refer to Subheading 3.3 diagram for detailed loading instructions.

**Notes**:

- All procedures are to be done at ROOM TEMPERATURE (22–27 °C). Do not use a heated stage.
- Have all necessary materials, tools, and equipment ready and handy before starting procedure.
- CryoTip®(s) should be pre-labeled with patient information and assembled with connector and syringe (for loading) prior to starting procedure. To protect the ultrafine tip from damage, keep it covered with metal cover sleeve until ready to load specimen(s).
- Where possible, select only the best quality embryos (2PN to blastocyst) for vitrification.
- The recommended CryoTip® capacity is a MAXIMUM of two specimens.
- Process only as many specimen(s) as will be loaded per CryoTip® at one time.
- Minimize exposure of specimens to light during equilibration in ES and VS solutions.
- Transfer specimens between drops using a minimal volume of medium.
- The timing for exposure to VS is CRITICAL. Maintain microscopic visualization of specimen(s) by adjusting focus as needed during rapid exposure to VS (specimens will float in the drop). Keep transfer pipette tip close to drop for quick manipulations. Count the seconds with timer during 30 s exposure to VS. Load, seal, and plunge the CryoTip® into LN2 within 110 s after transfer to VS drop.

### 3.3 CryoTip® Loading Protocol

It is important to note that all procedures must be performed at room temperature (22–27 °C).

As an added precaution during the preparation procedure, carefully examine each CryoTip® outside of the package. Prior to use, all CryoTip® should be examined under a suitable magnification (40× power) for possible damage (such as tip breakages or cracks) that may have occurred during transport or storage.

Have all necessary materials, tools, and equipment ready and easily accessible before starting procedure, including:

- Validated Impulse Heat Sealer (refer to Subheading 3.1).
- Load, seal, and vitrify within 80 s (within 110 s after initial exposure to vitrification solution.

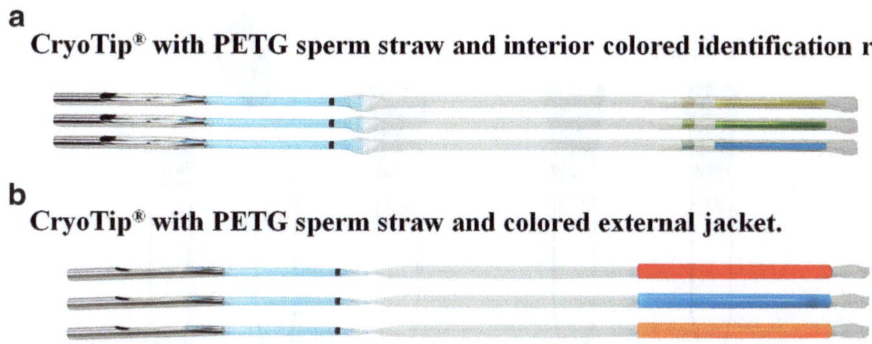


**Fig. 6** (**a**) CryoTip® with PETG sperm straw and interior colored identification rod. (**b**) CryoTip® with PETG sperm straw and colored external jacket

- Withdraw the plunger of the 25 μL Hamilton Gastight Luer Tip Syringe or equivalent aspiration tool (with connector attached) to the halfway position.
- Attach CryoTip® to connector and syringe (or equivalent aspiration tool).
- Gently slide up metal cover sleeve to expose the ultrafine tip of CryoTip®.
- Verify that CryoTip® are labeled with either a PETG sperm straw and interior ID rod (Fig. 6a) or external jacket (Fig. 6b).

**While viewing under the microscope: (Fig. 7)**

1. Aspirate vitrification solution (VS) to mark #1.
2. Aspirate specimen(s) and VS to mark #2.

**While viewing directly:**

3. Aspirate more VS to mark #3.
   NOTE: Specimen(s) MUST be between mark #2 and mark #3.

**Heat-seal both ends (use only recommended "impulse"-type heat sealers which have been previously validated):**

4. Disconnect the CryoTip® from the connector and syringe.
5. Seal #1 (Fig. 8): Set the impulse sealer to scale #3, holding the CryoTip® perpendicular to the sealer, place the ultrafine tip on the middle of the heating element, and brace your hand to ensure a steady seal. Gently push the handle of the heat sealer down to heat-seal the ultrafine tip end at or below the 1st mark. You will hear a beep and the light will turn off to indicate the seal is complete.
   - Slide cover sleeve down to protect the ultrafine tip.
   - Inspect the seal under the microscope to ensure a complete seal.

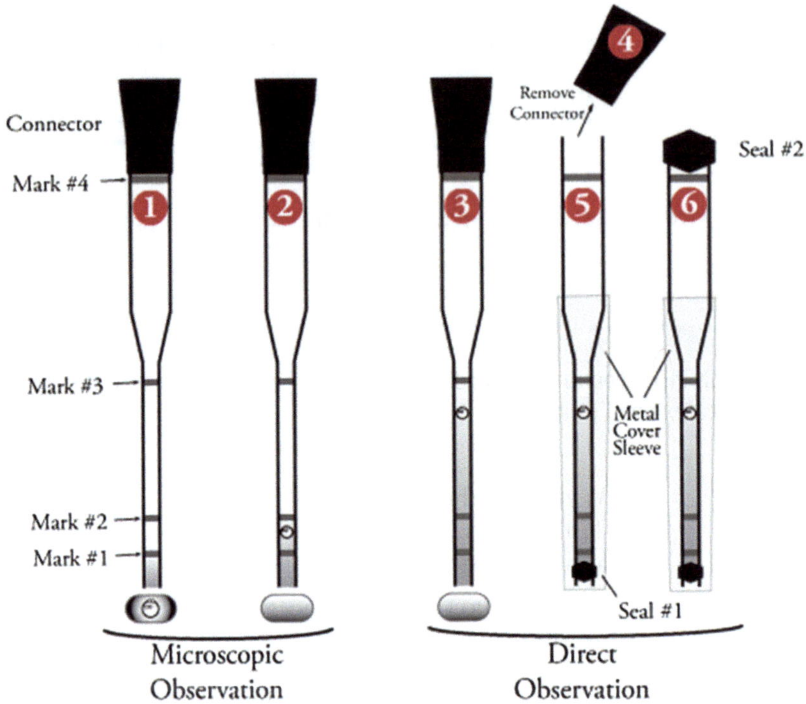


**Fig. 7** Step-by-step schematic of CryoTip® loading marks and specimen observation while viewing under a microscope (1 and 2) and while viewing the device directly (3–6)

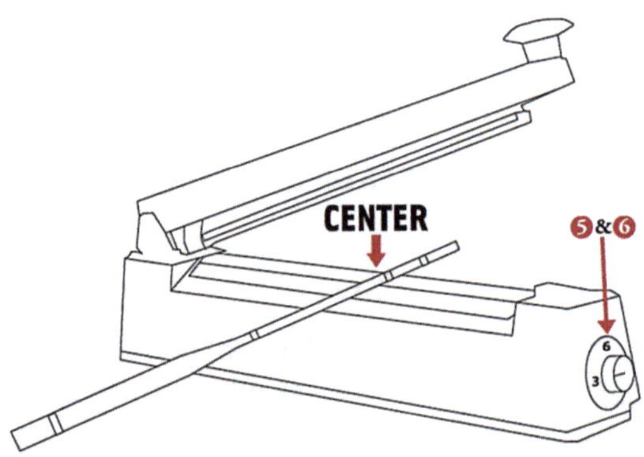


**Fig. 8** Schematic of impulse heat sealer and proper placement of the CryoTip® at time of sealing. The impulse sealer should be set to scale #3 for seal #1 and scale #6 for seal #2

6. Seal #2 (Fig. 8): Set the impulse sealer to scale #6, holding the CryoTip® perpendicular to the sealer, place the wide end on the middle of the heating element, and brace your hand to ensure a steady seal. Gently push the handle of the heat sealer down to heat-seal the wide end at or above the 4th mark. You will hear a beep and the light will turn off to indicate the seal is complete.
   - Inspect the seal under the microscope to ensure a complete seal.

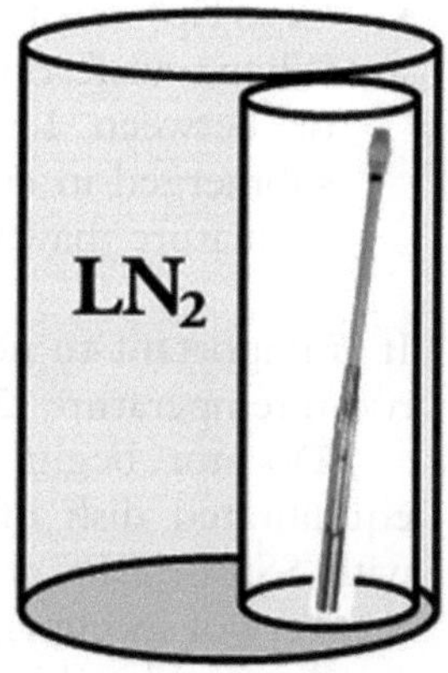


**Fig. 9** Schematic of CryoTip® plunged into liquid nitrogen ($LN_2$). After the metal cover sleeve is slid down over the sealed ultrafine tip of the CryoTip®, the device is plunged, metal cover sleeve side downward, into the $LN_2$

7. Plunge sealed CryoTip® into liquid nitrogen (LN2) (Fig. 9). Ideally, place into an LN2-filled goblet for long-term storage.

Notes:

- Have all necessary materials, tools, and equipment ready and easily accessible before starting procedure (e.g., LN2-filled holding reservoir, CryoCane and goblets, heat sealer, tongs or forceps, pipette or syringe).
- CryoTip® should be pre-labeled with patient information and assembled with connector and syringe (for loading), prior to starting vitrification procedure. To protect the ultrafine tip from damage, keep it covered with metal cover sleeve until ready to load specimen(s).
- The recommended CryoTip® capacity is a MAXIMUM of two specimens.
- The timing for exposure to VS is CRITICAL.

  Load, seal, and plunge the CryoTip® into LN2 within 110 s after initial exposure to VS drop.
- Rapid and controlled loading of the CryoTip® is essential and requires a secure seal between the CryoTip®, connector, and 25 μL Hamilton syringe or equivalent.

  Carefully attach the connector to the Hamilton syringe ensuring that there is a secure seal. Next, insert the wide end of the CryoTip® gently into the connector until you have achieved a slight seal.
- The recommended heat settings of the Impulse Sealer may vary in different heat sealers. The appropriate settings for heat sealing both ends of the CryoTip® should be determined by following the Subheading 3.1.

- CryoTip® must remain submerged in $LN_2$ until ready to thaw. When transferring CryoTip® from $LN_2$-filled holding reservoir, or between $LN_2$ storage tanks, CryoTip® should remain submerged in an $LN_2$-filled goblet to prevent uncontrolled/premature thawing in air.

### 3.4 Simplified Warming Protocol for CryoTip® (Oocytes and Embryos)

It is important to note that all procedures must be performed at room temperature (22–27 °C).

Do not begin warming procedure until you have a pre-equilibrated dish of appropriate culture medium supplemented with SSS or DSS at 20 % (v/v) or HSA at 12 mg/mL for final recovery of specimen(s).

Assisted hatching of embryos is highly recommended prior to embryo transfer.

Have all necessary materials, tools, and equipment ready and easily accessible before starting procedure.

1. Select CryoTip®(s) to be warmed from LN2 storage and quickly transfer to LN2-filled holding reservoir in preparation for warming procedure.
2. Place LN2-filled holding reservoir close to 37 °C water bath (minimum volume 500 mL) and microscope for subsequent rapid manipulation.
3. To set up warming dish, aseptically dispense (as shown in diagram):
   - One (1) 50 μL drop of TS
   - One (1) 50 μL drop of DS

   NOTE: For oocytes, dispense a minimum of 100 μL of TS.
4. Quickly remove CryoTip® from LN2 and within 1 s fully immerse the device in the 37 °C water bath (>500 mL) and swirl gently device for 3 s. Swirling the device is critical to ensure the most rapid warming rate (+24,000 °C/min) (Fig. 10).
5. Remove CryoTip® from the water bath and promptly remove metal cover sleeve from device by firmly grasping the lower end of the cover sleeve and pulling away from the CryoTip®. Gently wipe away any water with a sterile dry tissue ensuring the ultrafine tip of the device is dry.
6. Using sterile medical grade sharp scissors make Cut #1 below seal at wide end of CryoTip®.
7. Withdraw the plunger of the 250 μL Hamilton syringe (with connector attached) to the halfway position. Gently attach CryoTip® to connector and syringe.
8. Place ultrafine tip end over the prepared warming dish and quickly make Cut #2 above the seal at the fine end.

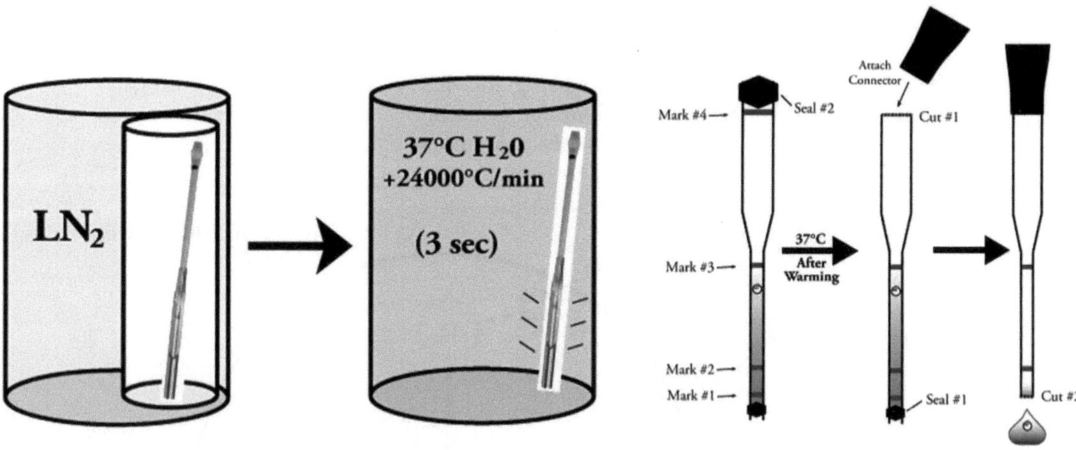


**Fig. 10** Schematic of CryoTip® warming procedure in a 37 °C water bath. Sharp sterile scissors are used to cut heat seals and contents are dispensed carefully with a syringe and connector

9. While visualizing under the microscope, dispense contents of CryoTip® as a small drop directly adjacent to TS drop. Once you visualize the specimen(s), touch the CryoTip® contents drop to TS drop with end of CryoTip® to mix. Set timer for 1 min and leave undisturbed (Fig. 11).
10. Transfer specimen(s) to DS, for 4 min. Gently pipette specimens once to ensure complete rinse in DS.
11. During the 4 min exposure in DS, aseptically dispense two (2) 50 μL drops of WS (WS1, WS2).
12. Transfer specimen(s) to WS1 and then WS2 for 4 min each undisturbed.
13. Transfer warmed OOCYTE(S) to pre-equilibrated culture medium with 20 % (v/v) protein supplement or 12 mg/mL for recovery (2–3 h to allow time for spindle reformation) prior to subsequent manipulations.

**There are two options for warmed EMBRYO(S):**

(a) For immediate transfer to patient: transfer EMBRYO(S) to pre-equilibrated "transfer" medium containing 20 % (v/v) protein supplement or 12 mg/mL.

(b) For further culture: transfer EMBRYO(S) to pre-equilibrated culture medium containing 20 % (v/v) protein supplement or 12 mg/mL for a 4 h recovery period. After recovery period, transfer EMBRYO(S) to culture medium with 10 % (v/v) protein and incubate accordingly until desired developmental stage has been reached for transfer to patient.

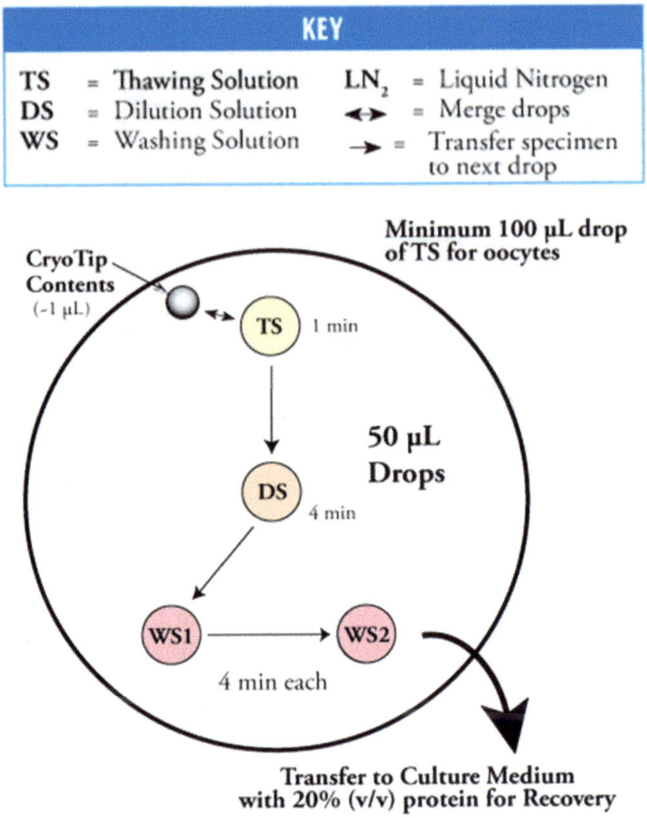


**Fig. 11** Schematic of solution distribution and method for embryo warming/thawing

Notes:

- Have all necessary materials, tools, and equipment ready and easily accessible before starting procedure.
- Pre-equilibrate a dish of appropriate culture medium supplemented with SSS or DSS at 20 % (v/v) or HSA at 12 mg/mL for final recovery of specimen.
- CryoTip®(s) must remain submerged in LN2 until ready to warm. When transferring CryoTip®(s) from LN2-filled holding reservoir, or between LN2 storage tanks, CryoTip®(s) should always be submerged in an LN2-filled goblet to prevent uncontrolled/premature warming in air.
- Use sterile medical grade sharp scissors.
- Set up warming dish with drops of solutions (*see* **step 3**) prior to removing CryoTip® from LN2.
- It is recommended to use a 250 µL Hamilton Gastight Luer Tip Syringe for the warming procedure.

- Use a 37 °C water bath with a minimum volume of at least 500 mL.
- Ensure that the plunger of the syringe has been withdrawn halfway prior to attaching the CryoTip® to the connector.
- Rapid and controlled dispensing of contents from the CryoTip® is essential and requires a secure seal between the CryoTip®, connector, and syringe (or pipette).
- Cut ultrafine tip end over the dish in case of premature dispensing of contents. Carefully dispense contents as a drop (ideally hanging on edge of CryoTip® before touching directly next to TS drop) to AVOID BUBBLES.
- Limit exposure to light while moving the specimens through the solutions.

*NOTE*: Following complete recovery (2–4 h post-warming), oocytes must be fertilized by ICSI due to zona hardening during vitrification.

## 4 Notes CryoTip® Loading, Sealing, and Dispensing

The following points should be considered when using the CryoTip®:

**Before Use:**

1. Inspect each CryoTip® under the microscope before use. Check to make sure there are no cracks, defects, or bends in the plastic. The front edge of the ultrafine tip end should be flat and straight. If any cracks, bends, or damage is apparent, DO NOT USE the CryoTip® as this may cause the device to burst resulting in the specimen being compromised.
2. Slide the metal cover sleeve up and down at least once. The slide tension of the cover sleeve should be firm but not score the plastic. Also, it should not be so loose that it slips up and down without assistance.
3. Connectors are reusable; we recommend you to replace the connector approximately every 30 CryoTip®(s).
4. Wipe and clean the Teflon sheet and silicone rubber of the heat sealer with 70 % ethanol to remove any debris. Debris on the heat sealer may prevent a complete heat seal.
5. Recommended heat sealer is the impulse heat sealer. Each heat sealer will have its own unique characteristics. It is up to the operator to determine exactly which settings are best for their sealer. Practicing with multiple CryoTip®(s) is essential for determining these settings (Subheading 3.1). You can heat-seal the same CryoTip® several times when validating your heat sealer; just make the next seal above the first and so on.

Approximate optimal settings for sealing each end of the CryoTip® are:

- Ultrafine tip end—low setting ~3.
- Wide end—high setting ~6: The settings can be marked on the specific heat sealer so that the same respective settings are used every time for sealing the corresponding ends (fine and wide). Alternatively, you can validate two heat sealers, one for the ultrafine tip end and one for the wide end. This method allows you to use a separate sealer for the fine end and the wide end to eliminate the need for adjusting heat settings between seals. Correct and complete sealing is key to achieving good recovery rates from the CryoTip®.

6. Heat sealers should be re-validated annually.

**Loading:**

1. We recommend that you use a glass Hamilton syringe to load the CryoTip®. You can get these from www.hamiltoncompany.com. There are several sizes available but we recommend the 25 μL syringe; this size gives you adequate control to easily load the CryoTip® with <1 μL of medium.
2. Pull back the syringe plunger slightly before attaching the CryoTip® to prevent any capillary action. Hold the CryoTip® steady and rotate the syringe slightly while pushing the CryoTip® into the connector to get the best seal; continue applying pressure until you see approximately 1 cm of the large end breach of the seal between the CryoTip® and connector. Be sure that you push the CryoTip® into the connector just enough so that the connector holds the CryoTip® in place, **do not push the CryoTip® in too far**.
3. Hold the metal cover slip from the ultrafine tip end and slide it up toward the connector. Ensure that the metal cover sleeve does not go into the connector; this will make it difficult to achieve a good fit between the connector and CryoTip®.
4. When loading the embryo(s) in the device, it is advised to increase your field of magnification but "zooming out" and using your peripheral vision to view the embryos moving up to mark #3.

**After Loading:**

1. Always ensure that when sealing the CryoTip® your hand is steady and braced. Movement during sealing can lead to stress on the seal and result in a failure.
2. After heat sealing always visually check the seals using the microscope. Check the small end to be sure the seal is flat and melted without bends or kinks. Check the large end to be sure

the plastic has a complete seal. If the heat seal is sufficient, the imprint of the brown plastic coating of the sealer should be visible on the blue plastic of the CryoTip®, and there should be no cracks along the sealed edge of the large end radiating down toward the small tip.

3. If the large end of the CryoTip® does not seal entirely, you need to seal it again. Make the second seal just below the first seal.
4. Always adjust the heat settings to accommodate the large end (~6) and small end (~3) of the CryoTip®.
5. Slide the metal cover sleeve completely down to the fine end until it stops to protect it during storage.
6. Plunge into LN2 and swirl for 3 s to maximize cooling rate.

**During Warming:**

7. One of the most critical steps to warming is to achieve a rapid warming rate (+24,000 °C/min). To maximize the warming rate with the CryoTip®, move the CryoTip® from LN2 to a large 37 °C water bath (>500 mL) within 1 s. Swirl CryoTip® for 3 s in water bath and continue with the recommended protocol for warming.
8. A pair of sharp scissors and a quick "snap" action when cutting away the ultrafine tip end are essential to prevent the contents of the CryoTip® from being accidentally dispensed onto the scissors. If the content is expelled immediately or onto the scissors, try not to connect the Hamilton syringe onto the connector too hard. This may create unwanted pressure in the CryoTip®.
9. If the freezing solution is not expelled from the device or you can visualize bubbles within the CryoTip® before cutting, the seals have not been correctly made (usually the seal on the wide end of the device).
10. Have your TS solution set-up and move your specimens into the solution as quickly as possible.
11. Specimens should be transferred to equilibrated culture medium containing 20 % protein for complete recovery after warming.

**Trouble Shooting:**

1. If cracks occur upon plunging, then most likely the seals were not complete before plunging. Check that the heat sealer platform is clean and sealing consistently, that the temperature setting is correct for each end of the CryoTip® (giving a firm, melted seal), and that the user is using the sealer correctly (push

the handle firmly, without crushing to seal, let up after the beep).

2. When just learning how to use the device, you can seal and unload several CryoTip®(s) prior to implementing the system with specimens (oocytes and embryos). Load solution, seal, plunge into liquid nitrogen, remove, plunge into water, and unload. Make sure you can do this repeatedly without getting bubbles in the CryoTip® or finding empty CryoTip®(s). If CryoTip® is empty then it wasn't sealed correctly.

## Acknowledgment

The authors would like to thank Dr. Joe Conaghan (Pacific Fertility Center) and Irvine Scientific® for providing images used in this appendix.

### References

1. Bielanski A, Nadin-Davis S, Sapp T, Lutze-Wallace C (2000) Viral contamination of embryos cryopreserved in liquid nitrogen. Cryobiology 40(2):110–116
2. Martino NA, Dell'Aquila ME, Cardone RA, Somoskoi B, Lacalandra GM, Cseh S (2013) Vitrification preserves chromatin integrity, bioenergy potential and oxidative parameters in mouse embryos. Reprod Biol Endocrinol 11(1):27
3. Popwell JM, Dougall KM, Conaghan J (2014) Clinical pregnancy and implantation rates of warmed, biopsied and re-vitrified blastocysts (W-CCS). Fertil Steril 101(2):e20
4. Tanaka M, Takahasi A, Seida K, Yoshida A (2008) Consideration of a single sperm freezing preservation method using a cryotip. Fertil Steril 90:S292
5. Valbuena D, Póo ME, Aguilar-Gallardo C, Martinez S, Cobo AC, Pellicer A, Simón C (2012) Comparison of Cryotip vs. Cryotop for mouse and human blastomere vitrification. Fertil Steril 97(1):209–217
6. Kuwayama M, Vajta G, Ieda S, Kato O (2005) Comparison of open and closed methods for vitrification of human embryos and the elimination of potential contamination. Reprod Biomed Online 11(5):608–614
7. Kuwayama M, Ieda S, Zhang J, Kato O (2005) The CryoTip method: aseptic vitrification of oocytes and embryos. Fertil Steril 84:S187
8. Kuwayama M, Vajta G, Ieda S, Kato O (2005) Vitrification of human embryos using the CryoTip TM method. Reprod Biomed 11:608–614

# Appendix E: Rapid-i™: Closed Vitrification Device by Vitrolife

## Mark G. Larman

## Abstract

Cryopreservation of gametes and embryos is a growing technique in numerous reproductive fields including human-assisted reproduction. With improved understanding of embryo physiology and optimized culture conditions, there are now more embryos than ever to vitrify for potential use in subsequent cycles. Many gametes and embryos have been cryopreserved in open systems, but there are concerns with regard to contamination from the liquid nitrogen and also cross-contamination between patients' germplasm. The development of the Rapid-i™, a closed vitrification device that does not use direct contact with liquid nitrogen during vitrification or subsequent storage, will be discussed as well as clinical protocols for human oocytes and embryos.

**Key words** Closed device, Cryopreservation, Contamination, Liquid nitrogen, Rapid-i, Vitrification

## 1 Introduction

Embryos vitrified on open devices and stored in liquid nitrogen are vulnerable to viral contamination [1]. In the embryology lab, liquid nitrogen is not stored under sterile conditions, so it is not surprising that bacterial and fungal contamination have been reported [2, 3] and that there are concerns with regard to the safety of gametes and embryos during vitrification and storage in liquid nitrogen. Methods to minimize contamination concerns have been discussed [4–6], but ultimately the optimal method to prevent or minimize contamination during vitrification and subsequent storage in liquid nitrogen is to use a closed system that does not require direct contact with liquid nitrogen.

A proof of principle paper was published by Larman and colleagues [7], demonstrating that an open device (Cryoloop) could be modified so that it did not require direct liquid nitrogen contact during vitrification. Cryovials were held in a polystyrene tray in such a way that liquid nitrogen around, but not inside the vial, cooled the air inside the vial to around −190 °C. Vitrification in

Zsolt Peter Nagy et al. (eds.), *Cryopreservation of Mammalian Gametes and Embryos: Methods and Protocols*, Methods in Molecular Biology, vol. 1568, DOI 10.1007/978-1-4939-6828-2_25, © Springer Science+Business Media LLC 2017

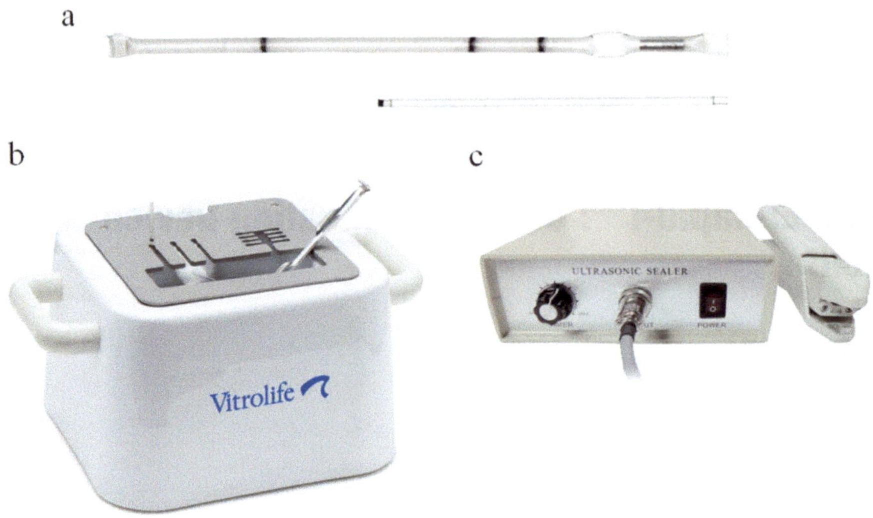


**Fig. 1** The Rapid-i is a closed vitrification device that requires no direct contact with liquid nitrogen. (**a**) It is composed of a weighted storage straw (RapidStraw) that is sealed at one end and a plastic rod (Rapid-i), which holds the embryos. (**b**) The Rapid-i is then inserted into the RapidStraw that is held in the SmartBox containing liquid nitrogen. The air inside the straw is supercooled to around −190 °C causing instantaneous vitrification. (**c**) The top of the straw is then sealed with the ultrasonic sealer

supercooled air was compared to the standard direct contact protocol using mouse pronuclear oocytes. Survival and embryo development, including cell differentiation, were comparable between the two techniques. The manufacturers of cryovials, however, only recommend storage in the vapor phase as they are not leakproof. Because cryovials could not provide a hermetic seal, the principle of supercooled air was applied to an in-straw device, the Rapid-i (Fig. 1a–d). The Rapid-i underwent significant preclinical testing. It was developed using mouse embryos and is capable of vitrifying mouse pronuclear oocytes with a 100 % survival rate. The subsequent embryo development, cell number, and embryo viability (assessed by embryo transfer) were not affected when compared to sibling non-vitrified embryos [8].

More recently the Rapid-i has been compared clinically to two open systems (Cryoloop and Cryotop) for human embryo vitrification [9, 10]. Embryo development and clinical outcomes with the Rapid-i were equivalent to both open systems. The results from these two studies demonstrated that a closed system such as the Rapid-i can vitrify human embryos with the same efficacy as open systems. It was thought that vitrification of human oocytes required the high cooling rates only provided by direct contact with liquid nitrogen. As with embryos, it appears that human oocytes can also be vitrified with the Rapid-i. In one ongoing clinical study, over 500 oocytes have been vitrified and warmed with the Rapid-i with a survival rate of 94 %. Following ICSI, the fertilization rate was 76 %.

Blastocyst transfer resulted in a 49 % ongoing pregnancy rate, with five healthy live births reported so far [11].

## 2 Materials

The Rapid-i can be used with any vitrification/warming media. For the purpose of this chapter, the procedure will be explained using Vitrolife's vitrification/warming media for blastocysts (RapidVit Blast and RapidWarm Blast). Two other kits are provided for vitrification/warming of oocytes and cleavage-stage embryos (Rapid-Vit/Warm Oocyte and Cleave, respectively). The protocols are similar with adjusted timings. Please consult each product insert for the specific instructions.

### 2.1 Vitrification

RapidVit Blast.

Rapid-i Kit.

SmartBox.

Ultrasonic sealer.

Dewar containing liquid nitrogen.

Cryocane and goblet.

Cryo-label or cryomarker for patient identification.

Multi-well dish such as Vitrolife's 5-well dish.

Pipettor.

Sterile pipette tips (1 mL and 20 μL).

Stereomicroscope with heated stage.

Embryo handling tool.

Timer.

### 2.2 Warming

RapidWarm Blast.

SmartBox.

Cutters.

Needle nose tweezers.

Dewar containing liquid nitrogen.

Cryocane and goblet containing patients' Rapid-i with embryos.

Multi-well dish such as Vitrolife's 5-well dish.

Pipettor.

Sterile pipette tips (1 mL).

Stereomicroscope with heated stage.

Embryo handling tool.

Timer.

## 3 Methods

Outlined below is the method for vitrifying and warming human blastocysts with the Rapid-i and RapidVit Blast and RapidWarm Blast kits. A video of the protocol can also be found on Vitrolife's website.

### 3.1 Vitrification of Blastocyst-Stage Embryos

1. Label the multi-well plate with the patient identification.
2. Place 0.5–1 mL of each of the solutions for vitrification in separate wells of the labeled multi-well dish, e.g., Vitri 1 solution in well number 1, Vitri 2 solution in well number 2, and Vitri 3 solution in well number 3.
3. Place the lid on the multi-well dish and allow the solutions to warm to 37 °C (*see* **Note 1**).
4. While the vitrification solutions are warming, collect the remaining items mentioned above in Materials Subheading 2.1.
5. Place the SmartBox on the lab bench close to the microscope. Ensure the bench used for the procedure can tolerate liquid nitrogen spills.
6. Fill the SmartBox with liquid nitrogen up to 1 cm from the box's rim and place the lid on top of the box (*see* **Note 2**).
7. Check that the Rapid-i Kit packaging is intact and that the use by date has not expired. Open the package using aseptic technique.
8. Label the exact number of RapidStraws needed with the patient's identification. Place the label below the top, black mark on the RapidStraw.
9. Once the vitrification solutions have warmed to 37 °C, move them on to a heated stereomicroscope stage that will maintain this temperature.
10. Transfer the embryos into Vitri 1 using a pipette (*see* **Note 3**). The embryos can remain in the solution for 5–20 min (*see* **Note 4**).
11. Move an appropriate number of embryos into Vitri 2 (*see* **Note 5**). Start the timer. The embryos are exposed to this solution for a total of 2 min.
12. When 30 s remains, make a 20 μL droplet of Vitri 3 on a nontoxic surface, preferably a culture dish (*see* **Note 6**).
13. When 10 s remains, begin collecting the embryos. Transfer the embryos in a minimal volume of Vitri 2 to avoid dilution of the Vitri 3 droplet.
14. Transfer the embryos into the 20 μL droplet of Vitri 3, and let them remain in this solution for 45 s. To allow proper exposure

**Fig. 2** The device is loaded with the embryos and vitrification medium by pipetting the embryos (or in this case 100 μm beads) directly into the 50 nL hole. The small volume and high viscosity of the vitrification solution mean that the embryos remain securely in place

of the embryos to Vitri 3, move the embryos around two–three times within the droplet.

15. When 5–10 s remains, collect the embryos in a minimal volume and place them onto the Rapid-i.
16. Bring the end of the pipette holding the embryos close to the wall of the hole on the Rapid-i. Gently expel the embryos and Vitri 3 until the hole has been filled (Fig. 2a–c).
17. Pick up the Rapid-i and place it in the RapidStraw (*see* **Note 7**). The total time from placing the embryos into Vitri 3, onto the Rapid-i, and then into the RapidStraw should be around 45 s.
18. Immediately seal the open end of the RapidStraw with the ultrasonic sealer.

### *3.2 Warming of Blastocyst-Stage Embryos*

1. Label the multi-well dish with the patient identification.
2. Place 0.5–1 mL of each of the solutions for warming in separate wells of the labeled multi-well dish, e.g., Warm 1 solution in well number 1, Warm 2 solution in well number 2, and Warm 3 solution in well number 3.
3. Place the lid on the multi-well dish and allow the solutions to warm to 37 °C (*see* **Note 1**).
4. While the vitrification solutions are warming, collect the remaining items for warming mentioned above in the Materials Subheading 2.2.
5. Place the SmartBox on the lab bench close to the microscope. Ensure the bench used for the procedure can tolerate liquid nitrogen spills.
6. Fill the SmartBox with liquid nitrogen up to 1 cm from the box's rim and place the lid on top of the box.
7. Move the cryocane and goblet containing the RapidStraws into the liquid nitrogen in the SmartBox. Remove one RapidStraw, without leaving the liquid nitrogen, and make sure the Rapid-Straw is secure in the lid and that the lower end touches the bottom of the SmartBox.

8. Warm the RapidStraw with your fingers around the black mark to get a better view of the black tab on the Rapid-i. Hold the RapidStraw well above the black mark, and cut it around 3 mm above the black end of the Rapid-i. Do not lift the RapidStraw from the lid, and make sure it stays upright in the liquid nitrogen.
9. Lift the Rapid-i (using needle nose tweezers) out of the Rapid-Straw just enough to enable you to grasp the end with your fingertips. Then quickly (as fast as possible, preferably less than 2 s), but carefully, remove the Rapid-i from the RapidStraw, and plunge the tip and hole of the Rapid-i into the Warm 1 solution immediately (*see* **Note 8**).
10. While watching under the microscope, gently move the Rapid-i back and forth. Verify that the embryos are free in Warm 1 and then remove the Rapid-i.
11. Embryos remain in this solution for 2 min.
12. Transfer the blastocysts into Warm 2 and let the blastocysts remain in the solution for 3 min.
13. Transfer the blastocysts into Warm 3 and let the blastocysts remain in the solution for 5–10 min.
14. Move the embryos into appropriate culture or embryo transfer medium.

## 4 Notes

1. Directly measure the temperature of the medium in the multi-well dish and adjust the temperature of the stage warmer to achieve 37 °C. The solutions are warmed to 37 °C to maintain the oocytes and embryos at a physiological temperature. This is particularly important for the oocyte since exposure to room temperature will facilitate microtubule depolymerization, leading to meiotic spindle disassembly [12]. Having a protocol that uses 37 °C also means that the incubation times are shorter, resulting in the oocytes and embryos spending less time out of the incubator.
2. It is critical to check that the level of liquid nitrogen in the SmartBox is sufficient to cover, at the very least, the lower half of the straw. This will ensure that during the vitrification and warming procedures, the embryos remain vitrified. The hole in the Rapid-i where the embryos are placed has a volume of around 50 nL, so it is easy for such a small volume to inadvertently warm if the Rapid-i and RapidStraw are not handled appropriately.
3. A finely pulled glass pipette that has a diameter slightly bigger than the oocyte/embryo allows them to be moved between the

solutions and into the hole of the Rapid-i in a minimal volume. The capillary action of the pipette can be broken by introducing air bubbles in the column of medium. This is simply achieved by rapidly moving the tip of the pipette in and out of the medium while applying light suction.

4. It is advised to culture embryos in media that have been supplemented with 0.125–0.5 mgL/mL hyaluronan [13, 14]. All of Vitrolife's culture and vitrification/warming media contain hyaluronan, which has been shown to increase cryotolerance [14–18].
5. Due to the density of Vitri 2, the oocytes and embryos will float to the top of the solution. It is advisable to collect the oocytes and embryos and replace them to the bottom of the drop to facilitate quick collection for transfer to Vitri 3.
6. As with Note 5, due to the density of Vitri 3, the oocytes and embryos will float. Since the time the oocytes and embryos spend in Vitri 3 is critical, a 20 μL drop of the Vitri 3 is created. This restricts the movement of the oocytes and embryos and allows them to be quickly mixed within the solution and gathered.
7. Using supercooled air results in a cooling rate that is much slower (1200 °C/min) than direct contact (>10,000 °C/min), but it is still sufficient to vitrify the vitrification medium (Fig. 3a, b).
8. Often the focus on the development of minimal volume devices and direct contact with liquid nitrogen was to increase cooling rates, but it appears that the warming rate is actually more

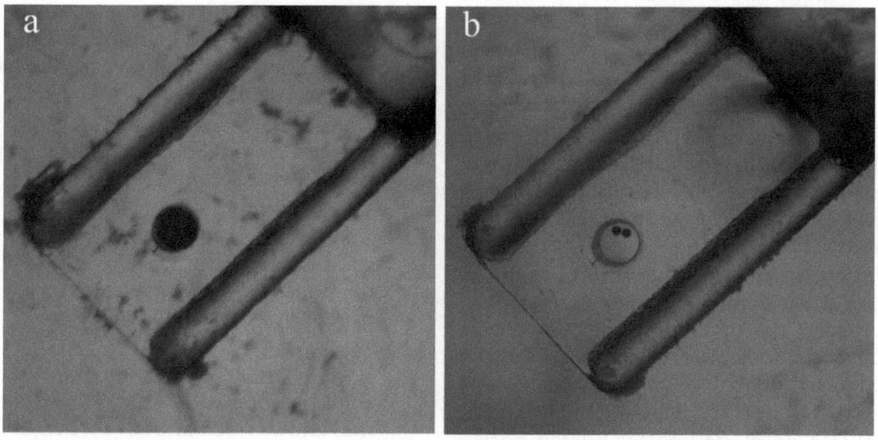


**Fig. 3** To demonstrate that the cooling rate with the Rapid-i permits Vitri 3 to vitrify, the RapidStraw was cut under liquid nitrogen, and an image was taken while the Rapid-i remained submerged. (**a**) The Rapid-i hole filled with Vitri 1 (no cryoprotectant) freezes, becoming opaque. (**b**) The Rapid-i hole filled with Vitri 3 vitrifies and remains translucent

critical. Two papers by Peter Mazur's group demonstrated the critical importance of warming rate in the survival of mouse oocytes [19, 20].

## References


1. Bielanski A, Nadin-Davis S, Sapp T, Lutze-Wallace C (2000) Viral contamination of embryos cryopreserved in liquid nitrogen. Cryobiology 40:110–116
2. Bielanski A, Bergeron H, Lau PC, Devenish J (2003) Microbial contamination of embryos and semen during long term banking in liquid nitrogen. Cryobiology 46:146–152
3. Morris GJ (2005) The origin, ultrastructure, and microbiology of the sediment accumulating in liquid nitrogen storage vessels. Cryobiology 50:231–238
4. Bielanski A, Vajta G (2009) Risk of contamination of germplasm during cryopreservation and cryobanking in IVF units. Hum Reprod 24:2457–2467
5. Bielanski A (2012) A review of the risk of contamination of semen and embryos during cryopreservation and measures to limit cross-contamination during banking to prevent disease transmission in ET practices. Theriogenology 77:467–482
6. Larman MG, Hashimoto S, Morimoto Y, Gardner DK (2014) Cryopreservation in ART and concerns with contamination during cryobanking. Reprod Med Biol 13:107–117
7. Larman MG, Sheehan CB, Gardner DK (2006) Vitrification of mouse pronuclear oocytes with no direct liquid nitrogen contact. Reprod Biomed Online 12:58–61
8. Larman MG, Gardner DK (2011) Vitrification of mouse embryos with super-cooled air. Fertil Steril 95:1462–1466
9. Desai NN, Goldberg JM, Austin C, Falcone T (2013) The new Rapid-i carrier is an effective system for human embryo vitrification at both the blastocyst and cleavage stage. Reprod Biol Endocrinol 11:41–49
10. Hashimoto S, Amo A, Hama S, Ohsumi K, Nakaoka Y, Morimoto Y (2013) A closed system supports the developmental competence of human embryos after vitrification. J Assist Reprod Genet 30:371–376
11. Machac S, Hubrinka V, Larman M, Koudelka M (2013) A novel method for human oocyte vitrification with a closed device using super-cooled air. Fert Steril 100(sup3):S108
12. Larman MG, Minasi MG, Rienzi L, Gardner DK (2007) Maintenance of the meiotic spindle during vitrification in human and mouse oocytes. Reprod Biomed Online 15:692–700
13. Gardner DK, Rodriegez-Martinez H, Lane M (1999) Fetal development after transfer is increased by replacing protein with the glycosaminoglycan hyaluronan for mouse embryo culture and transfer. Hum Reprod 14:2575–2580
14. Lane M, Maybach JM, Hooper K, Hasler SJ, Gardner DK (2003) Cryo-survival and development of bovine blastocysts are enhanced by culture with recombinant albumin and hyaluronan. Mol Reprod Dev 64:70–78
15. Palasz A, Alkemade S, And Mapletoft RJ (1993) The use of sodium hyaluronate in freezing media for bovine and murine embryos. Cryobiology 30:172–178
16. Stojkovic M, Kolle S, Peinl S, Stojkovic P, Zakhartchenko V, Thompson JG, Wenigerkind H, Reichenbach HD, Sinowatz F, Wolf E (2002) Effects of high concentrations of hyaluronan in culture medium on development and survival rates of fresh and frozen-thawed bovine embryos produced in vitro. Reproduction 124:141–153
17. Balaban B, Urman B (2005) Comparison of two sequential media for culturing cleavage-stage embryos and blastocysts: embryo characteristics and clinical outcome. Reprod Biomed Online 10:485–491
18. Dattena M, Mara L, Bin TAA (2007) Lambing rate using vitrified blastocysts is improved by culture with BSA and hyaluronan. Mol Reprod Dev 74:42–47
19. Seki S, Mazur P (2009) The dominance of warming rate over cooling rate in the survival of mouse oocytes subjected to a vitrification procedure. Cryobiology 59:75–82
20. Mazur P, Seki S (2011) Survival of mouse oocytes after being cooled in a vitrification solution to −196 °C at 95° to 70,000 °C/min and warmed at 610° to 118,000 °C/min: a new paradigm for cryopreservation by vitrification. Cryobiology 62:1–7

# Appendix F: Quinn's Advantage Embryo Freeze Kit

## Kiri Beilby and Patrick Quinn


## Abstract

Despite a large focus on the use of vitrification to cryopreserve embryos in recent years, there are still arguments for the use of slow freezing for the cleavage-stage embryo. Having said this, there are lessons to be learned from the process of vitrification that could be applied to slow freezing to improve post-thaw survival and ultimately clinical pregnancy rates. Specifically, increasing the concentration of sucrose in the freezing solution from 0.1 to 0.2 M and subsequently increasing the sucrose concentrations in thawing solutions could prove beneficial. The use of vitrification warming solutions in the thawing of slow-frozen embryos may also be an option that not only improves survival but also streamlines product purchasing and protocols within the laboratory.


**Key words** Embryo, Cryopreservation, Slow cooling, Cryoprotectants, Thawing

## 1 Introduction

The use of assisted reproductive technologies to treat infertility is increasing and so is the need for the effective cryopreservation of oocytes and embryos. With the industry developing and evolving, a greater proportion of high-quality embryos per treatment cycle can be expected. Subsequently, the chance of pregnancy from each embryo produced is increasing, resulting in a trend toward single embryo transfer (SET) [1] and more "spare" embryos per treatment cycle requiring preservation. Furthermore, a growing interest in preimplantation genetic diagnosis (PGD) and encouraging results following the transfer of embryos during natural cycles [2] have placed additional demands of the need for cryopreservation. One could argue that effective cryopreservation has allowed us to develop the aforementioned techniques, which are now dependant on good preservation technique for success. Finally, reports that children born from the transfer of cryopreserved embryos do not have the same likelihood of having a low birthweight as children born from fresh IVF embryo transfers add additional interest to this technology [3–5].

Zsolt Peter Nagy et al. (eds.), *Cryopreservation of Mammalian Gametes and Embryos: Methods and Protocols*, Methods in Molecular Biology, vol. 1568, DOI 10.1007/978-1-4939-6828-2_26, © Springer Science+Business Media LLC 2017

Numerous animal models and then early human embryo clinical studies have led to the commercial development of the cryopreservation products available, including Quinn's Advantage Embryo Freeze (QAF) Kit, supplied by ORIGIO, a CooperSurgical Company, Måløv, Denmark. The purpose of this chapter is to provide a technical explanation of this product in relation to its composition, and how to use it most effectively. A comparison between slow freezing and vitrification will be made, where the use of this product and protocol are placed in context with more current cryopreservation options; their merits and shortcomings are discussed. Finally, the relationship between freezing and thawing products and protocols (rapid thawing and rapid warming) will be considered. This chapter will focus on, but not be limited to, the physiology of the cleavage-stage embryo.

## 2 Materials

### 2.1 Product Description

Quinn's Advantage Embryo Freeze (QAF) Kit is for the effective freezing of pronuclear and cleavage-stage embryos. Developed in the 1990s, the kit supplies cryodiluents containing the permeating cryoprotectant propanediol (PROH, 1.5 M) and the non-permeating cryoprotectant sucrose (0.1 M). The PROH acts to support the cell membrane as it transitions from a relatively fluid state to a more ridged one, protecting the cells against osmotic shock by binding to electrolytes both inside the cytoplasm and surrounding it and acting as a partial substitute for the absence of water. The sucrose promotes the osmotic dehydration of cells, prior to cooling, and this minimizes the formation of damaging intracellular ice crystals [6].

The solutions are HEPES buffered to maintain a stable pH during cryopreservation procedures that take place outside of a high $CO_2$ atmosphere [7]. The addition of 12 mg/mL human serum albumin (HSA) has been used to provide a similar protein concentration to that in 20 % (v/v) human serum that has been successfully used for cryopreservation of human cleavage-stage embryos and zygotes [6].

The suggested freezing program specifies that embryos are taken from 37 °C to −6 °C at a rate of 2 °C/min, initially dehydrating the cells. This step is followed by a seeding period where embryos are held at −6 °C for 10–15 min before further cooling at a rate of 0.3 °C/min down to −35 °C. The embryos are then transferred to liquid nitrogen for storage.

### 2.2 Cryodiluent Development

The cryoprotective effect of glycerol in the preservation of semen was one of the first insights gained into the diluent composition required for the successful cryopreservation of mammalian cells [8]. While the mechanisms of this permeating cryoprotectant were yet to be elucidated, a new avenue of investigation had been opened.

The successful cryopreservation of cleavage-stage embryos came later with the validation of dimethyl sulfoxide (DMSO) as a more effective cryoprotectant, allowing mouse embryos to be frozen and thawed and resulting in pregnancies following embryo transfer [9]. The method was adapted from the mouse to numerous other species (reviewed by Whittingham; [10]) and ultimately paved the way for Trounson and Mohr [11] to conduct the first successful cryopreservation of human cleavage-stage embryos. The investigators compared the use of both glycerol and DMSO as permeating cryodiluents, but only DMSO produced a viable pregnancy, indicating the low permeability of cleavage-stage embryos to glycerol as was first suggested. The pregnancy obtained with DMSO was unfortunately terminated at 24 weeks due to development of septic microbial infection after premature membrane rupture.

The use of permeating cryoprotectants was explored further throughout the 1980s, with glycerol cited in the successful freezing of blastocysts [12] and the use of propanediol in the successful freezing of zygotes [13]. Propanediol would later also become the cryoprotectant of choice for freezing cleavage-stage embryos in combination with sucrose to increase intracellular dehydration before freezing [13, 14]. It is this cryoprotectant combination that is used in the Quinn's Advantage Embryo Freeze Kit and will be discussed in the next section of this chapter. More recently, the introduction of ethylene glycol as a permeating cryoprotectant was reported, demonstrating some positive results [15]. However, this method has yet to be adopted commercially.

Vitrification, the ultra-rapid cooling (and subsequent warming) of cells using much higher concentrations of both permeating and non-permeating cryoprotectants, was later introduced as a potential method of embryo cryopreservation [16]; the first pregnancies and deliveries in humans were achieved a few years later [17, 18]. Presently the most common method of cryopreservation for oocytes and blastocyst, but perhaps not yet for the cleavage-stage embryo. Reviewed by Edgar and Gook [19], the available evidence supporting the use of slow cooling versus vitrification for the cleavage-stage embryo was critically assessed and concluded that from the comparative studies available, it could not be shown that one method was outperforming the other as a measure of survival rate. While several authors have published poorer outcomes following the use of slow freezing compared to vitrification [20–23], evidence exists that a modification in the level of impermeable cryoprotectant (sucrose) concentration produces results comparable to those best achieved using vitrification [19].

The idea of modified cryoprotectant concentrations in slow freeing will be explored further in the following section. However, this observation does raise concerns regarding future medium composition, development, and commercialization. It seems obvious that the publication of evidence supporting a change in product composition should be rapidly adopted by media producers, yet the reality of conducting clinical assessments and producing the documents required by regulatory bodies worldwide is both timely and expensive. Gone are the days of rapid adaptation of products to keep pace with the speed of scientific observation. While there is no doubt that a high level of industry regulation is required in the field of reproductive medicine, to not only ultimately protect patients but also support those taking decisions in patient treatment, it can be a hindrance to incremental, albeit important progressions. The role of commercial companies in media production allows for scientists and clinicians working in the field to essentially "outsource" product quality control, including mouse embryo testing, sperm survival assays, endotoxin screening, sterility testing, and pH validation, just to name a few. Both customer preference and local governance have seen the majority of clinics worldwide shift from producing media in-house to purchasing products from a number of well-established suppliers.

### 2.3 Quinn's Advantage Embryo Freeze Kit

While the initial successes in embryo cryopreservation were achieved using DMSO [11], most commercially available products today contain PROH as the permeating cryoprotectant. This development stems from work published by Lassalle et al. [14] and Testart et al. [13] demonstrating the benefits of PROH. Following commercialization, this diluent composition was widely accepted, and this is today reflected in the numerous publications that exist concerning the quality of cleavage-stage embryos following slow freezing in a 1.5 M PROH and 0.1 M sucrose composition (reviewed by Edgar and Gook; [19]).

When critically assessing the published work available on slow-frozen embryos in a 1.5 M PROH and 0.1 M sucrose solution, these components combined will produce ≈70–80 % survival rates upon thawing in a corresponding warming solution, with ≈50 % of blastomeres fully intact. Of these fully intact embryos, a similar implantation can be achieved compared to equivalent fresh embryos (reviewed by Edgar and Gook; [19]). In this respect, improving the number of embryos that have intact blastomeres following thawing would subsequently improve clinical outcomes. The removal of lysed blastomeres has also been suggested to improve clinical rates of pregnancy following the recovery of damaged, cryopreserved embryos [24].

The major components in QAF are discussed below. How these components could be adjusted to enhance product performance is also considered.

#### 2.3.1 Permeating Cryoprotectant: Propanediol

As culture is prolonged, the quantity and quality of embryos available for transfer or cryopreservation are reduced. Therefore, early work has focused on cryopreserving 2-, 4-, and 8-cell cleavage-stage embryos as a means of having more material remaining following the additional insult of freeze-thawing. However, more reliable culture systems, a greater emphasis on single embryo transfer (SET), and the need for blastocyst culture to compliment advances in PGD/ PGS have meant the development of cryopreservation techniques through all stages of embryo development. Initial observations made suggesting probable differences in the permeability of different developmental stages (oocytes vs. cleavage-stage embryos vs. blastocysts) to different cryoprotectants has led to the development of different products for the cryopreservation of different stage embryos [19].

With regard to cleavage-stage embryos, the comparison of 1.5 M DMSO to 1 M glycerol showed DMSO to be the more effective cryoprotectant [11]. However, survival rates of post-thaw human embryos were disappointing, and it was hypothesized that by further extending the preseeding equilibration time and allowing for greater dehydration of cells, survival rates could be improved [7]. Further to this, partial dehydration of embryos at room temperature using a medium containing PROH and sucrose was suggested [25].

The use of PROH in place of DMSO lies in the more toxic and less stable properties of DMSO which has not been used on a regular basis for human embryo cryopreservation by slow cooling [26].

#### 2.3.2 Non-permeating Cryoprotectant: Sucrose

Although the initial sucrose concentration recommended by Quinn [6] was 0.2 M, the final concentration decided on for commercialization was 0.1 M. This lower concentration of sucrose was used previously (13, 14) and resulted in good outcomes. Subsequently it has been reported that 0.2 M sucrose gave better results than when only 0.1 M sucrose was used in the cryopreserved solution [27] and even 0.3 M sucrose gives better results than 0.1 M sucrose [28]. The better results using increased concentrations of 0.2 M and 0.3 M in the cryodiluent with 1.5 M PROH have probably been due to the greater amount of dehydration at the higher concentrations of sucrose [27, 29, 30].

#### 2.3.3 Protein

The role of protein in both embryo culture and cryopreservation diluents is often rigorously debated, and there have been strong observations/opinions expressed following reflection within isolated clinical setups and systems. The primary role of protein in traditional in vitro cell and tissue handling protocols was to prevent adhesion to glass and plastic ware, hence minimizing the potential mechanical damage suffered by cells during processing [31].

However, the closer inspection of proteins using modern technologies has revealed additional contributions of protein supplements to culture including high levels of amino acids and transition metals [32]. While the effects of these undefined components on the pre-implantation embryo remain to be elucidated, the addition of protein in large volumes, such as those in cryopreservation media, should be kept in mind.

Supplementation of media with a large protein molecule decreases surface tension, subsequently reducing the tendency of the embryo to float or adhere to plastic or glass and also inactivating heavy metals and other toxins. However, macromolecules derived from serum are also a possible vector for viral infection of embryos. The low but associated risks of disease transmittance between patients and the human protein source are still referenced as a cause of concern, despite the strict screening of materials used. For these reasons, recombinant protein products have been commercialized and are available to consumers worldwide [33, 34]. Unfortunately, the costs associated with these products are often a deterrent to clinics.

At the other end of the spectrum, we have the so-called complex protein sources, including SPS®, LGPS®, and SSS®, all containing an approximate mix of 85 % HSA and 15 % α- and β-globulins. While less defined, there is evidence that these protein sources may produce better rates of embryo production in culture and also provide greater support to embryos and oocytes that have undergone the freeze-thaw process [31]. However, the optimal type and dose rate of protein supplement in cryopreservation media remains to be elucidated.

The role of hyaluronate (HA) should also be mentioned here as an additional or alternative macromolecule component of cryopreservation media. This has been demonstrated as beneficial in bovine [35], murine [36], and human studies [37]. There is an indication that HA could have an effect on embryo cryotolerance, but further evidence would be needed to take greater inference on its role [38].

#### 2.3.4 Warming

The reverse process takes place when it comes to warming cryopreserved embryos—permeating cryoprotectants are removed from the cells, and rehydration takes place. Good technical descriptions of the process can be found in countless reviews and book chapters [6, 39]. With regard to the composition of thawing medium, several media with successively lower cryoprotectant concentration were first used to create a concentration gradient that would draw permeating cryoprotectants out of the cells in an osmotic stepwise manner. As the embryo moves between each medium, there is an initial increase in cellular volume as a result of isotonic equilibration, followed by a decrease in volume as the cryoprotectant leaves the embryo. Providing the initial volume increase does not exceed the maximum volume of the cells, the embryo should remain structurally undamaged [6].

Using both a permeating and non-permeating cryoprotectant has also been considered and forms the basis of existing cryoproducts available on the market today (including the ORIGIO Embryo Freeze Kit, also supplied by CooperSurgical).

The use of a non-permeating cryoprotectant alone to draw the permeating cryoprotectant out of the cytoplasm was introduced by Leibo and Mazur [40] and has largely replaced the use of permeating cryoprotectants in thawing diluents. This method allows for a better controlled rehydration process, with the gradual movement of permeating cryoprotectants out of the cells of the embryo, resulting in a more gradual swelling of cells. To illustrate with a practical example and in context of this chapter, for embryos thawed following freezing in the Quinn's Advantage Embryo Freeze Kit, the corresponding solutions recommended for use consist of two sucrose-containing media (0.5 M and 0.2 M, for successive rehydration). Importantly, the cryoprotectant concentrations of these solutions must be higher than the freezing diluent's concentration to ensure an effective osmotic gradient is established [41, 42].

With the adoption of vitrification as common clinical practice, an altered warming protocol has been required to account for the much higher levels of permeating cryoprotectants used in vitrification cooling media. With these media commonly containing approximately 1 M of non-permeating cryoprotectant [43], this concentration is higher than those found in both vitrification cooling and slow-freezing media. As such it has been proposed that these media can be used to thaw the countless slow-frozen oocytes in storage today, regardless of the media or protocol used for cryopreservation [44]. Subsequent testing of this "universal" thawing protocol has demonstrated nice results with regard to oocyte survival rates and post-thaw in vitro embryo development. The findings of this study indicate that rapid warming diluents that accompany slow-freezing products may become a thing of the past, simplifying the media requirements, reducing the cost, and standardizing the thawing of all cryopreserved oocytes. It has been suggested that higher sucrose levels in thawing diluents could also aid the rehydration of slow-frozen cleavage-stage embryos [45]. Further evidence for this was provided by Kojima et al. [46] for the thawing of pronuclear stage embryos using vitrification warming solutions.

The studies mentioned above indicate the usefulness of sucrose in the dehydration of oocytes/embryos prior to cooling and its benefit for rehydration after thawing/warming. There have been some studies of other mono- and disaccharides, for example, trehalose and galactose, for use in these processes because of their lower density than sucrose solutions which would assist in a better handling of the oocytes/embryos [47].

## 3 Methods

Both cryopreservation using Quinn's Advantage Embryo Freeze Kit and thawing using Quinn's Advantage Thaw Kit will be described because effective results are dependent on the proper use and performance of the kits for both cryopreservation and thawing. Complete versions of both procedures are available online [48, 49].

### 3.1 Quinn's Advantage Embryo Freeze Kit

1. Prepare solutions containing 0.5 M and 1.0 M PROH by diluting the stock solution of 1.5 M PROH with the freeze/thaw diluent solution which contains no cryoprotectants.
    (a) To prepare the 0.5 M PROH solution, add 0.3 mL of 1.5 M PROH to 0.6 mL of diluent solution.
    (b) To prepare the 1.0 M PROH solution, add 0.6 mL of 1.5 M PROH to 0.3 mL of diluent solution.
2. Embryos are pipetted at 37 °C into the 0.5 M PROH solution for 5 min, then into the 1.0 M PROH solution for 5 min, and finally into 1 mL of the 1.5 M PROH solution for 10 min. They are then transferred to 1 mL of 1.5 M PROH + 0.1 M sucrose freezing medium and then pipetted into straws containing this same solution. The straws are then sealed. They are held at 37 °C in the 1.5 M PROH + 0.1 M sucrose solution for a total of 5 min before cooling is initiated.
3. As an alternative, the embryos can be placed directly into the 1.5 M PROH solution for 10 min before transfer to the 1.5 M PROH + 0.1 M sucrose freezing medium. This procedure should be retained for use with good-quality, embryos whereas the complete procedure is described in 2. above provides less stress on embryos which can be detrimental to embryos of lesser quality.
4. Vials, e.g., 1.2 mL plastic cryovials, can be used in place of straws, but we have found that straws are easier to handle and provide somewhat better results.

### 3.2 Cooling Protocol

Embryos in straws or cryovials are taken from a starting temperature of 30 °C to −6 °C at 2 °C/minute. They are then seeded manually using a pair of metal tweezers or a metal rod that have been immersed in liquid nitrogen and held therein for sufficient time to cool to at least −100 °C. The tweezers are applied to the outside of the straw or cryovial at a site away from where the embryos are located, for example, above the fluid in the straw. Ice crystals will form in the thin layer of solution on the inside of the straw and move down into the cryosolution column containing the embryos.

After holding the embryos at −6 °C for a total of 10–15 min after seeding, the straw is cooled at about 0.3 °C/min to around −35 °C. They are then quickly transferred to a storage tank of liquid nitrogen.

### 3.3 Thawing Protocol

1. If straws have been transferred to liquid nitrogen after being slow cooled to between −30 °C and −37 °C, they should be thawed rapidly (at least 275 °C/min) so that intracellular ice is swiftly dispersed. This will help minimize cellular damage by ice crystals. The easiest way to achieve this is to initially hold the straw in air for 30–40 s and then immerse it in a water bath at 30 °C–35 °C until the ice has fully melted.
2. Thaw only one cryocontainer at a time. Transfer the liquid contents of the thawed solution to a dry dish and quickly locate the embryos. Pick the embryos up in a minimal amount of solution and transfer them first to 3 mL of 0.5 M sucrose thawing medium at 37 °C for 10 min followed by transfer to 3 mL of 0.2 M sucrose thawing medium at 37 °C for 10 min using a new transfer pipette for each procedure to minimize carry-over of cryoprotectant from one solution to the next.
3. The embryos are then washed through seven drops of freeze/thaw diluent solution at 37 °C. This can be done by placing seven drops, each of 100 uL, under sterile oil in a large culture dish. The embryos are placed in each drop and thoroughly washed by gently pipetting up and down several times over a period of about 1 min before being transferred to the next drop. A new transfer pipetted should be used after the first drop, but the same pipette can be used for subsequent transfers. After the sixth washing drop, the embryos can be transferred to the seventh drop and held up to 30 min at 37 °C before transfer or placed into culture.

## 4 Notes

1. The best-quality embryos should be chosen for cryopreservation. Poor-quality embryos give poor results and vice versa.
2. It is important to make sure that embryos are well mixed with the cryoprotectant solutions. This can be accomplished by pipetting the embryos up and down in the solution several times after adding them to the cryoprotectant solution [48].
3. It is recommended that the solutions are covered with sterile oil for tissue culture during use to minimize evaporation of water and a subsequent change in osmolality of the solutions [48].

4. Embryos are loaded into straws containing 1.5 M PROH + 0.1 M sucrose freezing medium by using an embryo handling pipette and observing the procedure under a dissecting microscope [48].
5. Using the thawing protocol wherein straws to be thawed are held in air for 30–40 seconds followed by immersion in water at 30–35 °C allows for any liquid nitrogen that may have entered the straw through an imperfectly sealed plug to be blown off before the straw is placed in the water bath. Little loss of straws or their contents occurs with this technique [49].

## References


1. Sullivan EA, Zegers-Hochschild F, Mansour R, Ishihara O, de Mouzon J, Nygren KG, Adamson GD (2013) International Committee for Monitoring Assisted Reproductive Technologies (ICMART) world report: assisted reproductive technology 2004. Hum Reprod 28:1375–1390
2. Levron J, Yerushalmi GM, Brengauz M, Gat I, Katorza E (2014) Comparison between two protocols for thawed embryo transfer: natural cycle versus exogenous hormone replacement. Gynecol Endocrinol 30:494–497
3. Aytoz A, Van den Abbeel E, Bonduelle M, Camus M, Joris H, Van SA, Devroey P (1999) Obstetric outcome of pregnancies after the transfer of cryopreserved and fresh embryos obtained by conventional in-vitro fertilization and intracytoplasmic sperm injection. Hum Reprod 14:2619–2624
4. Shih W, Rushford DD, Bourne H, Garrett C, McBain JC, Healy DL, Baker HW (2008) Factors affecting low birthweight after assisted reproduction technology: difference between transfer of fresh and cryopreserved embryos suggests an adverse effect of oocyte collection. Hum Reprod 23:1644–1653
5. Vergouw CG, Kostelijk EH, Doejaaren E, Hompes PG, Lambalk CB, Schats R (2012) The influence of the type of embryo culture medium on neonatal birthweight after single embryo transfer in IVF. Hum Reprod 27:2619–2626
6. Quinn P (1995) Cryopreservation of embryos and oocytes. In: Keye WR, Chang RJ, Rebar RW, Soules MR (eds) Infertility: evaluation and treatment. W. B. Saunders Company, Philadelphia, pp 821–840
7. Quinn P, Kerin JF (1986) Experience with the cryopreservation of human embryos using the mouse as a model to establish successful techniques. J In Vitro Fert Embryo Transf 3:40–45
8. Polge C, Parkes AS (1949) Revival of spermatozoa after vitrification and dehydration at low temperatures. Nature 164:666
9. Whittingham DG, Leibo SP, Mazur P (1972) Survival of mouse embryos frozen to −196 degrees and −269 degrees C. Science 178:411–414
10. Whittingham DG (1980) Principles of embryo preservation in Low Temperature Preservation in Medicine and Biology, (Ashwood-Smith MJ, Farrant J (eds) Pitman Medical, Tunbridge Wells, pp. 65–83
11. Trounson A, Mohr L (1983) Human pregnancy following cryopreservation, thawing and transfer of an eight-cell embryo. Nature 305:707–709
12. Cohen J, Simons RF, Edwards RG, Fehilly CB, Fishel SB (1985) Pregnancies following the frozen storage of expanding human blastocysts. J In Vitro Fert Embryo Transf 2:59–64
13. Testart J, Lassalle B, Belaisch-Allart J, Hazout A, Forman R, Rainhorn JD, Frydman R (1986) High pregnancy rate after early human embryo freezing. Fertil Steril 46:268–272
14. Lassalle B, Testart J, Renard JP (1985) Human embryo features that influence the success of cryopreservation with the use of 1,2 propanediol. Fertil Steril 44:645–651
15. Chi HJ, Koo JJ, Kim MY, Joo JY, Chang SS, Chung KS (2002) Cryopreservation of human embryos using ethylene glycol in controlled slow freezing. Hum Reprod 17:2146–2151
16. Rall WF, Fahy GM (1985) Ice-free cryopreservation of mouse embryos at −196 degrees C by vitrification. Nature 313:573–575
17. Barg PE, Barad DH, Feichtinger W (1990) Ultrarapid freezing (URF) of mouse and human preembryos: a modified approach. J In Vitro Fert Embryo Transf 7:355–357
18. Gordts S, Roziers P, Campo R, Noto V (1990) Survival and pregnancy outcome after

ultrarapid freezing of human embryos. Fertil Steril 53:469–472

19. Edgar DH, Gook DA (2012) A critical appraisal of cryopreservation (slow cooling versus vitrification) of human oocytes and embryos. Hum Reprod Update 18:536–554
20. Balaban B, Urman B, Ata B, Isiklar A, Larman MG, Hamilton R, Gardner DK (2008) A randomized controlled study of human Day 3 embryo cryopreservation by slow freezing or vitrification: vitrification is associated with higher survival, metabolism and blastocyst formation. Hum Reprod 23:1976–1982
21. Rama Raju GA, Haranath GB, Krishna KM, Prakash GJ, Madan K (2005) Vitrification of human 8-cell embryos, a modified protocol for better pregnancy rates. Reprod Biomed Online 11:434–437
22. Rezazadeh VM, Eftekhari-Yazdi P, Karimian L, Hassani F, Movaghar B (2009) Vitrification versus slow freezing gives excellent survival, post warming embryo morphology and pregnancy outcomes for human cleaved embryos. J Assist Reprod Genet 26:347–354
23. Vutyavanich T, Sreshthaputra O, Mongkolchaipak S, Wongtra-ngan S, Piromlertamorn W (2008) Slow programmable and ultra-rapid freezing of human embryos. J Obstet Gynaecol Res 34:457–463
24. Nagy ZP, Taylor T, Elliott T, Massey JB, Kort HI, Shapiro DB (2005) Removal of lysed blastomeres from frozen-thawed embryos improves implantation and pregnancy rates in frozen embryo transfer cycles. Fertil Steril 84:1606–1612
25. Renard JP, Babinet C (1984) High survival of mouse embryos after rapid freezing and thawing inside plastic straws with 1-2 propanediol as cryoprotectant. J Exp Zool 230:443–448
26. Renard JP (1985) The Cryopreservation of mammalian embryos. In: Testart JP, Frydman R (eds) Human In vitro fertilization. Elsevier, Amsterdam, pp 201–208
27. Edgar DH, Karani J, Gook DA (2009) Increasing dehydration of human cleavage-stage embryos prior to slow cooling significantly increases cryosurvival. Reprod Biomed Online 19:521–525
28. Wood MJ, Mollison J, Harrild K, Ferguson E, McKay T, Srikantharajah A, Bell L, Bhattacharya S (2011) A pragmatic RCT of conventional versus increased concentration sucrose in freezing and thawing solutions for human embryos. Hum Reprod 26:1987–1996
29. Fabbri R, Porcu E, Marsella T, Rocchetta G, Venturoli S, Flamigni C (2001) Human oocyte cryopreservation: new perspectives regarding oocyte survival. Hum Reprod 16:411–416
30. Jericho H, Wilton L, Gook DA, Edgar DH (2003) A modified cryopreservation method increases the survival of human biopsied cleavage stage embryos. Hum Reprod 18:568–571
31. Meintjes M, Chantilis SJ, Ward DC, Douglas JD, Rodriguez AJ, Guerami AR, Bookout DM, Barnett BD, Madden JD (2009) A randomized controlled study of human serum albumin and serum substitute supplement as protein supplements for IVF culture and the effect on live birth rates. Hum Reprod 24:782–789
32. Morbeck DE, Paczkowski M, Fredrickson JR, Krisher RL, Hoff HS, Baumann NA, Moyer T, Matern D (2014) Composition of protein supplements used for human embryo culture. J Assist Reprod Genet 31:1703–1711
33. Bavister BD, Kinsey DL, Lane M, Gardner DK (2003) Recombinant human albumin supports hamster in-vitro fertilization. Hum Reprod 18:113–116
34. Bungum M, Humaidan P, Bungum L (2002) Recombinant human albumin as protein source in culture media used for IVF: a prospective randomized study. Reprod Biomed Online 4:233–236
35. Stojkovic M, Kolle S, Peinl S, Stojkovic P, Zakhartchenko V, Thompson JG, Wenigerkind H, Reichenbach HD, Sinowatz F, Wolf E (2002) Effects of high concentrations of hyaluronan in culture medium on development and survival rates of fresh and frozen-thawed bovine embryos produced in vitro. Reproduction 124:141–153
36. Gardner DK, Rodriegez-Martinez H, Lane M (1999) Fetal development after transfer is increased by replacing protein with the glycosaminoglycan hyaluronan for mouse embryo culture and transfer. Hum Reprod 14:2575–2580
37. Valojerdi MR, Karimian L, Yazdi PE, Gilani MA, Madani T, Baghestani AR (2006) Efficacy of a human embryo transfer medium: a prospective, randomized clinical trial study. J Assist Reprod Genet 23:207–212
38. Lane M, Maybach JM, Hooper K, Hasler JF, Gardner DK (2003) Cryo-survival and development of bovine blastocysts are enhanced by culture with recombinant albumin and hyaluronan. Mol Reprod Dev 64:70–78
39. Chian R-C, Quinn P (2009) Fertility cryopreservation. Cambridge University Press, Cambridge, United Kingdom
40. Leibo SP, Mazur P (1978) Methods for the preservation of mammalian embryos by freezing. In: Daniels Jr J (ed) Methods in mammalian reproduction. Academic Press, New York, pp 179–201

41. Vanderzwalmen P (2012) Adjustment of cryoprotectant exposure time to counteract reduced cooling rates. Reprod Biomed Online 24:684
42. Vanderzwalmen P, Zech NH, Ectors F, Stecher A, Lejeune B, Vanderzwalmen S, Wirleitner B (2012) Blastocyst transfer after aseptic vitrification of zygotes: an approach to overcome an impaired uterine environment. Reprod Biomed Online 25:591–599
43. Alpha Scientists in Reproductive Medicine (2012) The Alpha consensus meeting on cryopreservation key performance indicators and benchmarks: proceedings of an expert meeting. Reprod Biomed Online 25:146–167
44. Parmegiani L, Bertocci F, Garello C, Salvarani MC, Tambuscio G, Fabbri R (2009) Efficiency of human oocyte slow freezing: results from five assisted reproduction centres. Reprod Biomed Online 18:352–359
45. Wale P, Jericho H, Edgar DH (2004) Comparison of rehydration protocols for human frozen embryos. Hum Reprod 19:1–225
46. Kojima E, Fukunaga N, Nagai R, Kitasaka H, Ohno H, Asada Y (2012) The vitrification method is significantly better for thawing of slow-freezing embryos. Fertil Steril 98:S124
47. McWilliams RB, Gibbons WE, Leibo SP (1995) Osmotic and physiological responses of mouse zygotes and human oocytes to mono- and disaccharides. Hum Reprod 10:1163–1171
48. ORIGIO (2015) Quinn's Advantage Embryo Freeze Kit. http://www.origio.com/documents/5r34pdjkyb9-5638_1.pdf. Accessed 21 Apr 2015
49. ORIGIO (2015) Quinn's Advantage Thaw Kit. http://www.origio.com/documents/9iocwl8fr-5640.pdf. Accessed 21 Apr 2015

# Appendix G: Vitrification of Blastocysts Using VitriBlast™ and ThermoBlast™: Nidacon

**Anna Niläng Laessker, Thorir Hardarsson, Ann-Sofie Forsberg, Tetsunori Mukaida, and Paul V. Holmes**

## Abstract

This appendix describes the vitrification of blastocysts using VitriBlast™ (VBK) and ThermoBlast™ (TBK) from Nidacon, Mölndal, Sweden. The technique used and the reason for not including DMSO in the medium at the production stage, but including it separately in the kit, and the importance of collapsing the blastocyst prior to vitrification will be explained and described.

**Key words** Vitrification, Blastocysts, DMSO, Ethylene glycol, Cryopreservation, Collapsing, VitriBlast™, ThermoBlast™, Nidacon

## 1 Introduction

The original formulation was developed by Lane et al. 1999 at which time the cryo-loop was also introduced [1]. During the cryopreservation of all gametes and embryos, they are at risk of injury from chilling ice crystallization, the toxicity of the cryoprotectants, extracellular ice, intracellular ice, fracture damage, osmotic swelling, and shrinkage. To obtain high rates of survival, all these problems must be circumvented and/or minimized.

A critical factor in all cryopreservation is to avoid ice crystal formation within the cells. It has been shown that the most effective way of achieving this is by using vitrification (glass transition) [2, 3]. One key factor is the vitrification media that involves the choice of cryoprotectants and the production method. An addition key factor is the technique used during the vitrifying process.

### 1.1 Cryoprotectants

For preventing intracellular ice from forming, rapidly cell membrane permeating cryoprotectants, like for instance, ethylene glycol (EG) and DMSO, are suitable since their rapid permeation character by the cryoprotectant is essential.

Zsolt Peter Nagy et al. (eds.), *Cryopreservation of Mammalian Gametes and Embryos: Methods and Protocols*, Methods in Molecular Biology, vol. 1568, DOI 10.1007/978-1-4939-6828-2_27, 

Ficoll is included as a non-permeating polymer (macromolecules promote vitrification), in particular since it has high solubility, low viscosity, and low toxicity. It has been shown that even large amounts of Ficoll are virtually nontoxic; consequently, a large amount of polymer occupies a significant proportion of the solution volume. Therefore, its inclusion increase the proportion of permeating cryoprotectant per water volume. This may be one mechanism that promotes vitrification of the solution.

The solutions contain not only EG and Ficoll but also sucrose, a non-permeating small sugar which has considerable osmotic effect. The inclusion of sucrose is effective at reducing the apparent toxicity of EG because sucrose promotes the shrinkage of embryos, thereby restricting excess permeation of ethylene glycol and, thus, reducing its toxic effect [4–6]. In addition, sucrose helps prevent overswelling during the removal of the ethylene glycol after warming.

Rapidly permeating agents are also suitable to prevent osmotic swelling during removal of the cryoprotectants after warming, since the faster the diffusion of the intracellular cryoprotectant out of the cell, the lower the risk of osmotic overswelling.

Human serum albumin functions chemically as a buffer and a molecular carrier. Principally, it also enables handling of the blastocyst, which otherwise can easily stick to plastic and glass.

### 1.2 DMSO vs. PrOH

It is often debated whether DMSO or propanediol (PrOH) is the optimal cryoprotectant to use. DMSO has often been criticized for being toxic and harmful. This is debatable since DMSO toxicity is low compared to EG and PrOH. Also, its superiority over PrOH when used in vitrification has been reported in numerous publications. Kartberg et al. (2008) found that, when adding DMSO to a medium containing PrOH and EG, the cell membrane damage did not occur as quickly as without [7]. Duus et al. (2008) concluded that vitrification of blastocysts using DMSO gave a higher percentage of top-quality blastocysts when compared to PrOH [8].

### 1.3 DMSO Added Just Prior to Use

A vital feature of the Nidacon vitrification solutions is the fact that DMSO is *not* included in the media and needs to be added to the medium just before use. DMSO and ethylene glycol are included as additives in the vitrification kit.

DMSO oxidizes human serum albumin (hSA) by the creation of disulfide bridges. This reaction is slow, but the longer the two substances are combined, the more bridges are created and dimers or oligomers of hSA would be created, thereby impairing its function. hSA could undergo considerable conformation changes and not effectively function as an excipient [9, 10].To avoid this risk, DMSO is not included in the media but supplied as an additive in the kit.

### 1.4 Artificial Collapse of Blastocysts Prior to Vitrification

In the instructions for use, "collapsing" the blastocyst prior to the vitrification process is recommended.

Although the solutions used in both the vitrification and the warming processes are important, there are other factors that will affect the survival rate during the vitrification process. A blastocyst consists of trophoblast cells surrounding a fluid-filled cavity, the blastocoele. The likelihood of ice crystal formation in the blastocoele during the vitrification process is directly proportional to the fluid volume and inversely proportional to its viscosity and the cooling rate. A decrease in blastocyst survival rate after vitrification has been noticed when the volume of the blastocoelic cavity and fluid is increased. This increased blastocyst volume can lead to an insufficient permeation of cryoprotectants into the cavity, thereby permitting ice crystal formation during the cooling step and reducing the post-warming survival. Vanderzwalmen et al. [11] found that the post-warming survival rate increased dramatically after artificial shrinkage of the blastocyst, and they suggested that the blastocoelic fluid is the source of injury, probably due to the higher likelihood of ice crystal formation, the ice crystals inducing mechanical damage to the cells.

When the blastocyst is exposed to the vitrification medium, it most often shrinks after some time. The shorter time the blastocyst is exposed to the medium, the better due, to the risk of injury from the chemical toxicity of the cryoprotectants. It has been observed that human blastocysts are probably less permeable to cryoprotectants since they shrink more slowly than mouse and bovine blastocysts [12]. Therefore, there is a greater risk for ice crystal formation inside the human blastocyst, and this will cause cell damage. There again, there are two ways to alleviate this damage, either increase the duration time in the vitrification medium or artificially shrink the blastocyst in order to shorten the exposure time to cryoprotectants [11] and their toxicity.

Artificial shrinkage has been used when performing preimplantation genetic diagnosis (PGD) since the embryos are biopsied prior to vitrification. In a report by Zhang et al. [13], the biopsied blastocysts showed a higher survival rate compared to their non-biopsied controls. Son et al. [14] examined the effects on blastocyst survival and subsequent pregnancy rate after induced collapse of the blastocoele before vitrification of human blastocysts; they concluded that the implantation and clinical pregnancy rates were increased by the artificial shrinkage program, and the pregnancy outcome was similar to that obtained with the transfer of fresh blastocysts. Also, Hardarson et al. [15] showed that artificial shrinkage prior to vitrification significantly improved survival rates of human blastocysts.

In conclusion, induced blastocyst collapse is a method that has been used with good results since 1998. The method is well documented and has been routinely used by Fertilitetscentrum in Gothenburg with excellent results (personal communication, T. Hardarson). Obstetric outcomes have also been followed up in 2010, and no adverse effects were found in the children born [16].

## 2 Materials

### 2.1 Vitrification Components

1. VitriBlast™ kit (containing solutions 1–3 and additives: EG and DMSO).
2. Sterile pipettes.
3. Culture dishes (NUNC 4-well).
4. Device for vitrification (*see* **Note 1**).
5. $CO_2$ incubator.
6. Stopwatch or timer.
7. Liquid nitrogen reservoir.
8. Liquid nitrogen.

### 2.2 Warming Components

1. ThermoBlast™ kit (Containing solutions 4–6).
2. Sterile transfer pipettes.
3. Culture dishes (NUNC 4-well).
4. Disposable gloves.
5. $CO_2$ incubator.
6. Stopwatch or timer.

## 3 Methods

### 3.1 Vitrifying Blastocysts Using VitriBlast™ (VB)

Work on a heated stage at all times when manipulating the blastocyst. Do not let the blastocyst remain exposed to the microscope light during incubations.

1. Remove the DMSO and EG bottles from refrigerator 1 h prior to use or the day before, and let the DMSO liquefy in RT. They can also be placed in the incubator prior to use (*see* **Note 2**).
2. Label a 4-well culture dish with the patient ID and each well with each solution number.
3. Prepare the 4-well culture dish by adding 1 mL of VitriBlast 1 (VB1) to the first well.
4. Add 850 μL of VitriBlast 2 (VB2), 75 μL of DMSO, and 75 μL EG, respectively, to the second well. Mix thoroughly (*see* **Note 3**).

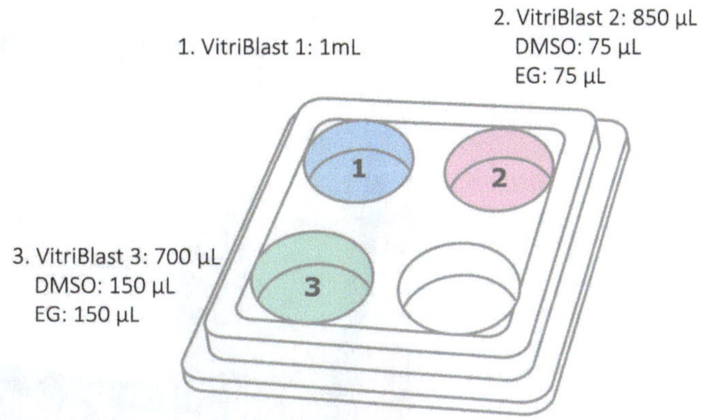


**Fig. 1** Preparation of culture dish for vitrification

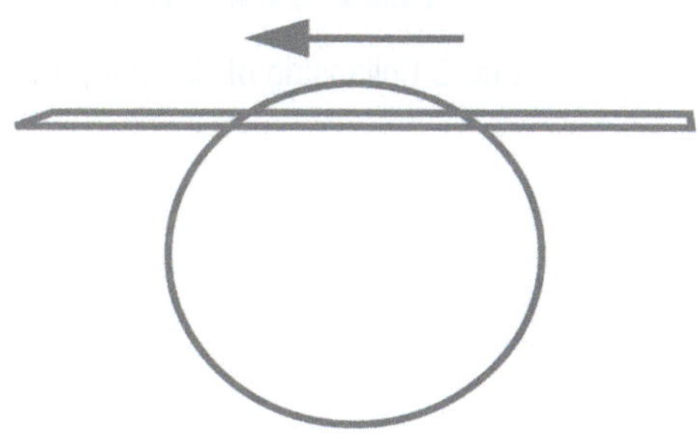

**Fig. 2** Collapsing of blastocyst using a pipette

5. Add 700 µL of VitriBlast 3 (VB3), 150 µL of DMSO, and 150 µL EG, respectively, to the third well. Mix thoroughly (Fig. 1).
6. Incubate at 37 °C in 5–6 % $CO_2$ for 30 min (*see* **Note 4**).
7. Collapse the blastocyst by either laser or by pipette tip under microscope. Punctuate as far from the inner cell mass (ICM) as possible (*see* **Note 5**, Figs. 2–4).
8. Transfer the blastocyst to VB 1, and incubate 1.5–2 min on the heated stage (*see* **Note 6**).
9. Transfer the blastocyst to VB 2, and incubate for EXACTLY 2 min on the heated stage (*see* **Notes 6** and 7, Fig. 5).
10. During the 2 min incubation, prepare 2 × 10 µL droplets of VB 3 in the middle of the dish. Illustration (*see* **Note 8**, Fig. 5).
11. At the correct time, move the blastocyst by aspirating VB 3 into the pipette tip, collect the blastocyst from solution 2 in the second well, and transfer to solution 3 in the droplet (Fig. 6).
12. The blastocyst must remain in VB 3 for 30–45 s, including the time on the device.
13. Place the blastocyst onto the vitrification device, leaving the smallest volume possible of the VB 3 (*see* **Note 9**, Fig. 7).

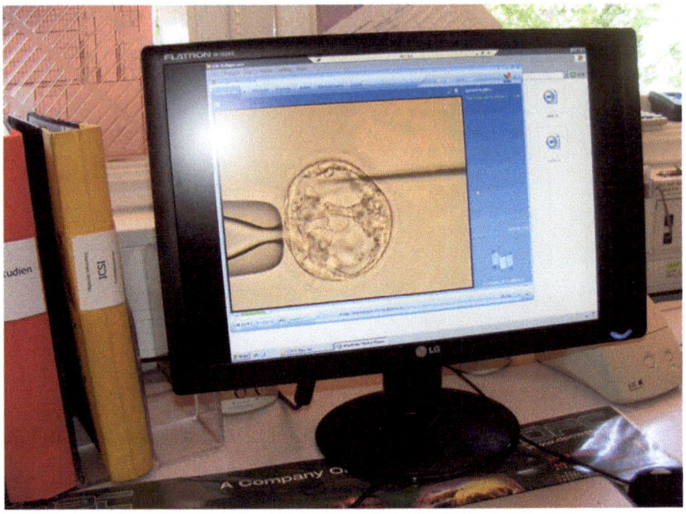

**Fig. 3** Collapsing of blastocyst using a pipette

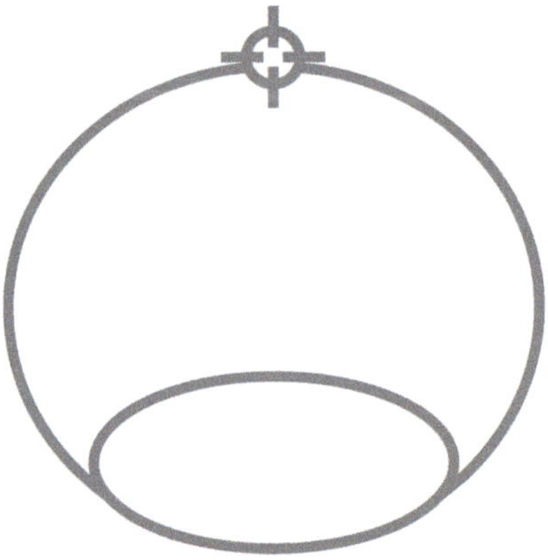

**Fig. 4** Collapsing of blastocyst using laser

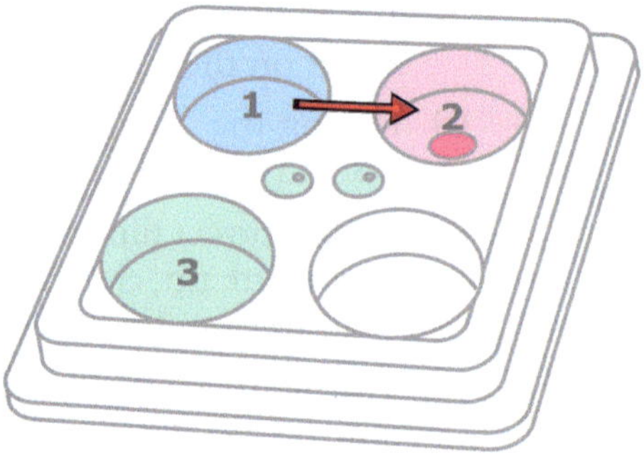


**Fig. 5** Transferring the blastocyst from solution 1 to solution 2

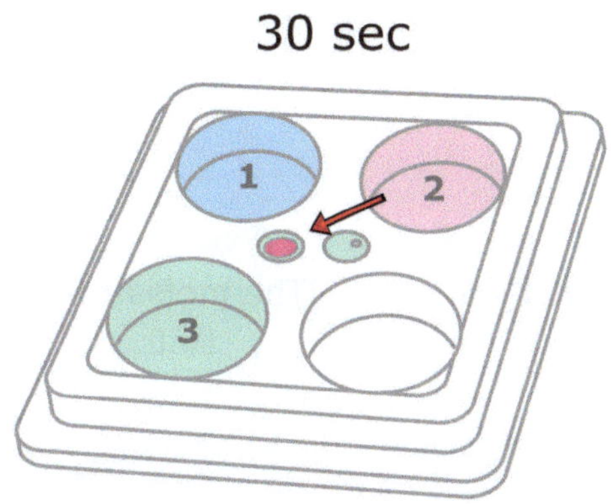


**Fig. 6** Transferring the blastocyst from solution 2 to 10 μl of solution 3

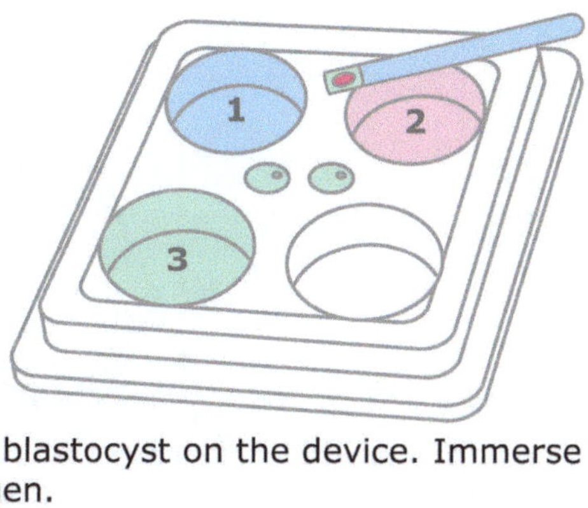


**Fig. 7** Placing the blastocyst on vitrification device

14. Plunge into liquid nitrogen.
15. Transfer to storage (*see* **Note 10**).

### *3.2 Warming of Vitrified Blastocysts Using ThermoBlast™ (TB)*

1. Label a 4-well culture dish with the patient ID and each well with each solution number.
2. Prepare the 4-well culture dish by adding 1 mL each of solution 4 and 5 to the first two wells. Add 1 mL of solution 6 the third (Fig. 8).
3. Incubate at 37 °C in 5–6 % $CO_2$ for 30 min.
4. Immerse the part of the device, containing the blastocyst under the surface of solution 4. Allow the blastocyst to fall off. Identify its presence in the well, and incubate for 2 min on heated stage (*see* **Note 11**, Fig. 9).
5. Transfer the blastocyst to solution 5 and incubate 3 min (*see* **Note 12**, Fig. 10).

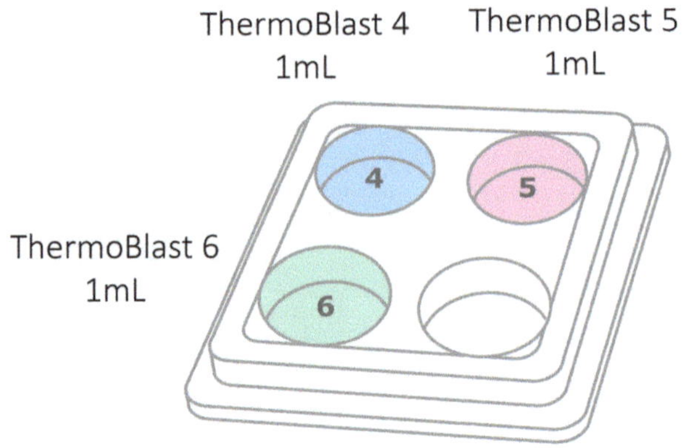


**Fig. 8** Preparation of culture dish for warming

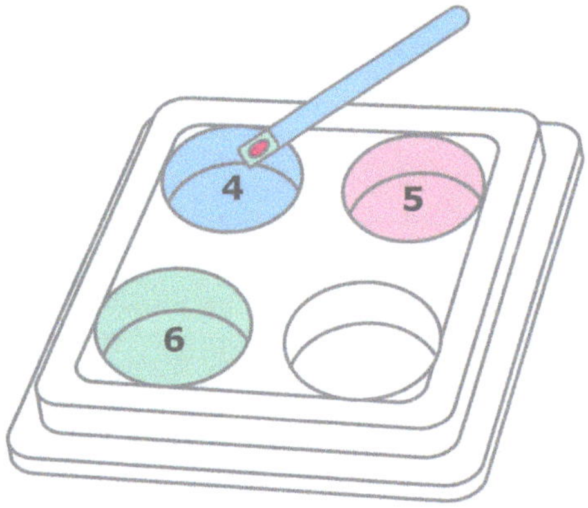


**Fig. 9** Immersing the blastocyst in warming solution 4

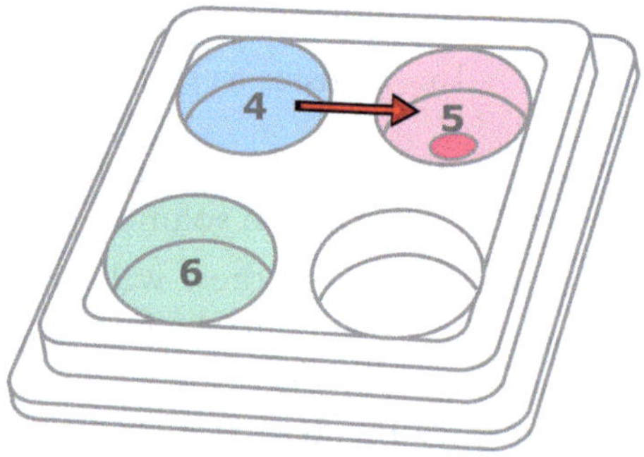


**Fig. 10** Transferring the blastocyst from solution 4 to solution 5

6. Transfer the blastocyst to solution 6 and rinse quickly (*see* **Note 13**, Fig. 11).
7. Incubate in 37 °C in 5–6 % $CO_2$ for 5 min.

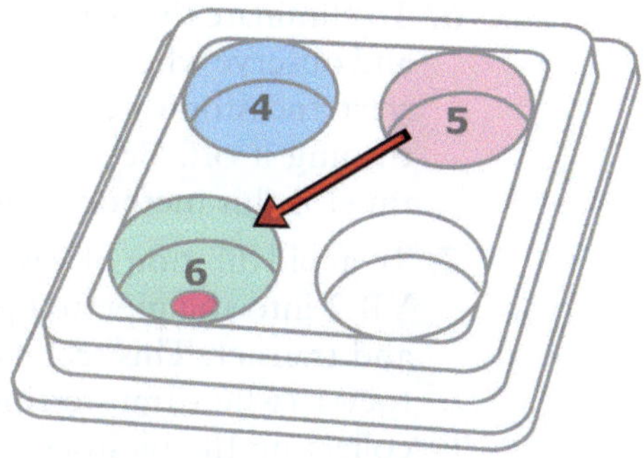


**Fig. 11** Transferring the blastocyst from solution 5 to solution 6

8. Transfer to culture medium.
9. For correct evaluation, wait 1–4 h before transfer, in order to allow the blastocyst time to re-expand (*see* **Note 14**).

## 4 Notes

1. VBK is compatible with most devices. The most important factor is to use the type of device that feels safe and easy to use.
2. DMSO is solid below +18 °C and needs to be above this temperature to reach liquid form. If there is a shortage of time, it can be warmed in the hand or warmed in the incubator.
3. The EG and DMSO can be premixed in the bottles (VB 2 and VB 3) and stored in the refrigerator for up to a week. However, it is important to know that the volumes in the VB 2 and 3 are not EXACTLY 10 mL; there is always a little surplus. Therefore, the volumes in the VB 2 and 3 must be measured and the surplus discarded prior to adding the EG and DMSO in order to achieve the correct concentrations.
4. Do not incubate for longer than 60 min. Any longer incubation can cause the solutions to become more viscous.
5. Collapsing the blastocyst will improve the results. If laser is used, shoot as far from inner cell mass (ICM) as possible, and ensure both the zona and the trophectoderm are breached (Fig. 4).

   If an ICSI pipette or other sharp instrument is used, puncture right through the trophoblast cell layer into the blastocoele, and be sure to puncture as far possible from the ICM (Fig. 3).

   The pipette should be inserted at the 1 O'clock position and exit through the blastocyst at the 11 o'clock position (Fig. 2).

   Collapsing is optional when vitrifying early blastocysts with smaller blastocoele cavities.

6. To eliminate the stress of the beeping noise, start the stopwatch and observe when the correct time is approaching. It is easier to start the stopwatch and let it run toward desired time than turning it off. Be aware of the risk of forgetting to watch the timer if this method is used.
7. To avoid dilution of VB 2 when transferring from VB1, aspirate VB 2 into the pipette tip and, thereafter, collect the blastocyst and transfer. This technique applies to all transfers of the blastocyst (when transferring from 2 to 3, aspirate VB3 prior to collecting the blastocyst in VB2).
8. Finding the blastocyst in the droplet is easier than in the well. Having 1 mL of VB 3 in the well will ensure correct pH and avoid evaporation, which would occur for the 10 μL drop.
9. The second drop is for coating the loop, if this is the device used. If another device is used, the second drop is not necessary. The well with the 1 mL is needed to achieve correct pH and temperature at incubation.
10. Ensure that the device with the vitrified blastocyst does NOT come in contact with air during transfer to storage in liquid nitrogen.
11. The 2 min includes the time locating the blastocyst.
12. Do not collect solution from TB5 when collecting from TB4. This is to allow the blastocyst to sink in TB5 which makes it easier to identify.
13. Do not collect solution from TB6 when collecting from TB5. This is to allow the blastocyst to sink in TB6 which again makes it easier to identify.
14. If the blastocyst has not re-expanded after 4 h, the chance of re-expansion is low.

## References

1. Lane M, Forest KT et al (1999) Live births following vitrification of hamster embryos using a novel containerless technique. Theriogenology 51:67
2. Loutradi K, Kolibianakis E et al (2008) Cryopreservation of human embryos by vitrification or slow freezing: a systematic review and meta analysis. Fertil Steril 90(1):186–193
3. Mukaida T, Takahashi K (2007) Vitrification of blastocysts using the cryoloop technique. In: Tucker M, Liemberman J (eds) Vitrification in assisted reproduction, a user's manual and trouble-shooting guide. Informa, UK, pp 219–237
4. Kasai M, Niwa K et al (1980) Survival of mouse embryos frozen and thawed rapidly. J Reprod Fertil 59:51–56
5. Kasai M, Nowee K et al (1983) Protective effect of scrose on the survival of mouse and rat embryos stored at 0 °C. J Reprod Fertil 68:377–380
6. Kasai M, Edashige K (2007) Vitrification in animal reproduction: vitrification of embryos using conventional straws with an ethylene glycol-based solutions. In: Tucker M, Liemberman J (eds) Vitrification in assisted reproduction, a user's manual and trouble-shooting guide. Informa, UK, pp 75–85

7. Kartberg A-J, Hambiliki F et al (2008) Vitrification with DMSO protects embryo membrane integrity better than solutions without DMSO. RBM Online 17(3):378–384
8. Duus A, Lundin K (2008), Vitrification of human blastocysts with either DMSO or PrOH as a cryoprotectant. In: Poster presented at the 24th annual meeting of the European society of human reproduction and embryology (ESHRE), Barcelona, 6–9 July 2008
9. Abedini A, Singha G et al (2006) Recovery and purification of highly aggregation-prone disulfide containing peptides: application to islet amyloid polypeptide. Anal Biochem 351:181–186
10. Pogorelyi, V. K. Barvinchenko V. N et al (1988) Solvation effects in human serum albumin radioulysis in the presence of dimethyl sulfoxide. V. K. Teoreticheskaya i Eksperimental'naya Khimiya 26 107–110. Translated 1990
11. Vanderzwalmen P, Bertin G et al (2002) Births after vitrification at morula and blastocyst stages: effect of artificial reduction of the blastocoelic cavity before vitrification. Hum Reprod 17(3):744–751
12. Mukaida T, Oka C et al (2006) Artificial shrinkage of blastocoeles using either a micro-needle or a laser pulse prior to the cooling steps of vitrification improves survival rate and pregnancy outcome of vitrified human blastocysts. Hum Reprod 21 (12):3246–3252
13. Zhang X, Trokoudes K et al (2009) Vitrification of biopsied embryos at cleavage, morula and blastocyst stage. RBM Online 19 (4):526–531
14. Son W-Y, Yoon S-H et al (2003) Pregnancy outcome following transfer of human blastocysts vitrified on electron microscopy grids after induced collapse of the blastocoele. Hum Reprod 18(1):137–139
15. Hardarson T et al (2006) Vitrification and warming human blastocysts by use of a laser to artificially induce blastocyst collapse prior to vitrification. Acta Obstet Gynecol Scand 86:119–120
16. Wikland M, Hardarson T et al (2010) Obstetric outcomes after transfer of vitrified blastocysts. Hum Reprod 27(7):1699–1707

# Index




Zsolt Peter Nagy et al. (eds.), *Cryopreservation of Mammalian Gametes and Embryos: Methods and Protocols*, Methods in Molecular Biology, vol. 1568, DOI 10.1007/978-1-4939-6828-2, © Springer Science+Business Media LLC 2017

MIX
Papier aus verantwortungsvollen Quellen
Paper from responsible sources
FSC® C105338

If you have any concerns about our products,
you can contact us on
**ProductSafety@springernature.com**

In case Publisher is established outside the EU,
the EU authorized representative is:
**Springer Nature Customer Service Center GmbH**
**Europaplatz 3, 69115 Heidelberg, Germany**

Printed by Libri Plureos GmbH
in Hamburg, Germany